ROUTLEDGE LIBRARY EDITIONS: URBANIZATION

Volume 2

ECONOMIC GROWTH AND URBANIZATION IN DEVELOPING AREAS

ECONOMIC GROWTH AND URBANIZATION IN DEVELOPING AREAS

Edited by
DAVID DRAKAKIS-SMITH

for the IGU Commission on Third World Development

Routledge
Taylor & Francis Group
LONDON AND NEW YORK

First published in 1990 by Routledge

This edition first published in 2018
by Routledge
2 Park Square, Milton Park, Abingdon, Oxon OX14 4RN

and by Routledge
605 Third Avenue, New York, NY 10017

Routledge is an imprint of the Taylor & Francis Group, an informa business

British Library Cataloguing in Publication Data
A catalogue record for this book is available from the British Library

ISBN: 978-0-8153-8014-6 (Set)
ISBN: 978-1-351-21390-5 (Set) (ebk)
ISBN: 978-0-8153-7872-3 (Volume 2) (hbk)
ISBN: 978-1-351-22782-7 (Volume 2) (ebk)

Publisher's Note
The publisher has gone to great lengths to ensure the quality of this reprint but points out that some imperfections in the original copies may be apparent.

Disclaimer
The publisher has made every effort to trace copyright holders and would welcome correspondence from those they have been unable to trace.

Economic Growth and Urbanization in Developing Areas

Edited by David Drakakis-Smith
for the IGU Commission on Third World Development

ROUTLEDGE
London and New York

First published 1990 by Routledge
11 New Fetter Lane, London EC4P 4EE

Simultaneously published in the USA and Canada by
Routledge
a division of Routledge, Chapman and Hall, Inc.
29 West 35th Street, New York, NY 10001

Printed in Great Britain by Mackays of Chatham PLC, Kent

British Library Cataloguing in Publication Data

Economic growth and urbanization in developing
areas.
1. Developing Countries. Economic development.
Effects of urbanisation
I. Drakakis-Smith, D.W. (David William) *1942–*
330.9172'4

ISBN 0-415-00442-X

Library of Congress Cataloging in Publication Data

Economic growth and urbanization in developing areas / edited by David Drakakis-Smith.
p. cm.
Includes bibliographies and index.
ISBN 0-415-00442-X
1. Developing countries — Economic policy. 2. Urbanization — Developing countries. I. Drakakis-Smith, D.W.
HC59.7.E286 1989
338.9'009172'4–dc19 88-34312
CIP

Contents

Figures

Tables

Acknowledgements

The production of an edited volume is usually a team effort and this is no exception. I would like to offer my thanks to the following people for their invaluable contributions: Muriel Patrick for the maps and diagrams, Don Morris for photographic work, Pauline Jones and May Bowers for their typing.

However, as this book is the product of a conference held in Madrid by the IGU Working Group on Urbanization in Developing Countries, there are some acknowledgements to be made in this context too. I would like to express the thanks of all participants to Professor Aurora Garcia-Ballasteros and her colleagues at the Department of Geography in the Universidad Complutense de Madrid, not only for their excellent organizational arrangements, but also for the warm and friendly welcome they bestowed upon us. Second, our thanks go to Dr John Naylon, of the University of Keele, for arranging our accommodation, for acting as the group's unpaid translator, and for organizing so well our evening forays into old Madrid.

Introduction: urbanization, economic development and space

The chapters presented in this book comprise a selection of the papers presented at a meeting of the IGU Working Group on Urbanization in Developing Countries which was held in Madrid in August 1986, the theme of which was 'economic development and urbanization in the periphery and semi-periphery'. It is axiomatic to state that this is very broad, in both a thematic and spatial sense, and consequently the papers presented at the meeting covered a very wide range indeed. It is the task of this brief introduction to link these chapters both to one another and to the themes of the meeting in general.

The world economy has clearly undergone major changes since the early 1970s. Henderson and Castells (1987: 3-4) summarize these major changes as follows:

(i) Inflation has been controlled by fiscal and social austerity.

(ii) The social costs of labour have been reduced or held stagnant, enabling profits to rise.

(iii) Productivity and profitability have been increased by technological innovation and redundancy.

(iv) Restructuring of industry has occurred, particularly through privatization of public sector assets, with certain sections, regions and activities favoured over others.

(v) There has been rapid growth of the petty commodity sector, both legal and illegal, to provide economic and social sustenance to those outside the economic mainstream, but to the eventual benefit of the dominant sector.

(vi) Increasing internationalization of domestic/national economies has taken place, with accentuated favouring of regions, sectors, etc.

(vii) Core economies have attempted to stabilize prices of raw materials and energy.

Partly as a result of the impact of such economic changes, in both urban and rural areas, the scale of urbanization has accelerated over the last two decades. Towns and cities in the Third World are growing at three times the rate of those in developed countries and now times the rate of those in developed countries and now contain some 55 per cent of total world urban population. Moreover, this urbanization process is producing increasingly larger cities with one-third currently living in 'million' cities and one-quarter predicted to be resident in cities of more than four million by AD 2000 (Habitat 1987).

The nature of the urbanization process in the Third World is in general quite different from that experienced by developed countries, particularly at present. Most developing countries are still in the throes of a rapid and heavily concentrated urban growth, and relatively few have experienced suburbanization, let alone de-urbanization. The main problems common to the overwhelming majority of Third World cities are consequently those that result from massively accelerating demands for urban services of all kinds (housing, transport, health, and the like) from urban governments without the means or the will to respond.

However, no matter how desperate are these problems in general terms, it is also true that massive variations exist in the nature of the urbanization process in individual countries or regions. Such differences relate not only to rates of growth but also to the characteristics of the cities themselves, with the nature of their inhabitants and their built environments. Much of this variation is due to the myriad of cultural, economic, historical, and political features that characterize individual social formations, but the unevenness of the global economy is also an important factor. The consequence is not only the 'familiar' poverty and squalor of urban squatter settlements but also, in some 'favoured' cities, a degree of architectural, financial, transportational, and social sophistication matched by only very few western metropolises.

This is in itself, of course, a value judgement, since the afore-mentioned sophistication is western in its nature and is invariably and inevitably representative of an extensive level of penetration by global capitalism in its most advanced form. Clearly there is a strong link between economic development and urbanization, not only in the overall processes involved but also in the character of the end-product i.e. the built environment, its functions, and the life-style of its inhabitants.

The relationship between economic development and urbanization has long been recognized. Development theorists from both left and right have centred their examinations of change on the pivotal role played by the city. Geographical analysts have often preferred to express this relationship in a more spatial context. Indeed, for some, the context has become the objective of study per se, justifying the criticisms of 'spatial fetishism' or 'reification of space' which have then, unfortunatley, been attached to all development geographers.

More recently, many non-geographers have begun to recognize the importance of the spatial dimension to the anlaysis of uneven development, and core-periphery analysis has been absorbed almost too enthusiastically by dependency and world-system theorists. To many, however, core-periphery concepts have severe limitations as an analytical tool, leading inter alia to investigations into the validity of the notion of a semi-periphery. Several chapters in the present book directly address the conceptual issues involved in this discussion and it would be inappropriate to anticipate this debate here.

In effect, what has happened over the years in development theory is that space has shifted from being merely a formal framework for underdevelopment (Hinderink and Sterkenburg 1978), in Lefebvre's (1976) terms the context for development, to a situation in which it is seen as being created by changes in the political, economic, or social dimensions of development. However, for many development strategists this still means that space must be seen as a subordinate element, its unevenness being a consequence rather than a cause of changes in the political economy of a particular social formation.

Others, such as Slater (1986) and Soja (1984) have attempted to emphasize the importance of space as a more independent analytical and explanatory variable, stressing in particular the inherent spatial dimension to the transfer of value. Although this has formed the focus of several analytical texts in geography during the 1980s (one of the most comprehensive and underrated of which is Rimmer and Forbes 1983), only recently has this interpretation of the role of space in fluid rather than static terms been accepted by many non-geographical development analysts. A useful example, in this respect, is Henderson and Castells' (1987) distinction between 'a space of flows' and 'the space of places'.

In many ways, the city continues to be the focus of both spatial and social inequality, the irony being that whilst urbanization often accentuates national unevenness in development, it gives rise to urban units that contain even greater contrasts in wealth and well-being. There are many observers who would credit the state with a role in creating such inequalities by intervening on behalf of international capitalism. There is much truth in such observations, but in other ways the state struggles to cope with the consequences of uneven development and the upheavals consequent upon it.

The role of the state in influencing economic and urban development is an important issue running through the essays in this volume. The picture revealed is one of enormous complexity, not the least of which involves the various levels of state governments involved, namely national, sub-national, or supra-national, and how contradictory their multiple interventionist policies may be. As the various chapters reveal, however, the state ultimately plays an enormous role in shaping the nature of the impact of economic change on the city.

The consequence of the varied combinations of the macro-forces, briefly outlined above, and their impact on particular Third World social formations with their own historical patterns of development, produces an immense differentiation. Any attempt to understand the precise way in which external and local factors articulate must, therefore, be heavily dependent upon empirical research for access to the rich variety of information which is required.

Obviously this does not mean a regression to detailed descriptive research for its own sake, for the goal must be to ascertain how individual social formations develop in the complex context of the nested levels of the world, regional, national, and other local economies. Empirical investigation which is to be of real value (to development strategies, national or urban planners, etc.) can, therefore, often pose many difficult problems for its practitioners, such as the scale of the unit of analysis, the ways in which development theories or concepts can be blended into the investigation, or the practical use to which the information collected might be put (by allies or adversaries). It is in such a context that the contributions to this volume must be placed.

The opening chapter by David Smith and Roger Nemeth epitomizes the nature of the volume as a whole by taking a particular concept, discussing some of its theoretical dimensions, and attempting to test these against empirical evidence. Each of the chapters attempts a similar analysis, although the emphasis on ideas, theories, and empirical evidence varies considerably. Smith and Nemeth not only examine the 'slippery' concept of semi-peripheral urbanization against the realities of the contemporary Third World, in Southeast Asia and West Africa, but also in the historical context of North America. Their conclusion that semi-peripheral urbanization has been and is still dependent development reaffirms the utility of the concept as an analytical tool.

John Naylon, in chapter two, also pursues the notion of the semi-periphery in his examination of the development of Spain in general, and the city of Barcelona in particular, during the Franco regime. He charts the emergence of the country from semi-peripheral towards core status and details the impact of such economic change upon the development of Barcelona. As in chapter one, the evidence reveals clear parallels in the relationships between economic and urban change between the developed and developing semi-periphery, not least of which is the authoritarian nature of the regime under which this occurred. However, the freewheeling chaos even within important cities, such as Barcelona, is also very evident: a contradiction between the control of national and urban development which is also repeated throughout this volume.

Chapter three, by Alan Gilbert, raises several questions in his lucid overview of Latin American urban economic development which, he feels, are inadequately addressed in the current literature. His first goal is to assess whether Latin American cities have experienced the same types of change as those documented for Europe and East Asia; his second objective is to examine the impact of the recession of the 1980s on existing trends. To these ends Gilbert draws upon housing provision, service systems, and employment patterns to provide supporting evidence. His prognosis for the future of Latin American cities is one of enormous challenges.

The dimension of analysis changes amost completely in chapter four, in which Mei Ling Young and Kamal Salih investigate the response of urban households to the industrialization process. The scale of this conceptual bridge is enormous, ranging from global trends in industrialization to the adjustments made within the household to accommodate changes in local urban economies. The study draws heavily on anthropological approaches, and its scope and effectiveness illustrate clearly the extent to which the extremes of macro- and micro-analyses can be successfully fused together.

The household remains the unit of analysis for Jennifer Bryant's study of the impact of economic pressures on the composition of urban squatter settlements in Fiji, which comprises chapter five. Here the focus of investigation is the micro-economic adjustments which have begun to appear in response to the broader forces of world recession which, along with internal problems, have affected low-income urban residents in the capital city. The result has been a readjustment in the economic structure of such settlements, particularly in the increase of subletting, thereby affecting the nature of the built environment itself.

The built environment and its relationship with social, economic, and political events is also at the heart of David Drakakis-Smith's contribution to the volume in chapter six. The geographical focus reverts to the advanced semi-periphery, this time in Northern Australia, but the analytical thrust is somewhat different from previous chapters. Drakakis-Smith is attempting to illustrate the ways in which social and political processes can be just as influential as economic forces in shaping the built environment of individual settlements. This much is evident in some of the other studies, but chapter six more specifically addresses the issue.

The theme of the built environment also runs through the two following chapters, seven and eight. In the former, Charles Mather and Susan Parnell investigate the process of urban renewal in Soweto. In particular, they chart the emergence of what might be termed a middle-class housing market and highlight the political and economic forces - global, national, and local - which have given rise to this process. Their attention to the historical dimension is echoed in chapter eight, in which Johannes Augel investigates the economic considerations that form the backdrop to the changes that have occurred, and might yet take place, in the historic centre of the city of Salvador in Brazil.

Keith Beavon and Chris Rogerson once again take up the historical dimension as they pursue archival information on the evolution of hawking in Soweto in chapter nine. Their detailed review gives an unusual degree of depth to the analysis of changes in the nature of petty-commodity production and their impact on the social morphology of the city. It constitutes a fine example of research on the micro-relationships between economic and urban development. Chris Rogerson goes solo in the following chapter, which again investigates the 'informal sector', but this time he paints with a broader brush the role of the state in shaping the economics of the consumption patterns of Black South Africans and how this in turn has influenced the adaptation of the petty-commodity response.

It is the consumer sector of the urban economy which furnishes the link with the final chapter by Scott MacLeod and Terry McGee on changing urban food consumption and retailing patterns in Hong Kong. Their contribution illustrates the way in which the permeation of western values has resulted in nothing less than the industrialization of the food system in Hong Kong. They examine the production, circulatory and consumption spheres of urban food both at the macro- and micro-levels and clearly illustrate, once again, the ways in which the character of an individual city is shaped by a combination of global and local processes.

Examination of the integration of global, national, and local processes is a feature of all the chapters in this volume, with their geographical foci being fairly evenly spread across the periphery (Fiji, Latin America), and semi-periphery (Northern Australia, historical North America, and Spain) and the uncertain areas between (South Africa, Hong Kong, Brazil). In addition, there are certain focal themes, such as consumerism, the built environment, and the household economy, which appear and reappear to weave the other issues together into what is hopefully as reasonably coherent a volume as might be expected from any edited conference proceedings.

Chapter One

Dependent urbanization in the contemporary semi-periphery: deepening the analogy

David A. Smith and Roger J. Nemeth

INTRODUCTION

Urbanization is a dynamic process that is irrevocably intertwined with other aspects of economic, social, and political development. Classical social theorists such as Marx and Weber emphasized the profound social transformations that accompany the growth of cities. Twentieth-century urbanologists have also stressed the role of the city as both locus for, and product of, broader socio-economic changes. Hawley (1971: 3) argues that 'cities stand at the vortices of the currents and cross-currents of broad scale change that alters and re-constitutes societies'. Castells (1977: ix) sees cities as foci for epochal forces and conflict which generate 'increasingly explosive urban contradictions'. It is in the study of 'the urban question' that researchers confront 'the problematic of the development of societies' (Castells 1977: 7).

This view of cities and urbanization suggests two important themes. One is that research on urban structures and processes can potentially reveal a great deal about more general patterns of change in a society. Understanding the growth, form, and function of a society's cities is a critical component of efforts to unravel the complexity of its overall development trajectory. Conversely, an appreciation of the inherent interpenetration of urbanization and development highlights the importance of conceptualizing the growth of cities as eddies in wider currents of social transformation. Conceptualizations of city growth need to be grounded in broad-based theories of development. Models of urban structure and spatial dynamics frequently fail to explicitly do this, even though they rely on implicit assumptions about the generic nature and causes of social change. This becomes particularly problematic when these unexamined assumptions are of dubious value and are called into question by other social scientists. Clearly, a theoretical perspective that systematically explains the relationship between the dynamics of urbanization and wider processes of macro-structural change is analytically desirable.

The recent rise of an international political economy perspective on the 'dependent city' appears to offer this type of comprehensive, theoretically grounded approach to urbanization. The emphasis on locating cities and urban systems in circuits of global capitalism and the webs of the international commerce and geo-politics provides an exciting new framework for urban research (Walton 1981). Several authors have attempted to present schematically a world-system theory of comparative urban development (Walton 1982; Chase-Dunn 1984; Timberlake 1985). However, many of the implications of this perspective for particular types of cities, urban systems, and regions need further specification.

The idea of 'semi-peripheral urbanization' is one concept that has not been extensively explored. While the notion of a distinct urban dynamic in the intermediate stratum of the world system seems consistent with Wallersteinian theory, to our knowledge, no one has yet attempted to systematically discuss cities and urban growth in the semi-periphery. This is the goal of this paper; one which is fraught with pitfalls and perils. We hope, however, the theoretical pay-off justifies the particularly problematic nature of our effort.

We begin by summarizing the theoretical orientation which world-system analysts apply to urbanization. The next section grapples with the slippery, elusive concept of the semi-periphery. There follows a summary of patterns and processes of semi-peripheral urbanization, drawing on three historical case studies of colonial North America and three case studies of the contemporary Third World. This leads to a concluding discussion on the limits of the analogy of dependent urbanization and the generalizability and practical implications of a pattern of semi-peripheral urbanization.

THEORETICAL ORIENTATION

While it would be possible to delve deeper into the history of dependency theory, a key intellectual forebear of the world-system approach to the 'dependent city' and 'peripheral urbanization' was A.G. Frank. His 1969 description of world capitalist exploitation operating through 'a chain of constellations of metropolises and satellites' was polemical, but proved to be theoretically pregnant. Systematic explication of a theory of dependent urbanization was left to Castells (1977: 3). Arguing that cities can only be fully understood as products of the expansion of the capitalist world economy, he claims that

> the process of urbanization becomes, therefore, the expression of this social dynamic at the level of space, that is to say, of the penetration by the capitalist mode of production, historically formed in the western countries, of the remainder of the social formations at different technological, economic, and social levels.
>
> (Castells 1977: 44)

These pioneering formulations have been followed by a number of empirical studies applying the logic of the international political economy approach to comparative research on cities and urban systems (Walton 1977; Roberts 1978; Slater 1978; Friedmann and Wolff 1982; Gilbert and Gugler 1982; Timberlake 1985; Armstrong and McGee 1985; Meyer 1986; Smith and Nemeth 1986). Timberlake succinctly summarizes the main premise of this approach:

> Urbanization must be studied holistically - part of the logic of a larger process of socio-economic development that encompasses it, and entails systematic unevenness across regions of the world. The dependence relation is an important theoretical concept to pry into the ways in which the processes embodied in the world system produce various manifestations of this unevenness, including divergent patterns of urbanization.
>
> (Timberlake 1985: 10)

The penetration of the world economy into peripheral areas leads to a development dynamic which gives rise to a few relatively large cities which act as trade centres in the web of colonial or neo-colonial exploitation. The result is a process of urbanization which leads to urban primacy, regional inequalities, centralization of political and economic power within cities, and intra-urban ecological segregation and inequality (for theoretical summaries, see Walton 1982; Timberlake 1985: ch. 1). Chase-Dunn explains the role of the dependent city in the world system: 'Peripheral primate cities are nodes on a conduit which transmit surplus value to the core and domination to the periphery, while primate cities in the core receive surplus value and transmit domination' (Chase-Dunn 1984: 115).

The important point is that dependent urbanism conceived in this way not only leads to 'uneven' urban hierarchies and high levels of intra-urban inequality, but (using terminology from Hoselitz 1954) creates cities that are more likely to be economically 'parasitic' on the surrounding region than 'generative'.

The highly skewed class structures and large numbers of poor in dependent cities, in this view, are more than the result of 'too many' people or a sectoral 'misallocation' of labour.

> High rates of inequality and poverty in large primate Third World cities may be very undesirable for the masses of people who live in them and an obstacle to genuine national development. But when the role of the 'informal sector' in surplus extraction in the capitalist world economy is considered, it becomes clear that the poverty and inequality may be functional for the wider system.
>
> (D. Smith 1985: 211)

The large pool of under-renumerated, semi-proletarianized labour is a structural element of the peripheral city; one which allows goods and services to be produced cheaply (see Portes 1985a for a discussion of mechanisms). This, in turn, reduces the cost of high consumption by elites and directly and indirectly subsidizes wage costs for local formal sector firms. The paradox of 'peasants in cities' is resolved - the urban economy of the dependent city is a mechanism of 'unequal exchange' (McGee 1973, 1978). Portes claims that:

> The informal sector - a vast network of activities articulated with, but not limited to, remaining subsistence enclaves - has implications that go beyond the peripheral countries. Direct subsidies to

> consumption provided by informal to formal sector workers within a particular peripheral country are also indirect subsidies to core-nation workers, and, hence, means to maintain the rate of profit. Thus, through a series of mechanisms well hidden from public view, the apparently isolated labour of shanty-town workers can be registered in the financial houses of New York and London.
>
> (Portes 1985a: 61-2)

Obviously a critical question for the present analysis is, to what extent does the image of unevenness, parasitism, and inequality apply to urban centres in the semi-periphery? In what sense are they also 'dependent cities'? What differentiates them from peripheral cities?

To begin even tentatively to answer these questions, we need to be clear about what the semi-periphery is, what its salient characteristics are, and what role it purportedly plays in the world economy. That is the object of the next section.

But before delineating what constitutes the semi-periphery and attempting to conceptualize how it fits into a theory of dependent urbanization, it is useful to highlight some differences between the world-system formulation and more conventional views of urbanization and development. Many of the images of this process in the regional science and urban geography/sociology literature rely on modernization theory assumptions about social change at a time when this old approach has become discredited in the sociology of development.

One key problem with the orthodox perspective is its emphasis on a bipolar conception of progress manifested in 'the urban-rural duality thesis' (Slater 1986: 9-10). The international political economy view emphasizes the complex interpenetration of city and countryside. Developmentalist perspectives have emphasized cities' potentials as centres for innovation, opportunity, and political transformation (Friedmann 1978: 82). World-system analysts are much less sanguine about the potential for autonomous change in dependent cities, emphasizing instead the structure of domination emanating downward from the global level (Walton 1982). When Third World cities are plagued by poverty and economic stagnation, orthodox explanations focus on 'excessive migration' and 'population maldistribution' (World Bank 1985: 96 cited by Slater 1986:

10; see also Gugler 1982), whereas those who emphasize the dependent nature of these cities conceptualize the problems not as 'spatio-demographic' (Slater 1978), but as embedded in, and perhaps even functional for, the capitalist world economy (C. Smith 1985).

A final difference concerns the conventional approach's use of stages and phases to describe models of urban development. The widely accepted regional science theory of urban concentration as 'polarization-polarization reversal' (Richardson 1977, 1980) is a blatant example of an uncritical adoption of the modernization theory assumption that societies pass through universal sequences. This implicit developmentalism is so pervasive that it occasionally creeps into analyses which are sensitive to the key role that links to the global economy play in urbanization in the less-developed world (see Meyer 1986). World-system researchers are extremely sceptical about all models of national development which suggest universal phases, particularly when these schemes imply that the ultimate stage is some sort of modern prosperous, balanced, core-like pattern.

The strongest argument against models incorporating a stage theory of cities in development is empirical. Put simply, the pattern of macro-structural change in Third World societies in recent years does not square with the assumptions of developmentalist approaches. Clearly, the urbanization process in regions like Southeast Asia, Africa, and Latin America belies the image of cities as dynamic generators of economic and social development (McGee 1967; Portes and Walton 1976; Gugler and Flanagen 1978). The urban dynamic in these places is very different from the growth of cities in the west. This realization led to the formulation of such concepts as 'pseudo-urbanization' (McGee 1967) or 'dependent urbanization' (Castells 1977) to indicate the basic divergence in trajectories. Now, as more detailed regional and national studies of urban processes are available, and social scientists become more sensitive to the nuances of particular histories, it is equally obvious that these dichotomous categories do not begin to do justice to the diverse experiences of city growth in these societies either.

Armstrong and McGee highlight the variations in processes such as capital accumulation and urbanization and point out that:

> the interaction with the global expansion of capitalism of particular elements of historical experience, political institutions, physical resources, economic structures, and socio-cultural relationships within each society will result in highly specific chemistries of development for each society.
>
> (Armstrong and McGee 1985: 32)

We agree with this conjunctural view of development and would argue that a complete understanding of the details and mechanisms of urban processes is only attainable through in-depth case studies. However, in this chapter we have followed the traditional social science urge to generalize about urbanization (see Gilbert and Gugler 1982, for a discussion of this predilection). This prompts us to seek a pattern of semi-peripheral urbanization in the hope that it may help to explain some of the diversity.

THE SEMI-PERIPHERY: WHAT IT IS AND WHY IT IS IMPORTANT

While eschewing 'stage theory', world-system analysts are not 'historical particularists' (Lenski 1976). The structure of the international system as a whole remains fairly stable over time, despite changing roles for its regional and national sub-units. Within the long spans of time defined by various periodization schemes (see Wallerstein and Hopkins 1977; Frank 1978), historically bounded generalizations can be made about matters such as the mechanisms of unequal exchange, the forms of labour control, or the structure of political domination. Most important for present purposes, however, is the idea that while the development processes follow different laws of motion in the core, periphery, and semi-periphery, within the world-system strata similar dynamics of social change are discernible (Chirot 1977; Wallerstein and Hopkins 1977; Evans 1979b).

During the past two decades, an enormous amount has been written on peripheral urbanization. The meaning of antonymous terms such as dependent/autonomous, metropolis/satellite, and core/periphery has been debated, researched, and continually refined. From this literature, a consensus has begun to emerge regarding the character, structure, and consequences of urban systems within countries occupying these polar structural positions in the world economy. Unfortunately, the same cannot be said for the semi-periphery. Few concepts in dependency/world-system theory have remained as elusive and misunderstood as this one, and even fewer have (until recently) received such scant attention from urban researchers. But this paucity of research does not necessarily mean that the concept of the semi-periphery is irrelevant to our understanding of the interplay between the world economy and processes of urbanization (see also chapter six).

Beginning in the sixteenth century and continuing up to the present day, the semi-periphery has played a unique and vital role in shaping the modern world economy (Wallerstein 1974a, 1976, 1980). In this section, we will indicate how an understanding of the semi-periphery can contribute to the study of Third World urbanization. We begin by delineating some of the essential characteristics of the semi-periphery and discussing how they are likely to affect urbanization. Next, we propose some hypotheses on how urbanization in the semi-periphery and periphery differ. In the next section, we examine some case studies to evaluate the usefulness of our ideas.

In his seminal work on the origins of the modern world economy, Wallerstein (1974a: 349) locates the semi-periphery 'between core and periphery on a series of dimensions'. Although this definition lacks sufficient operational criteria to identify specific semi-peripheral states (Arrighi 1985), it does provide a starting point for discerning the essential nature of the concept. Being located between core and periphery does not mean that the semi-periphery is simply a middle point on some continuum between core and periphery. Nor does it mean that the semi-periphery is just a residual category which includes any country that does not conveniently fit into either of the other positions. The value of Wallerstein's conceptualization is that the semi-periphery is seen as a unique and 'necessary structural element in the world economy' (1974a: 349). The semi-periphery comprises a distinct structural position in the international division of labour and the particular role it plays is essential to the maintenance of the world capitalist system as a whole.

The semi-periphery differs politically and economically from the core or periphery. Because they play the dual role of both core and periphery in the international economy, the internal productive activities of semi-peripheral countries are more evenly divided than those of nations in either of the other structural positions (Chase-Dunn 1978). The nature of foreign investments and the possibility for 'dependent development' in the semi-periphery is also different. Transnational corporations are more willing to invest in manufacturing industries, and to transfer technology designed to promote domestic industrialization, in countries where there is an industrial base able to absorb the new technology and adequate internal markets to consume the products of the new industries. Thus, it is in the semi-periphery that economic growth and the transformation of a more diversified industrial structure is most likely to occur. Empirically, studies of non-core countries find that foreign investments of transnational corporations help to diversify domestic industry and advance technologies in semi-peripheral countries (Gereffi and Evans 1981; Yang and Stone 1985).

These economic functions, however, must be viewed within a political framework which recognizes the particular state structures of semi-peripheral nations. In the world economy, the semi-periphery acts as a buffer between core and periphery; a role that reduces the likelihood of social revolution in the latter (Wallerstein 1974a). The more evenly mixed core-periphery economic functions of semi-peripheral nations mean that state policies can more directly influence the accumulation of capital within them. Indeed, Wallerstein argues that 'semi-peripherality is important because it points to a concentration of state-oriented political activity by major internal (and external) economic actors' (Wallerstein 1985: 35). Similarly, Chirot identifies semi-peripheral societies as those 'organized into states powerful enough to play an international role; [ones that] could not be pushed around as easily as peripheral states' (Chirot 1986: 93). While in general agreement with Wallerstein and Chirot, Chase-Dunn and Robinson (1977) argue that the particular nature of state structures depends on the mobility of the country. Although upwardly mobile countries that rely on alliances with the core tend to develop rightist state regimes (e.g. South Korea or Brazil), those attempting autonomous development tend to form regimes left of centre (e.g. China or the Soviet Union). In either case, semi-peripheral countries 'tend to employ more state-directed and state-mobilized development policies than do core countries' (Chase-Dunn and Robinson 1977: 472).

Despite this apparent agreement over the conceptual meaning of the semi-periphery, disagreements persist over the specific countries occupying this intermediate position in the world economy. According to Wallerstein, the semi-periphery includes

> the economically stronger countries of Latin America: Brazil, Mexico, Argentina, Venezuela, possibly Chile, and Cuba. It includes the whole outer rim of Europe: the southern tier of Portgual, Spain, Italy, and Greece; most of Eastern Europe; parts of the Northern tier such as Norway and Finland. It includes a series of Arab states: Algeria, Egypt, Saudi Arabia, and also Israel. It includes in Africa at least Nigeria and Zaire and in Asia: Turkey, Iran, India, Indonesia, China, Korea, and Vietnam. And it includes the old white Commonwealth: Canada, Australia, South Africa, possibly New Zealand.
> (Wallerstein 1976: 465)

Chirot, on the other hand, argues that today 'only a few societies remain purely peripheral in the old sense of that term' and that 'the distinctions between the two categories of non-core societies have become so blurred that such labels as peripheral and semi-peripheral are dated and can be applied to the present only with great care' (Chirot 1977: 197-81). On the other hand, findings from several empirical analyses suggest that the tripartite division of the world economy is useful in explaining differences and similarities among countries (Evans 1979a; Gereffi and Evans 1981; Nemeth and Smith 1983). So is there really a semi-periphery, that is, an intermediate position within the world economy which differs fundamentally from either of the other positions?

Recently, several quantitative studies have attempted to locate countries in structural positions within the world economy (Steiber 1979; Snyder and Kick 1979; Breiger 1981; Arrighi 1985; Nemeth and Smith 1985; Smith and White 1986). Although these attempts vary greatly in the criteria used and types of analyses employed, they have been successful in identifying a fairly consistent and well-defined core and periphery. Interestingly, some of the more recent research finds evidence of stratified (or differentiated) layers within the semi-periphery (Nemeth and Smith 1985; Smith and White 1986; Kick 1986). The essential point of these studies is not whether there are two, or more, semi-peripheries, but that countries do occupy this distinct structural position in the global economy. If nations can be meaningfully differentiated by world-system position, and these strata play identifiable roles in the world economic system, one would expect both industrialization and urbanization to vary systematically by structural position.

Assuming the above argument makes theoretical sense, we are still faced with the question: can differences in urbanization and urban development be discerned between non-core countries? Recently in a large, quantitative, cross-national study of urban systems, Fiala and Kamens (1986) found that, generally, the periphery had much higher rates of population primacy than the semi-periphery. D. Smith (1984) found similar patterns. These studies of general patterns of primacy are supported by a growing body of regional urban studies which highlight key differences in urbanization between Third World countries. In the following section, we will examine a pattern of urbanization which appears to be common to much of the semi-periphery. Our analysis will compare contemporary upwardly mobile countries from East Asia and Africa with the experience of colonial North America. The objective of this comparative analysis is to identify common patterns among these disparate case studies, and to evaluate the effect of semi-peripherality on their urban development.

CONTEMPORARY REGIONAL ANALYSES

The dependency/world-system approach suggests that countries' processes of urbanization and urban development are intrinsically related to the roles they play in the world economy. This proposition has received at least partial support from both large, cross-national studies and in-depth case studies of many Third World countries. This section focuses on the comparison of contemporary urban patterns in two underdeveloped regions: East Asia and West Africa. These geographically distinct regions were chosen for analyses because of their disparate cultural and historical experiences. By selecting these contrasting regions, our intent is first, to account for regional and cultural explanations of any similarities and differences in urbanization patterns, and second, to supplement the extensive work by dependency theorists on Latin American urbanization.[1] Our analysis provides comparisons between the peripheral urbanization pattern (common to East Asia and West Africa) and the urbanization of semi-peripheral countries in each region. We also are able to compare the urban development of Nigeria before and after its recent ascendance into the semi-periphery.

Although East Asia is a region of varied urban trajectories, one of its sub-regions (Southeast Asia) is similar to West Africa in a number of ways. The southern nations of East Asia are, like those of West Africa, amongst the poorest and least urbanized in the world. Both regions have GNP/per capita levels substantially below the average for their continents and for most other major global regions (Population Reference Bureau 1985). Nevertheless, city growth in both areas has been quite rapid since the Second World War. Indeed, at a rate exceeding 4 per cent per annum, their urban growth exceeds the reported average for all 'less developed' regions (United Nations 1980: Table B). However, because rural populations have grown even more rapidly, Southeast Asia and West Africa remain two of the least urbanized regions of the world (Population Reference Bureau 1985). In fact, with less than one-fourth of their populations counted as city dwellers, both areas are far less urban than almost every other country on their continents.

Although their city populations have grown rapidly in the last three decades, they have done so in a very uneven and concentrated fashion. Countries in each region have large portions of their urban populations concentrated in one (usually the capital) city, and lack a system of secondary urban centres (McGee 1967, 1976; Rosser 1973; D. Smith 1984). This primate urban pattern, while common to much of the Third World, is particularly pronounced in Southeast Asia, where five countries have capital cities at least five times the size of their next largest cities. Although urban primacy is not quite as high in West Africa, nearly all countries in the region have primacy ratios above 3.5 and rising (United Nations 1980). Although demographic concentration is only a rough approximation of 'functional primacy' (London 1980), other social and economic indicators confirm that each region suffers from urban bias Todaro and Stilkind 1981). Of course uneven growth is not limited to inter-city comparisons. Within the large cities of each region, one finds many of the same problems of rapid growth common throughout the Third World. In-depth case studies of urban areas in both regions reveal problems of squatter, or spontaneous, settlements with inadequate housing and basic infrastructural necessities, high rates of unemployment and underemployment, large disparities in incomes and life chances, and large informal sector economies (McGee 1967, 1976; Hance 1970; Cohen 1974; Hollnsteiner and Lopez 1976; Sethuraman 1977; Riddell 1978; D. Smith 1984).

It seems remarkable that two regions as different geographically, ethnically, and culturally as Southeast Asia and West Africa should share a general pattern of urbanization. But the traits common to both characterize 'dependent urbanization'. Although countries in both regions generally share this common pattern, there are notable exceptions. These 'deviant' cases may not, however, be anomalies. Instead they fit the pattern expected of countries undergoing dependent development. They too are explainable within a dependency/world-system approach to urbanization if we use the idea of the semiperiphery. Below we describe some of the salient characteristics common to these deviant cases, and later attempt to explain this type of urbanization as a consequence of the role each country plays in the global economy.

Regional exceptions

East Asia is a region of varied urban patterns. The urban experience of South Korea and Taiwan stands in sharp contrast to the Southeast Asian pattern described above.[2] Both of these countries, while lagging behind Japan in overall urbanization, have experienced tremendous city growth since the Second World War and currently count well over one-half of their population as urban (Population Reference Bureau 1985). Although their populations have become increasingly urban, they have not become more primate. Along with the growth of their capital cities there has been a concomitant growth of a system of secondary cities. Indeed, primacy indices indicate that neither could be considered primate by world-wide standards, and that each is slightly less primate than several core nations (Mills and Song 1979; D. Smith 1984). Nor have these countries had the degree of urban poverty and economic inequality that plagues much of Southeast Asia and other regions experiencing rapid urban growth. This, in part, has been possible because of the exceptional economic growth each has experienced - averaging near double-digit economic growth over the last decade and enjoying GNP per capita figures well above the norm for 'less developed countries' (World Bank 1982; Population Reference Bureau 1985). In contrast to many developing countries, South Korea and Taiwan have maintained a relatively equal income distribution in addition to rapidly increasing production. This rapid and relatively equitable economic growth, resulting largely from labour-intensive export industrialization, has increased opportunities and reduced the time needed for new migrants to assimilate into the urban labour market (Mills and Song 1979; D. Smith 1984).

Similar to South Korea and Taiwan, Nigeria represents a 'deviant case'. With a GNP per capita nearly one-third higher than the average for West Africa, Nigeria has been one of the fastest-growing economies in Africa (Population Reference Bureau 1985). Likewise, Nigeria's system of relatively even-sized cities stands in stark contrast to the extremely high levels of primacy for other West African nations (Gugler and Flanagan 1978). Indeed, its near log-normal city-size distribution makes it exceptional for most of Africa. Comparatively, Nigeria's urban migration also appears to be far better balanced, with regional urban centres growing at a rate commensurate with Lagos. Recent migration flows, instead of leading to increasing primacy, have instead pushed the country toward an inter-regional equilibrium (Amin 1974: 83). Limited case studies suggest that Nigerian cities also differ in the degree to which they exihibit the characteristics of dependent urbanization. For example, in a study of Kano, the second largest city, Lubeck (1977) argues that relatively moderate growth, a predominance of local and national investment in growing industrial production, and an absence of sprawling shanty towns differentiate this city from others in dependent countries. Nigerian cities, however, continue to have far larger tertiary sectors and greater intra-urban inequalities than might be expected of a semi-peripheral country.

How are we to explain the urban patterns of South Korea, Taiwan, and Nigeria? Given their diverse geographical and cultural experiences, what factors can account for their deviation from the predominant urban pattern found in their respective regions? And to what extent can their differentiated or changing roles within the world economy account for the similarities they share? To begin to answer these questions, we offer recent quantitative evidence which indicates that each is part of the upwardly mobile semi-periphery (Snyder and Kick 1979; Nemeth and Smith 1985; Smith and White 1986). In addition to these statistical analyses, major monographs by dependency/world-system theorists consistently place them in an intermediate stratum in the world economy. For example, all three appear on Wallerstein's (1976) list of countries in the semi-periphery (op. cit.). Similarly, Chirot classifies them as part of a group of countries that 'are potentially major economic and political powers' in 'the new semi-periphery' (Chirot 1977: 213). Moreover, Evans (1979b) suggests that they are experiencing dependent development; a process which is driving each towards semi-periphery status. Thus, these countries may deviate from the peripheral patterns of Southeast Asia and West Africa, at least partly, because they perform a different role in the world economy. This statement, however, explains little about how this intermediate role has helped shape their urban configurations.

The historical dimension

To get at the mechanisms leading to the patterns of urbanization and development in these semi-peripheral countries, we need to compare their historical experiences. The following is a very brief synopsis of several detailed historical/structural analyses of the urban development of Nigeria, South Korea, and Taiwan (see Nemeth and Smith 1983; Smith and Nemeth 1986; D. Smith 1984, 1985). Our intent here is not to provide a thorough explanation of each country's urban history, but rather, to attempt to synthesize some of their common experiences and to suggest how these might account for the similarities.

Our search for significant common features begins by noting their pre-colonial pasts. Unlike most countries in their regions, by the time these countries were incorporated into the capitalist world economy they had developed agrarian societies and had experienced a long history of indigenous urban development (Nemeth and Smith 1983; D. Smith 1984). The pre-colonial period of each country witnessed the emergence of a system of relatively large urban centres based on an incipient geographical division of labour and linked together by long-established trade routes. Indeed, evidence from this early period indicates that many of these early commercial and administrative centres had attained a size comparable to the largest pre-industrial European cities. These distinct pre-contact urbanization and development patterns were reinforced by the particular pattern of incorporation into the world economy experienced by each society. The concatenation of this historical heritage and the dynamics of the international economy have propelled each of these countries towards the assumption of a semi-peripheral role. This status in the world economy has influenced the political economy of urbanization within each society.

While Nigeria possessed a network of large settlements prior to European penetration, colonialism and the peripheralization process which it generated promoted skewed urban growth and lead-city primacy. Nigeria's first sustained contact with European core powers came with the rise of the Atlantic slave trade in the seventeenth century. Because it came to serve as an entrepôt for the expanding slave trade, Lagos grew rapidly. Its growth is partly accounted for by the internal strife experienced by other major ports and Britain's ability to control the coastal territory surrounding Lagos. This provided European merchants the 'stable' political conditions they sought and provided Lagos with an important commercial advantage (Mabogunje 1968; A. Hopkins 1973). With the advent of the steamship in the 1870s, foreign merchants were able to control oceanic shipping, and in so doing, gained an additional advantage over the indigenous merchant class. The power of this foreign merchant class is represented in the relatively few development projects that were attempted - especially the building and extension of railways and port facilities. These developments in the transportation system served to funnel increasingly greater amounts of commerce and people toward Lagos (D. Smith 1984). Indeed, decisions as to where to lay rail lines often neglected important indigenous trade routes and seemed almost arbitrary to contemporary Nigerians (Mabogunje 1968). There is little doubt that these decisions served the interests of a foreign merchant class located in Lagos and worked to the detriment of other existing trade centres (ibid.).

It was not until the early twentieth century, however, that Lagos came to dominate Nigeria economically and politically. In fifty years, it grew more than five-fold, and by 1950 its population was in excess of a quarter million. By 1950 it was the national capital, the centre of national, European, and American financial institutions in Nigeria, and the destination of floods of migrants seeking well-paid jobs (D. Smith 1984). With an already bloated tertiary sector and stark inequalities between its urban classes,

Lagos had begun to exhibit the intra-urban inequalities found in many Third World cities.

Korea and Taiwan escaped colonialism until the late nineteenth century, and the Japanese imperialism which these societies endured was quite different from earlier European versions. With its victory in the Russo-Japanese War in 1905, Japan embarked upon an expansionist policy in Northeast Asia by colonizing first Taiwan and then Korea. During this period Japan was emerging as a major core power and its colonial policies reflected this fact (Chirot 1977). With the elimination and/or co-optation of indigenous elites, Japan initially developed both colonies as suppliers of raw materials and as markets for finished goods. Following a renewed policy of imperialism in the 1930s, however, Korea and Taiwan became avenues for Japan's penetration of the Asian continent and essential to its preparations for war (Chirot 1977; Cumings 1981). This required heavy investment in developing the infrastructure of both countries (Cumings 1981; Nemeth and Smith 1983). The extension and development of the existing transportation/communication network became a top-priority policy. The origins of a system of modern roadways, rail lines, port facilities, and telecommunications connecting all of the major urban centres can be traced to the Japanese colonial period in both countries (ibid.).

As a result of this heavy investment, manufacturing output in both colonies grew rapidly during the 1930s and by 1941 their conversion to wartime economies had been completed. The diversification of their economies resulted in industries locating near the factors of production and served to reinforce regional specialization and inter-regional dependencies. It should be stressed that these colonial developments were unrelated to indigenous Korean or Taiwanese entrepreneurship - which was limited primarily to small industries. Thus, while massive investment in both colonies transformed their economies to meet Japan's imperial designs, it maintained and even strengthened the existing urban structures by increasing the flow of people and commerce between cities. Demographic data from their early colonial periods indicate that a system of cities had already developed in each and, although Seoul and Taipei emerged as the largest and most important of these cities, neither exhibited the degree of dominance found in many contemporary peripheral countries.

Although the end of the Second World War brought

liberation from Japanese rule for Korea and Taiwan, each remained heavily dependent and dominated by non-indigenous powers. In South Korea, the US military government was in direct control and in Taiwan the Kuomintang Party held power. As a result of the complete domination by Japan, and the destruction created by the Second World War (and the civil war in Korea), indigenous elite classes were very weak in both countries (Cumings 1984; D. Smith 1985). Consequently, this allowed for the formation and continued growth of a relatively autonomous state (Koo 1982; Nemeth and Smith 1983). This does not mean that either state has ever been completely free from indigenous or core elites. Rather, it implies that the state, in actively directing the development of each country, has transcended (to a remarkable degree) the individualized interests of particular class fractions (Koo 1982; Cumings 1981; D. Smith 1984). For our purposes, the strong position of South Korea and Taiwan has meant the implementation of policies which allowed for more balanced urban and regional development.

Among these policies were the successful land reform initiatives which not only helped to equalize rural income distributions, but also depressed urban migration (Kim and Roemer 1970; Cumings 1984; D. Smith 1985). Furthermore, after initially pursuing a policy of import-substitution industrialization (ISI) during the 1950s, each has made the transition to an economy based on labour-intensive export industrialization (Nemeth and Smith 1983; Cumings 1984). This has proven to be a very successful strategy in attracting core capital to both countries. Given the relatively high costs of labour, and ageing technology in the core, transnational corporations have invested heavily in semi-peripheral countries with cheap, disciplined labour and political stability ensured by strong bureaucratic-authoritarian states (O'Donnell 1978; Cumings 1984). During the past two decades both South Korea and Taiwan have satisfied these criteria extremely well.

Export-led industrialization has helped to maintain the deconcentrated urban patterns and the generative nature of cities in each country. Under this type of industrialization it is efficient for plants to locate close to the factors of production and to utilize the advantages of regional resources. Conversely, in countries pursuing ISI, it is most profitable for industries to locate close to concentrated domestic markets (Roberts 1978). This is precisely what appears to have happened in Nigeria during the decades of the 1950s and 1960s.

Earlier we discussed Lagos' rapid growth during the early 1900s. It was not until after 1950, however, that the real metropolitan explosion occurred (Mabogunje 1968; D. Smith 1984). Between 1950 and 1976 Lagos experienced an astonishing ten-fold increase and attained a metropolitan population of over 2.5 million. Certainly, major factors accounting for this astronomical growth were the country's independence and the efforts made to gain greater control over its increasing trade deficit (Mabogunje 1968). Like many Third World nations, Nigeria pursued a policy of ISI during the 1950s and 1960s. This developmental strategy requires access to a highly trained and disciplined labour force and the existence of a large, concentrated consumer market (Roberts 1978). Typically, large primate cities fit these requisites and provide the ideal location for these types of industries. Not too surprisingly, the explosive growth of Lagos during this period has been attributed to the economic opportunities accompanying its industrialization (Mabogunje 1968).

While it is true that Lagos is an extremely large, fast-growing capital city with a large tertiary sector, it is not a primate city. Indeed, it was not until the mid-1960s that it became the largest city. Despite Lagos' rapid growth in recent years, Nigeria's primacy index is lower than that of most core countries (United Nations 1986). By 1980, four cities had populations of greater than 500,000 and the top ten cities had populations of over 200,000. The persistence of this urban system is, to an extent, a testimony to the momentum of institutional patterns within cities. Elites acting to protect and promote their own interests often preserve the status quo. Indeed, the indigenous class structures that developed in Nigeria's pre-colonial cities were conducive to local economic innovation and expansion and were able to survive and adapt to capitalist penetration and incorporation (A. Hopkins 1973; Lubeck 1977; D. Smith 1984). Nigeria's changing role within the world economy, however, must also be considered.

Nigeria's ascendancy into the semi-periphery has been noted by several writers (see earlier discussion). In perhaps the most serious attempt to come to grips with Nigeria's role in the world economy, Evans (1979b) argues that, while the country has only begun the process of dependent development, it is clearly in the transition to semi-periphery. Not only is its economy one of the strongest in Africa, but with an average 9 per cent increase over the last 10 years, it is also one of the fastest growing. Moreover, the Nigerian state has begun to take an active role in developing the nation's petroleum industry and its alliances with transnational corporations extracting and refining oil (ibid.). This localization of the economy can also be seen in the active participation of indigenous capitalists. At present, 40 per cent Nigerian ownership is required of all enterprises (Evans 1979b: 312).

In summary, Nigeria's present non-primate urban system resembles the semi-peripheral pattern found in South Korea and Taiwan. It exhibits, however, the bloated tertiary sector and enormous intra-urban inequality usually associated with peripheral urbanization. Given that Nigeria is a newly emerged semi-peripheral country, and has only recently begun the process of dependent development, we can speculate as to how this hybrid pattern will be affected by its changing role in the world economy.

Nigeria's large oil reserves are a major reason for its recent economic success.[3] The role of the Nigerian state in developing this valuable resource will help determine growth in various cities and regions. A transition to more national control of an increasingly closed economy has led to policies directed at the internal development of the nation (A. Hopkins 1973; D. Smith 1984). This development is likely to have profound effects on future urban growth. Given Lagos' rapid growth during the initial efforts of ISI, increasingly export-led industrialization is likely to result in an enhancement of the growth potential of other cities with regional production advantages (D. Smith 1984). To some extent, this is already happening with the development of cities in the less urbanized oil-rich eastern region of the country (Kirk-Greene and Rimmer 1981). Furthermore, the willingness of the state to implement national policies directed explicitly towards balanced growth could lead to a more even distribution of economic opportunities between cities and could maintain their generative nature for balanced economic development. The continued ascendancy of Nigeria into the semi-periphery would be likely to reinforce already existing counter-primacy forces.

What accounts for Nigeria's large and persistent intra-urban inequality - an aspect of its urbanization distinguishing it from South Korea and Taiwan? Although peripheral status is often associated with increased primacy, heightened inequality, and large informal sector economies, this does not mean that ascendancy into the semi-periphery automatically equates with a diminution of all of these processes. Indeed, some writers have argued that economic growth in the semi-periphery has often resulted in greater, rather than reduced, inequality (Barnet and Muller 1974; Chirot 1977).

The effect of economic growth on the level of intra-urban inequality is, at least partially, accounted for by the role informal sectors play in the national and international economy. By providing non-core nations with a large, cheap labour force, the informal sector subsidizes formal capitalist enterprises, and allows these countries to be more competitive in the production of goods in the world market (Portes 1983). Interestingly, Portes finds significant differences in the size and development of informal sectors among the semi-peripheral countries of Latin America. He attributes these dissimilarities to historical differences that have influenced the development of the state, the nature of class conflict, and more particularly, strategies for controlling cheap labour. Portes suggests that urban informal sectors might actually decline in those countries where authoritarian regimes succeed in keeping wages low by direct repression of labour movements (Portes 1983).

Similar to other upwardly mobile countries that rely heavily on aid and alliances with core powers, South Korea and Taiwan have developed strong authoritarian states which have consistently and openly repressed labour (Chase-Dunn and Robinson 1977; Cumings 1984). Perhaps the relative absence of urban informal sectors in these countries is a result of their states' ability to use direct political repression to keep wages down. Nigeria, on the other hand, has a much weaker state which has been fragmented by ethnic and regional divisions. The British colonial legacy left a government committed, at least formally, to political pluralism and electoral democracy. These factors combined to make the Nigerian state both less willing and less able to resort to open repression of labour to maintain depressed wage rates. Because of this, it is not likely that the counter-primacy forces associated with semi-peripheral status and an export-led industrialization will lead to a lessening of intra-urban inequality and the disappearance of informal sector activities in Nigeria.

Having used three contemporary examples to illustrate the distinct nature of semi-peripheral urbanization, we now switch our analysis to a historical comparison of two regional urban patterns found within an emerging semi-peripheral country during the seventeenth and eighteenth centuries. We will be paying particularly close attention to the roles each of these regions played in the world economy, and the effects this had on their respective urban development.

A HISTORICAL PERSPECTIVE ON SEMI-PERIPHERAL URBANIZATION

A hallmark of an analytically useful concept in comparative social science is its ability to highlight surprising similarities between seemingly disparate cases, as well as to discriminate between those which are not so obviously different. We have argued that urbanization in the contemporary semi-periphery (with emphasis on the upwardly mobile parts) is different in pattern and process from city growth in the periphery (or, for that matter, in other strata in the world economy). Basically, this illustrates the plausibility of a distinct trajectory of urbanization in countries playing this role in the present global division of labour.

However, both the international political economy approach, in general, and research on dependent cities, in particular, have emphasized the need to study change historically. Can the idea of semi-peripheral urbanization be fruitfully applied to other cases of dependent development? Are there similarities in general patterns? Can we locate and describe analogous processes and mechanisms of urban growth? Answering this type of question should allow us to deepen and delimit the analogy of the semi-peripheral city/urban system (Stinchcombe 1979).

Despite the historical slant of research on dependent cities, it has largely ignored one type of urbanization: differential patterns of city growth in colonial regions that later became parts of advanced core nations. A recent analysis of urban dynamics in colonial/early federal North America attempts to fill this lacuna in the literature (D. Smith, 1987). A basic premise of this argument is that New England and the Middle Colonies and the northern cities were beginning to assume the function of an upwardly mobile semi-periphery in the late eighteenth century, while the South and its ports were undergoing peripheralization (see also Chase-Dunn 1982).

For present purposes, it is not necessary to detail the causes of this pattern of differential development (see D. Smith, 1987, for a detailed discussion). However, it is important to demonstrate that the northern colonies were, in fact, semi-peripheral and did exhibit a distinct urban dynamic.

A.G. Frank puts forward the argument succinctly:

> The Northern-eastern colonies came to occupy a position in the expanding world mercantile capitalist system and the process of capital accumulation which permitted them to share in the latter as a sub-metropolis of Western Europe... This privileged position - not shared by others in the New World - must be considered as contributing crucially to the economic development of the Northeast during colonial time... and in the associated capital accumulation and concentration in northern cities.
>
> (Frank 1978: 61)

This privileged position of the North was based on trade and commerce. Northern ports, particularly the bustling entrepôts of Boston, New York, and Philadelphia, were becoming major mercantile centres engaged in a great variety of shipping to and from Europe, Great Britain, and the other New World colonies (Nettels 1952; Shepherd and Walton 1972). A quantitative analysis of the trade patterns of these ports in the years just prior to the American Revolution shows a large volume of intra-coastal trade, commerce with continental Europe, and much less two-way trade dependence on Britain than that exhibited by the colonies which became part of the US South (D. Smith, 1987). Data on the movement of commodities in the Atlantic economy show a large inflow of value into New England and the Middle Colonies. Shepherd and Walton (1972: 130-6) argue that these goods were paid through invisible earnings that accrued to the North, and particularly the region's port cities, from 'interest, insurance, and mercantile profits' made possible by the area's growing share of Atlantic shipbuilding and commercial co-ordination. Wallerstein claims that even before this period 'the northern colonists came to be competitors of English producers as shipbuilders, ship conveyors, and suppliers of provisions to the West Indies and

Europe' (Wallerstein 1980: 238). In short, the northern colonies were increasingly taking on core-like commercial productive activities which resulted in economic profiles which were distinctly semi-peripheral or semi-metropolitan (Chase-Dunn 1982).

The result was a social structure and political economy very different from the other New World colonies; one that would be conducive to 'relatively autonomous capitalist development in the nineteenth century' (Frank 1978: 22). Merchants and businessmen dominated urban life in the northern ports (see Baltzell 1980; Pease and Pease 1985). Historians also point to qualitative and quantitative evidence of a 'sturdy middle class of small proprietors and artisans' (McKelvey 1969: 17) in cities like Boston, New York, and Philadelphia (see also Bridenbaugh 1938, 1955; VerSteeg, 1975; R. Walsh 1970). The class configuration of these semi-peripheral centres contrasted sharply with that of port cities in primary-product exporting areas undergoing more thorough peripheralization. Colonial and early federal Charleston, South Carolina - dominated by a planter aristocracy, a large slave population, and a small, politically weak middle class - provides a clear contrapositive case.

What effect did this distinct semi-metropolitan class structure have on city form and function and urban growth? It has implications for each of the three key characteristics of theoretical importance: intra-urban inequality, the generative versus parasitic nature of cities, and the expansion and balance in the urban system. The effects of the semi-peripheral city's class structure on internal inequality are rather self-evident. While there was a good deal of poverty alongside wealth in the large colonial towns of New England and the Middle Colonies (see, for example, Pease and Pease 1985, on Boston's poor), it was qualitatively different from the proscribed economic and political inequality of the ports of the slave colonies.

The peripheral cities where urban slavery was pervasive also provide a sharp contrast to urban areas of the rising Northeast semi-periphery, when the cities' broad economic role in society is considered. Urban slavery in places like Charleston subsidized the conspicuous consumption and genteel life-styles of the planter elite (D. Smith, 1987). Free from the tight mercantile dependence on Great Britain, and actually encouraged by the mid-eighteenth century Navigation Acts to develop a level of economic self-sufficiency (Nettels 1952), the entrepreneurs and financiers of the northern ports amassed fortunes by investing in innovative industrial and commercial enterprises. Businessmen and merchants became both economically and politically powerful and pushed for pro-development policies (Pease and Pease 1985). The role of state policies in helping to create a climate conducive to business and commerce in many of these cities has been noted by others (Lunday 1984). By the middle of the eighteenth century, these cities were clearly growing loci for trading houses and 'factors', and closely conformed to the image of the generative city.

This push for growth extended beyond the city boundaries. Duncan and Lieberson (1970) emphasize the key role which urban business interests in the various northern ports played in competing for urban hinterlands for their own cities. Civic boosterism led to the extension of canals and railroads westward. Interior cities were founded by men closely tied to particular Atlantic cities with an eye towards capturing a greater share of the trade generated by continental colonialism. The result was the rise of an integrated network of large urban centres, centred on the Northeast and Midwest in the mid to late nineteenth century. This developing national system of cities is very different from the large primate headlinks in peripheral areas (Chase-Dunn 1984). Once again, the case of peripheral Charleston is instructive. Evidence suggests that the city's planter aristocracy provided scant support, and even a measure of antipathy, to efforts to build a railroad connecting the city to new cotton-producing areas in the West (D. Smith, 1987). Peripheralization and underdevelopment surely contributed to the almost complete lack of any real system of cities in the US South until the early twentieth century.

CONCLUSIONS AND CAVEATS

In the comparative analyses above, we have tried to indicate the distinct nature of semi-peripheral urbanization and, at least partially, to account for it. In conclusion, we would like to review briefly some of the more important points which can be gleaned from this study, and to suggest some of the limits of the analogy of dependent urbanization and the generalizability and practical implication of a pattern of semi-peripheral urbanization.

The argument advanced in this chapter is that the roles countries play in the capitalist world economy have very definite implications for the type of urban development they experience. Although the pattern of parasitic peripheral urbanization would appear to be ubiquitous in today's Third World, contemporary and historical cases examined above suggest that perhaps it is not universal. Indeed, our analysis has identified major aspects of semi-peripheral urbanization which deviate from what would be expected of non-core countries. Of the four cases examined, all had a much more even city-size distribution than their surrounding regions and peripheral countries in general. Furthermore, three of the areas had relatively low intra-urban inequality as well. The reasons for these deviations are not to be found in cultural or geographical variations, but rather in the political economy of semi-peripheral areas. In each of these areas the mix of economic activities moves towards increasing levels of core-like production, often located in dynamically growing cities. Local capitalist and urban-based elites are actively involved in this process and adopt a 'developmentalist' political stance, often advocating policies favourable to more growth-oriented balanced urbanization. In the contemporary Third World case, in particular, the political-economic aspect becomes most obvious as local business and government officials use the state to actively plan and direct industrialization and development through regulatory agencies and state enterprises. But the importance of the relatively autonomous, active, capitalist class is just as evident in the development of New England. In fact, while the fledgling US nation state remained weak in the early nineteenth century, local government at the municipal and state level were pursuing the type of proactive development policies characteristic of semi-peripheral governments in the twentieth century (Lunday 1984, cited by Bornschier and Chase-Dunn 1985: 49). In all the cases of semi-peripheral urbanization, government policies toward tariff and trade restrictions, infrastructural developments (especially regarding transportation/ communications systems), and the location of key industries, have opened the possibility of more balanced urban and regional economic and demographic growth.

The success of South Korea, Taiwan, and Nigeria in creating and maintaining a relatively balanced pattern of urban and economic growth appears to hold out promise for the rest of the Third World. However, while the Asian cases in particular are often seen as models to be emulated, there are limits to generalizability. The historical context and timing of their incorporation into the world economy, while not unique, limits the possibility of their balanced growth being reproduced in contemporary peripheral societies. Past cases, like the rise of the semi-peripheral United States, occurred in a much less tightly integrated world-system. Modern Third World societies have limited opportunities for the type of colonial expansionism which the North American frontier offered during the late eighteenth/early nineteenth century. The ascendancy of our three contemporary examples of dependent development and semi-peripheral urbanization also relies on historical antecedents which are not easily replicated.

Likewise, can the patterns of urbanization exhibited by these four examples be generalized to the entire semi-periphery? In a recent paper, Wallerstein asked much the same question:

> Are there significant differences among semi-peripheral zones? Obviously, in concrete detail there always are. But are the processes fundamentally different in Southern Europe from, say, Brazil and Mexico? And are there significant differences between states that are gliding 'downward' towards semi-peripherality (e.g. Great Britain) and those that are gliding 'upward' however measured?
>
> (Wallerstein 1985)

Obviously, there are many significant differences between the urban dynamics of semi-peripheral countries experiencing opposing forms of mobility. Some of the more obvious of these differences are their origins, historical contexts and nature of development (colonial vs. colonized), state strength in promoting balanced growth and development, the presence and extent of informal sector economies, and the entrenchment of powerful class factions. These and other distinctions provide ample warning not to extend the analogy of semi-peripheral urbanization willy-nilly to all countries occupying this status. Indeed, even among the three contemporary cases examined, we found important distinctions. Although there are concrete and important differences, it is interesting to note that there seem to be similarities as well. For example, a cursory examination of demographic data indicates that the five

Southern European countries often considered to be in the semi-periphery, all have low primacy index scores. A similar reading of data for semi-peripheral Latin American countries reveals that they too have relatively even city-size distributions. Perhaps a detailed comparative analysis of the urban structures of Southern European countries would reveal even more likenesses among them and other downwardly mobile countries of the semi-periphery. Indeed, given the importance of mobility in the world economy for the creation and relocation of capital, it is possible that this upward/downward dimension serves as an axis for two distinct urban patterns within the semi-periphery.

Finally, given the balanced and rapid growth of the contemporary cases examined, are we to expect that they are on the same path towards development that led the colonial United States to core status in the world economy? If the number of countries making the transition from the semi-periphery to core during this century is any indication (i.e. only Japan and the Soviet Union), the chances are not very good. Upwardly mobile countries like South Korea, Taiwan, and Nigeria are still experiencing, in varying degrees, dependent development. In other words, each is dependent upon core powers to provide the capital, technology, and markets that make their development possible. Although each of the three countries' growth is likely to exceed greatly the norm for the rest of the Third World, their leap to core status is unlikely to occur in the near future.

NOTES

1. The premise that there are fundamental differences in peripheral and semi-peripheral urbanization is also apparent, we believe, in Latin America. For example, a cursory examination of population data for 1980 reveals that of the Latin American countries which often are considered semi-peripheral, none has a four-city primacy ratio of greater than 1.75 (United Nations 1986). Ratios of this magnitude are substantially below levels that are considered primate, and are comparable to the semi-peripheral countries of Southern Europe.

2. Japan, the lone core country of East Asia, exhibits a third distinct urban pattern and huge socialist China represents a fourth. For a fuller description of the

configuration of its urban development, see D. Smith (1985).

3. The recent world oil glut and the dramatic decline in world petroleum prices obviously place Nigerian economic progress and upward mobility in the world system in grave doubt, and highlight the precarious nature of dependent development. The optimistic prognosis offered on this page assumes a recovery in oil prices and the continuity of development programmes underway in the late 1970s and early 1980s.

Chapter Two

'Anteroom to a madhouse': economic growth and urban development in Barcelona in the Franco era

John Naylon

'la comarca ha anat esdevenint, cada vegada mes, l'antesala del manicomi' ('the region has become, more and more, the anteroom to a madhouse') - Lluis Cassasas i Simo, 1977

INDUSTRIALIZATION AND URBAN GROWTH

Spain's emergence from the semi-periphery

The thirty-six years of General Franco's authoritarian rule, from 1939 to 1975, saw Spain's belated transformation from a rural-agricultural into an urban-industrial society. From the Stabilization Plan of 1959 until the onset of the economic recession of the mid-1970s (coinciding with the death of the dictator), Spain was one of the fastest-growing of the newly industrializing nations. In the decade of the 1960s (the years of the Spanish 'economic miracle') its economic growth rate was the third highest in the world after Brazil and Japan; between 1960 and 1972, gross industrial output increased thirteen times, a performance surpassed only by Japan among the OECD nations. By 1973, Spain had become the fifth ranking industrial nation in Western Europe and the twelfth in the world in the value of its gross national product.

This late industrialization can be correlated with a particular phase in the evolution of Spain's political economy. In the period prior to the end of the ancien regime at the end of the eighteenth century, Spain, along with the other Southern European countries, had lapsed from core to periphery status, politically and economically, as the Mediterranean had been superseded by the North Atlantic core. Throughout the nineteenth century, while Northwest Europe industrialized, Spain played the role of one of today's Third World countries, touched by the Industrial Revolution in only a handful of regions (Catalonia, the Basque Country, and Asturias) and was mostly exploited as a source of raw materials by the entrepreneurs of the European core. The political environment of this economic marginalization was constitutional turmoil and a failure to complete the democratization process. Oligarchic control by Spain's elites culminated, towards the end of the century, in the cynicism of turnismo - the ritual alternation in office of liberals and conservatives - while the inability of the system to admit the proletariat, or even the radical middle class, resulted in a perpetual state of tension and violence.

The twentieth century saw the appearance of dictatorial regimes (Primo de Rivera 1923-31; Franco 1939-75) which aimed to bring order out of political chaos, and at the same time to terminate Spain's economic dependency and to 'modernize' the nation. The latter would be accomplished by state capitalism in association with big business, both domestic and foreign. Spain would be 'developed' and 'modernized' but at the same time the privileges of the established elites would be preserved. This phase in the country's political and economic maturation corresponds to its emergence from the semi-periphery and its full incorporation, from the 1950s onwards, into the world economy. It is the arugment of this study that the super-urbanization which accompanied Spain's late industrialization not only reflects the core-periphery and cumulative causation processes which can be expected in the early stages of economic development, but also the structures, institutions, and attitudes which characterize the authoritarian stage of political evolution.

Spatial imbalances in development

Spain's transformation - in the structures of the economy and society; in the relationships between town and country; in the speed of communications; in new forms of industry, services, and ways of working the land; in new human and industrial location factors; in the importance, power and direction of banking, credit, insurance, and investment - took place in the absence of any effective or committed urban or regional planning. In the open Spanish economy, with free rein given to market forces, funds flowed from the periphery to a handful of established centres, particularly Madrid, Barcelona, and Bilbao. The advice of the World Bank in 1963, that Spain should forego the goal of greater regional equality and seek instead that of maximum economic growth, found ready acceptance in the philosophy and allegiances of the Franco regime which, unable to claim political legitimacy or popular support, sought prestige and justification in industrial achievement. It also suited Spanish big business, foreign investors, and the country's seven largest banks, which handle over 70 per cent of all financial credit and directly control 50 per cent of Spanish industry (Naylon 1987).

Well-intentioned but inadequate regional development programmes, based on hydraulic works, irrigation, land settlement, land consolidation, and industrial growth poles, failed to diminish spatial disparities, especially between the countryside and the town. Nor would an authoritarian centralist regime such as that of Franco have wished for a really effective regional policy, since this might have constructed take-off platforms for regional autonomist movements, the traditional bete noire of Madrid (Higueras Arnal 1980).

Urbanization processes

The populations of Spain's cities exploded in line with industrialization. The most visible physical manifestation of Spanish socio-economic modernization since the 1950s is the South American-scale mushrooming of the major urban centres (Gormsen 1984). This urban explosion can be seen as the product of first, Spain's incorporation into the world economy; and second, national policies pursued by a non-elected government allied to big business and able to disregard the problems, complaints, and aspirations of the less dynamic regions.

In contrast to the cities of long-industrialized North-west Europe, which are now at the stage of suburbanization and even counter-urbanization, Spanish cities have - up to the onset of the current economic recession - still been at the stage of urban development characterized by growth and centralization (thus, the 'inner city problem' of Northwest Europe has not yet affected Spain or the Mediterranean countries in general) (Gaspar 1984; Wynn 1984b). Cohen (1981) postulates world capitalism operating through a constellation of metropolises and their satellites, penetrating peripheral and semi-peripheral countries through a relatively small number of large cities, which play the dominant role in determining the patterns of spatially uneven development and the geographical transfer of value.

In the 1950s and 1960s, Madrid, Barcelona, and Bilbao became part of the global process of economic expansion. Industrialization, initially via import substitution, and later via export orientation, accumulated capital and concentrated economic power in these cities, which became responsible for structuring space for the control of production, distribution, and consumption, and for the reproduction of labour (Fiala and Kamens 1986; Harvey 1981). At the same time, urban growth itself generates profound social transformations through innovation and opportunity. Urban living conditions and types of employment mirror changes in the economy. City populations adopt global tastes in food, dress, and the built environment. A consumer society is created which resembles those in the already industrialized, highly urbanized countries of the world (Borja and Serra 1984).

Spain has a long history of indigenous urban development and a marked geographical division of labour; but the post-1959 'economic miracle' altered the nature of its urban centres, channelling the profits generated on domestic and overseas markets through the urban system and adding manufacturing to the traditional administrative, commercial, and communications functions (Precedo Ledo 1976). Import-substitution industrialization requires access to management resources; a substantial, skilled, and disciplined labour force; adequate infrastructures and services, including transport facilities; and a large, concentrated market - requisites provided by the primate cities of Madrid, Barcelona, and Bilbao (Harvey 1985; Timberlake 1985). The classical metropolitan dominance model, with its emphasis upon economies of agglomeration and urbanization, and upon risk avoidance, fits the Spanish experience well. 'Modern, urban industry killed off rural manufacturing employment, and the government's aping of French industrial growth-pole practice did nothing to ameliorate this polarization.

At the same time, failure to achieve meaningful or effective agrarian reform, or rural or regional development, produced a comparative impoverishment of the countryside and a diaspora of those with initiative and marketable skills to the expanding opportunities - in industry and in commercial and service activities, including tourism - in the cities (Ferrer and Precedo 1981). Rural landowners and other country folk with capital invested it in urban real estate and speculation, while the diffusion of 'modernization' processes and images to all regions sucked the peasantry out of their 'backward' villages and small towns and into a Northern-European life-style in the provincial and regional capitals (Lasuen 1974). Between 1960 and 1973, 4.2 million Spaniards left their homes, mostly in rural areas, to live and work abroad or in Madrid, Barcelona, or Bilbao. These three cities became the nerve centres of Spanish economic growth: they contained the nation's three stock exchanges and the headquarters of 75 out of the 100 biggest Spanish industrial firms; Spain's financial oligarchy and industrial bourgeoisie resided there; and there the decisions were taken about investment and expansion (Lasuen and Racionero 1971). By the mid-1970s one in every four Spaniards was living in the provinces of Madrid or Barcelona (Table 2.1) (Leira et al. 1976). To dwell in the city, with its amenities and ambiente, became synonymous with the transition from traditional to modern living. It was in the city that the opportunity was presented for the first time, to the majority of Spaniards, to ingress into the middle class through office employment in expanding government activities, industry or commerce. Hence, perhaps the most striking contrast within this nation of contrasts: the gulf between Spain's highly sophisticated cities and the primitive rural world of its small towns and villages.

Table 2.1 Spain: population growth in the main urban areas

	1950	1970	% change	National rate of change
Madrid	1,984,033	3,950,686	99.13	21.58
Basque coast	1,133,238	1,992,833	75.85	21.58
Barcelona	2,510,382	3,827,988	52.49	21.58

Centre-periphery and urban growth in Catalonia

The spatial disequilibrium which characterized Spain's socio-economic transformation are to be found nowhere more strikingly than in Catalonia; the nation's richest region. At the macro-economic level, the boom of the 1960s and early 1970s would appear to have confirmed the relative affluence and economic dominance of Catalonia over the rest of Spain (Sarda Dexeus et al. 1984). With only 6.3 per cent of the national territory, Catalonia contains 15 per cent of the Spanish population and enjoys nearly one-quarter of the national income; it accounts for over 25 per cent of Spain's net industrial output and provides 22 per cent of all exports. Catalan-based banks hold more than 19 per cent of all Spain's bank deposits and 23 per cent of industrial bank deposits. This prosperity was a powerful attraction for workers from less-developed parts of Spain such as Murcia and Andalusia; between 1950 and 1974 the population of Catalonia rose from 2 million to 5.5 million (Giner 1985). However, these generalities conceal the fact that during the 'economic miracle' years of 1964-73, over 70 per cent of the industrial investment in Catalonia was placed in the sub-regions of the Barcelones (Llarch et al. 1987), Maresme, Baix Llobregat, Valles Occidental, and Valles Oriental, whose centre is Barcelona (Figure 2.1); and that while 75 per cent of all new Catalan industrial employment in this period was created in and around Barcelona, 28 out of the remaining Catalan sub-regions received only 7.3 per cent of total industrial investment and no more than 10.4 per cent of new industrial jobs (Cassasas i Simo 1977; Rodriguez and D'Alos Moner 1978). In some parts of the Catalan periphery industrial employment hardly exists.

Demographically, too, Catalonia suffered an extreme form of macrocephaly: a shrivelled body supporting a bloated head (Vidal Bendito et al. 1980). Two-thirds of Catalonia lost population; among the rural comarcas, for instance, the Pallars Sobira had 68.6 per cent fewer inhabitants in 1975 than in 1857 (the date of the first modern Spanish census); other severe cases of decline were the Priorat (56.5 per cent), the Vall d'Aran (45.0 per cent), the Pallars Jussa (43.1 per cent), and the Conca de Barbera (39.0 per cent) (Miro et al. 1974). By contrast, the province, and more specifically the Metropolitan Area, of Barcelona attracted by far the greater part of population growth. In 1975, the Baix Llobregat showed a population increase of

Figure 2.1 The Catalonian sub-region

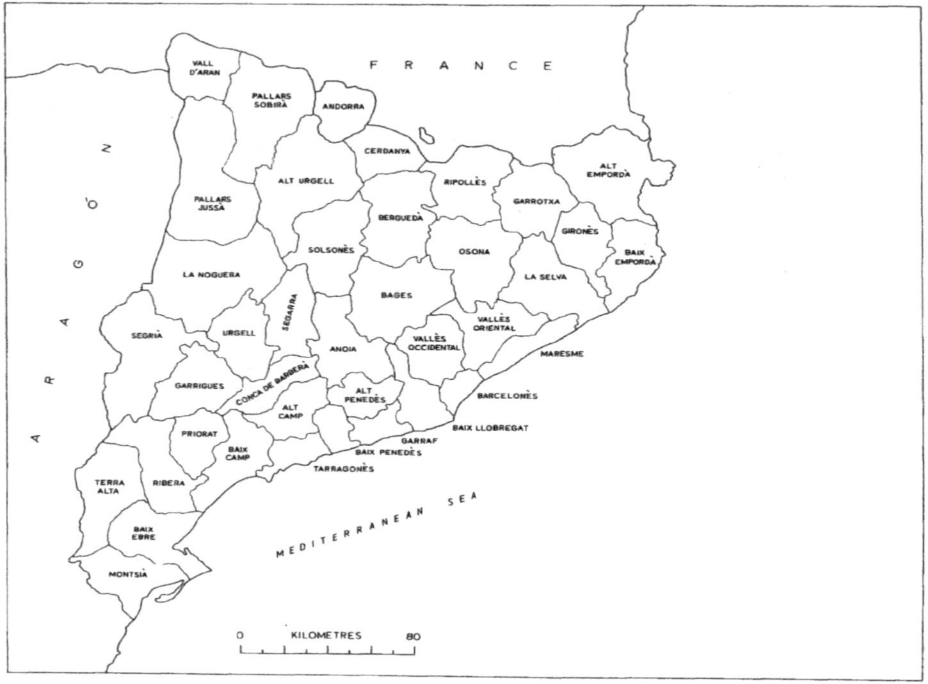

998.7 per cent over the 1857 census, the Valles Occidental 923.7 per cent, the Barcelones 909.9, the Valles Oriental 331.0, and the Maresme 221.2. In 1974, the province of Barcelona contained 4.25 million out of Catalonia's 5.5 million inhabitants or 77.3 per cent (Sabater 1977). Some parts of the Metropolitan Area had arrived at the almost complete saturation of living and working space; the population density in the Barcelones was 14,746 persons per square kilometre while the average density for Catalonia as a whole was only 32.3, falling in some parts of the periphery to as low as 10.6 (Pallars Jussa) and 4.3 (Pallars Sobira). The uncontrolled growth of Barcelona impoverished and debilitated most other towns and villages in Catalonia, as well as many elsewhere in Spain.

THE POLITICAL ENVIRONMENT OF URBAN GROWTH

The pressures and conflicts to be seen in Spain's regions and cities during the 'economic miracle' were not simply those likely to emerge in any nation experiencing very rapid industrialization and socio-economic change. The classic features of centre-periphery, cumulative causation, and metropolitan dominance were exaggerated in Spain's case by the short space of time over which economic growth took place and by the authoritarian nature of the regime in power from the end of the Spanish Civil War in 1939 to the death of General Franco in 1975. Spain had the misfortune to undergo its most rapid urban growth during a period of right-wing dictatorial rule, so that the distortions, crudities, and conflicts which would anyway have resulted from the uneven regional development process were exacerbated by the failure of the state to step in to cushion the impacts - for example, by redistributing income regionally and sectorally. Instead, the central power exhibited insensitivity and disregard for the needs of the working class generally and of migrants in particular, and made a serious situation worse by its prejudices, alliances, and favours, so that urban growth reached chaotic proportions (Teran 1978).

The interests of centralism

The regime which emerged victorious from the Civil War had among its prime political objectives the imposition of centralized authority upon Spain's varied regions and peoples, and the eradication of regional separatism and autonomy. This obsession had a direct effect upon urban planning in Catalonia. During the 1920s and 1930s there had been an imaginative effort to plan the future growth of Barcelona within the context of Catalonia as a whole. The idea, clearly inspired by Ebeneezer Howard, was first proposed in 1920 by the Societat Civica La Ciutat Jardi. During the 1920s, the Mancomunitat de Catalunya drew up a zonal scheme for Catalonia, based on regional resources and ignoring the official provincial divisions of 1833; but action had to wait until the Second Republic was established in 1931 and conferred a degree of autonomy on the Catalans.

In 1936, the newly constituted Generalitat de Catalunya (the autonomous regional government) promulgated a decree on the 'Territorial Division of Catalonia'; and in 1937 a conference was held on the 'industrial exploitation of the natural resources of Catalonia'. The tenor of these projects was a strong emphasis on assessing the resources of the whole of the region. The particular problems of Barcelona were recognized, but always with a view to preventing the excessive growth of the city, and in terms of an evolutionary Catalunya Ciutat (Catalonia City) involving the planning of the region as a single unit.

The Civil War abruptly truncated this line of development: the Generalitat was abolished, Catalonia ceased to exist as a separate entity, and its four provinces were once more administered from Madrid in the same way as the other forty-six in Spain. Academic and professional institutions were closed down or survived only clandestinely; regional planning criteria were forgotten; and there was a return to the 1833 provincial division and administration of the country, better suited in the eyes of the new regime in Madrid to the purposes of centralized control and repression. Planning perspectives in Catalonia shifted from a view of the region as a planning unity to the view that the motor of urban and industrial expansion should be Barcelona. Given the hostile attitude of the Franco government towards the potentially autonomist regions of Spain, to divorce the planning of the Barcelona conurbation from that of Catalonia as a whole was a foreseeable tactic. From this time onwards, planning for the growth of Barcelona would always be subordinated to the interests of the Madrid government and its political and business allies.

Barcelona in the Franco era

The shortcomings of planning legislation

In the aftermath of the Civil War, the interests of the national government were concentrated upon the targets of industrial development and economic autarchy, as a reaction against Spain's backward agrarian condition and 'colonial' dependency status. Partly for ideological reasons and partly to achieve economic self-sufficiency, the regime also attempted to maintain the population on the land and to inhibit internal migration. The essential conflict between efforts to promote economic growth, on the one hand, and negative attitudes towards migration and urban expansion, on the other, appears to have been ignored. There was no national urban planning policy and hardly any urban planning institutions. The Direccion General de Arquitectura, set up in 1939, was the only body with any competence in the urban planning field until 1949, when the Jefatura Central de Urbanismo was created; while the strong centralist structure of the administration frustrated efforts at local level. A number of Comisiones Superiores de Ordenacion Urbana and Planes de Ordenacion came into being, but always at provincial level and charged with rural as well as urban planning. There was no recognition given to the fact that Spain's urban problems called for special attention. Up to 1953, the official development plan for Barcelona was still the Cerda Plan of 1860.

It was not until the mid-1950s that the central government broke away from its inhibitions about internal migration and city growth, which were already uncontrollably under way. The arrival of the first United States' aid ended Spain's long post-war isolation and brought with it brighter economic prospects, improved transport facilities, and greatly increased migratory movements. Belatedly, the government created an institutional and legal superstructure of urban planning, and a series of housing programmes, which, under a more democratically responsible regime, might have beneficially influenced the growth of Barcelona and other large Spanish cities during the following two decades.

The most important measure was the 1956 Land and Urban Planning Act, which well illustrates what was to become a familiar type of legislation, being excellent in its conception and stated objectives, but of negligible practical application (Lasuen 1972). The Act introduced a five-level planning system of local plans, municipal development plans, sub-regional plans, provincial plans, and national plans, and set out the procedural steps for plan approval. At the same time, it proposed the creation in each municipality of a 'patrimony' of publically owned land for development purposes. In practice, although the Act formed the basis of Spanish urban planning up to the late 1970s (it was reformed in 1976), like other public sector measures during the Franco period, its impact was derisory and it proved completely unable to cure the major problem of speculation.

Outside a few main urban areas, very few plans were actually drawn up, although it was obligatory under the Act for every municipality to present its development plan. The majority of town councils proved incapable of producing plans because of their lack of initiative, experience, and technical knowledge; a concerted land-use plan in conurbations such as Barcelona (where independent municipalities often had conflicting interests) proved particularly elusive. The Act was aborted by the lack of technical teams and specialized administrative organizations, inadequate finance, an absence of inter-mininsterial co-ordination, inactivity on the part of those institutions which did exist, and by the breakdown of communication between the planning technocrats and the executive (the former proposing objectives which proved to be unrealistic or politically inconvenient in the eyes of the latter). Slow, inefficient, and hierarchical administrations delayed physical planning processes, so that impatient investors were encouraged to ignore procedural steps and even entire plans. At all stages, planning considerations took second place to those of economic growth and speculative profit (Teran 1982a, 1982b).

Five plans for Barcelona

The Catalan capital illustrates the climate of planning disorder particularly well. From 1953 onwards, five different plans were formulated to regulate the expansion of the metropolis; sometimes these were co-existant, often they were contradictory, and all were largely failures (Grupo 2c 1972).

Following the plans for Greater Paris (1939), Greater London (1940), and Greater Madrid (1946), in 1953 the Plan de Ordenacion de Barcelona y su Zona de Influencia defined a 'Greater Barcelona' within which urban development would be regulated (Borja et al. 1972). Commonly known as the plan comarcal, or sub-regional plan, the scheme embraced twenty-seven neighbouring municipalities and was to last almost fifty years, up to the year 2000 (Borja 1973). It was followed in 1959 by the Barcelona provincial plan and in 1966-8 by plans for a new administrative region, the Barcelona Metropolitan Area, comprising 162 municipalities with a combined area of 3,206 square kilometres (approximately half the province of Barcelona and 10.3 per cent of the area of Catalonia) and containing over 3.5 million people (almost 70 per cent of the total Catalan population). This time, the Barcelona conurbation was to be decentralized by accommodating overspill population, industry, and tertiary activities in inland growth centres. Considerable effort was put into the theoretical organization of the Metropolitan Area. Bureaucratic delays and conflicts, however, held up the publication of the new plan until 1972, by which time it was evident that the Metropolitan Area could only achieve decentralization if it were endowed with authority on a par with other administrative organizations, and were given its own adequate resources and its own political identity with legislative and decision-making powers. These it did not have and could not hope to have under the Franco government, as events soon proved.

In August 1974, during the summer recess and without prior consultation, the central government's Council of Ministers created a new Metropolitan Corporation of Barcelona which, the following year, superseded the Metropolitan Area of Barcelona and put the planning clock back twenty years. Ostensibly, the Decree-Law 5/1975 created an effective and officially recognized executive organization which responded to the spreading popular demand for reform of the 1953 sub-regional plan. In fact, from the 162 municipalities of the Barcelona Metropolitan Area, the planning area of the new Metropolitan Corporation was reduced to the twenty-seven municipalities of the 1953 plan - an area totally inadequate to meet urban development needs.

The true objective of Decree-Law 5/1975 was to forestall the setting up of an administrative unit (the Barcelona Metropolitan Area) which, by virtue of its size and socio-economic potential, might serve as a platform for the expression of Catalan regionalism. The Metropolitan Area would have overshadowed the national capital in population, economic power, and wealth; to the Madrid of General Franco, the creation of such a power base outside the direct control of the political nerve-centre was anathema. The Barcelona conurbation finally saw the end of the dictatorship and entered the democratic transition and the world economic recession with no development perspective more ambitious than a reform (published in 1976 as the General Metropolitan Development Plan) of the 1953 plan comarcal, which was already demonstrating its inadequacy by 1960 (Romaguera and Dot 1972).

Housing: a real-life game of Monopoly

It was in the fundamental and all-important sphere of housing that the Franco regime's priorities and lack of concern were best revealed (Wynn 1984a). The post-Civil War, 1940s house-building programme of the central government largely ignored 'Republican' Barcelona and addressed itself mainly to Madrid, to the 'Nationalist' cities of central and southern Spain, and to war-devastated towns and villages. Barcelona struggled, unaided, to absorb, house, and employ the rising tide of immigrants which flowed (despite government policies) from Spain's rural areas. At the end of the Civil War, Barcelona's housing shortfall was aproximately 20,000 dwellings: with the arrival of over 100,000 immigrants during the decade of the 1940s the shortage increased to 80,000 - yet up to 1950 only 15,000 new dwellings were built and many of these were for middle- and upper-middle-class residents. By the end of the 1940s, an estimated 26,000 people were living in squatter settlements on the slopes of the Montjuich and the Tres Turons hills; by 1954 this figure had more than doubled.

During the boom years of the 1950s and 1960s, one in seven of the Spanish population moved residence from one part of Spain to another, only to find an alarming lack of housing in the receptor cities. Yet a separate Ministry of Housing, given sole responsibility for urban planning decisions, was not created until 1957. This was followed in 1959 by the setting up of the Gerencia de Urbanizacion (after 1972 renamed the Instituto Nacional de Urbanizacion). An examination of the regime's housing legislation of the 1950s and early 1960s reveals not only a very tardy response to the urban housing crisis, but also a willingness to countenance the manipulation of the legislation for the untrammelled benefit of the private building sector (King 1971).

The earliest legislation (the 1954 Law of Limited Rent

Housing and the National Housing Plan of 1955) launched modest and inadequate programmes of state-owned housing for low-income families; but most of the millions of dwellings built during the 'economic miracle' (and which have given Spain the highest proportion of population in Europe living in high-rise blocks) were provided under the National Housing Plan of 1961-76, which placed the greater part of house-building in the hands of the private sector. The plan more than achieved its objective of erecting 4 million new dwellings by making half the target the responsibility of private builders, on their own terms. The other half of the housing programme, state-subsidized, was called Viviendas de Proteccion Oficial (VPO), or Officially Protected Housing; the subsidies took the form either of subsidized rates of interest or of concessionary loans. Some were developed by the state via local authorities or through the state housing agency (the interest rate of which was 5 per cent, repayable over 25 years), but most were developed by the private sector according to government guidelines. The interest rates on this latter type of state-subsidized housing were higher, and over shorter repayment periods (10-15 years); but it was this kind of housing which made up the greater part of the state programme.

The 1961-76 Housing Plan did not, however, solve the problem of homelessness, principally for two reasons. First of all, the scale of internal migration was far greater than was officially calculated. By the second half of the 1970s there were still 35,000 squatter dwellings on the outskirts of Madrid; even in the 1980s Spain still needs to build 250,000-310,000 dwellings per year if it is to eliminate homelessness by the 1990s. But the main shortcomings of the plan were flaws in the legislation and the emphasis on official stimuli to the private sector. This meant that property developers, operating in a free market, constantly tried to go up-market, building for the better-off strata of society. Moreover, all the housing acts of the 1950s and 1960s made it possible to sell and sublet state-subsidized housing.

During the 1960s (the main period of construction), there was a wholesale shift towards sale and away from renting by the public and private sectors alike; probably less than one-quarter of the total housing stock was for rent. Consequently, most housing, even state-subsidized housing (probably as much as 65-70 per cent of VPO), went not to the poorest but to the better-off, because of unrealistic selling prices and a failure to institute a means test until the early 1980s. This situation explains the high proportion of lodgers in Spanish cities - not only the result of pressure on dwelling space, but also a necessary means of helping to repay a mortgage; in the mid-1960s one-fifth of all working-class families in Barcelona was living in someone else's (occupied) apartment (Candel 1964). There was also, naturally, a political significance in the situation: workers' minds were concentrated on paying their mortages rather than going on strike. Housing policy was part of the embourgeoisement of Spanish society - creating a petit bourgeois, property-owning nation with a stake in the prosperity and stability of the regime.

Planning and reality

At the technical level, Spain in general and Barcelona in particular have kept abreast of developments in planning theory. Spanish technocrats have consistently produced excellent plans to solve the nation's spatial problems. Most of the schemes for Barcelona's urban development contained admirable provisions: to avoid speculation; to prevent Barcelona becoming one immense sprawl between the Besos and the Llobregat; to foster an orderly and balanced growth; to channel industry and population to other urban nuclei. New spatial administrative frameworks were to be created. The conurbation was to be zoned into as many as thirty-seven land-use classes, with a given area in each municipality to be allotted to each function. Population and building densities were to be reduced by intra-provincial communications, improved traffic circulation, stipulations about building densities and other norms, and decongestion into zones of preferential development. There was even a deep social commitment in some of the plans to subject private interests to the common good and to endeavour to achieve the maximum well-being and effective functioning of the area as a unit of residence, work, consumption, and services. Concern was expressed for the quality of urban living; for the provision of sufficient recreational space, infrastructures, and services; for sociological aspects of city life, including the integration of immigrants into Catalan society; and for the protection of the environment and the cultural and artistic patrimony. The reality was quite different.

Barcelona in the Franco era

Non-democratic institutions

A widely applicable lesson of Barcelona's practical experience is the irrelevance of sophisticated projects when local authorities are not democratically representative and when they lack the funds and the human resources to translate proposals into practice. The way in which the Barcelona conurbation grew during the Franco dictatorship bore little relation to the various plans formulated from 1953 onwards, and their irrelevance is explained by the nature of the institutions responsible for administering them.

Between 1936 and June 1977, Spain held no democratic central government elections, and it was not until April 1979 that the country held its first full municipal elections since 1931. In the authoritarian Francoist state central government bodies held sway in the urban planning commissions; planning teams, composed of technocrats of the central administration, were often in conflict with the municipalities; municipal mayors and presidents of provincial councils were nominated by the Home Office (Ministerio de Gobernacion); town councillors were elected by a very limited suffrage; and there was no popular participation in municipal affairs. The principle of hierarchy and tutelage built into the centralized Spanish politico-administrative system left local authorities with only residual powers over those issues which concerned them most, and even here they could be overruled by state bodies. Local authorities, for the generality of their functions, came under the Home Office, but for the purposes of urban development and planning they came under the Ministry of Housing - adding to the insecurity and lack of initiative which characterized them. Some of the most important elements in urban planning - for instance, the primary road network- were (and are) directly under the control of the central government and the provincial councils; as a consequence, notably less attention was paid to Barcelona's transport network than to that of Madrid.

Fragmentation of responsibilities

There was a conscious irony in the evident intention of the Madrid government to divide and rule the Barcelona conurbation by maintaining the fiction of the administrative and juridical independence of the municipalities which it comprised. Ostensibly to avoid 'gigantism', but in reality from motives of central government political control, none of the urban development plans for Barcelona annexed the municipalities adjacent to the city but simply defined a 'zone of influence' which would allow the city to export its problems of urban growth but not oblige it to share the burden of solving them. The vested interests of Barcelona city inevitably prevailed at the expense of the other corporations; at the same time, the actual execution of the plans' directives and norms was left in the hands of those same corporations who went ahead, in so far as they could, with their own partial plans. However, given the inadequacy of funds, equipment, technical know-how, and coercive powers to carry out the tasks of public works building, expropriations, communal service provision, and the like, the plans were inoperable from the start. Sometimes the plans were in direct contradiction; thus, whereas the 1953 plan comarcal aimed to potentialize Greater Barcelona, the 1959 provincial plan placed emphasis on the development of the rest of the province, and thus could not count on the collaboration of the city.

The plans were generally indicative only, in the sense of placing no obligations upon the public sector. The really big infrastructural elements depended on the goodwill of the central administration, which instead showed lack of interest and delayed the execution of such projects as the arterial road network, the Sabadell-Terrassa motorway, and the Tibidabo road tunnels, all of which were the responsibility of the Ministry of Public Works and vital to the spatial decentralization which was the main objective of the plans.

Notably missing was an effective, legal, administrative body with adequate powers of decision, capable of co-ordinating all the common services of the relevant municipalities and of integrating all planning steps into a general plan with executive force for the whole region. When population and industrial growth were concentrated mainly in Barcelona city, the situation could be handled by the single municipal authority which, up to 1936, had in its hands powers of initiative, control, administration, and execution of urban planning. After the Civil War, when growth began to spill over adjacent municipal boundaries, control became extraordinarily complicated. Not only was responsibility for urban planning fragmented between municipalities, with each authority sovereign (albeit only in theory) in its own territory, but the plurality of bodies involved, and the disparity of their powers, turned planning into a conflict of interests. Within the Metropolitan Area, for example, functions were divided between the Comision de Urbanismo y Servicios Comunes de Barcelona y Otros Municipios, responsible for the twenty-seven municipalities of the 1953 sub-regional plan, and the Comision Provincial de Urbanismo, embracing the remaining municipalities of Barcelona province and also those of the Metropolitan Area which were not included in the 1953 plan; both these bodies had equal rank, but depended on the Direccion General de Urbanismo in Madrid. Similarly, in the housing sphere, central state organizations such as the Syndical Housing Authority (attached to the Ministry of Labour) and the National Housing Institute (attached to the Ministry of Housing) conducted their operations alongside the local public housing authorities and private foundations.

Financial dependency and bankruptcy

The financial situation of Spanish cities warrants special mention. In a rational socio-political situation, local authorities invigilate and mediate the quality of urban life. One of the major constraints upon Spanish urban authorities performing this function has been their state of virtual bankruptcy. Financially, as in almost every other respect, Spain's municipalities have depended upon Madrid (Medhurst 1973a); yet as urban populations have mushroomed and town councils have been called upon to provide an ever wider range of services, municipal budgets have dwindled. In 1930, Spanish local authorities accounted for 23.6 per cent of all central government spending; by 1970 this proportion had shrunk to 18.3 per cent, and by 1979 to 9 per cent. The corresponding figure for Italy in 1979 was 32 per cent and for the United States almost 45 per cent. Any absolute increase in the cities' budgets was more than swallowed by by population increases.

In the early 1970s, it was estimated that the arrival of each new immigrant in Barcelona called for an investment of not less than 400,000 pesetas (about £2,580) in adequate service provision; but in 1970 the city budget of 5,420 million pesetas worked out at only 3,111 pesetas per inhabitant (little more than £20). Collectively, Spain's town halls reached the end of 1979 with a combined debt of some 400 billion pesetas (£2.9 billion). Barcelona was the worst affected, with debts in excess of 60 billion pesetas (£437 million). In these circumstances, the authorities could not fulfil their normal responsibilities. They could barely cope with roads, paving, drainage, lighting, and similar basic necessities, while anything which was not absolutely urgent or within their most specific brief - housing, education, health, social assistance - was relegated to second place and could be covered only by extraordinary budgets, by borrowing from the Banco de Credito Local (which had limited funds and could not fully meet Barcelona's needs because this would have left nothing for the other municipalities), by public debt issues, by international loans, or by the private sector. During the Franco period the situation was not helped by the size of the municipal bureaucracy, its thirty-six years of unchecked graft and administrative incompetence, and the spending of scarce resources on the city centre, the better-off residential districts, and costly projects such as the Tibidabo tunnels, costed at 27.3 billion pesetas (£202 million) in 1967 and abandoned in the early 1970s.

Central control of public finance and tax revenue has traditionally been extremely strong in Spain (up to the advent of the post-Franco regional autonomies) (Medhurst 1973b). In the last years of the Franco regime, 92.6 per cent of tax income went on public spending, the allocation of which was decided in Madrid; 49.1 per cent of town council resources and 67.0 per cent of provincial council resources were allotted by the central administration, which furthermore retained a degree of control over the way this money was spent.

Spain's fiscal system in general has remained anachronistic in comparison with the country's level of socio-economic development. While the public spending sector of most advanced OECD countries equals 30-40 per cent of their GNP, in Spain in the early 1970s it amounted to only 20 per cent. Similarly, although progressive income tax and other flexible automatic taxes provide 40 per cent of government income in the United Kingdom, in Spain they provided less than 1 per cent. The Spanish state's historic inability, or unwillingness, to raise revenue by taxing the rich has an obvious relevance to the provision of public services. Spain's antiquated taxation system has reflected the enduring inequalities of her society, favouring the well-to-do at the expense of the poor. Spain has traditionally relied on indirect taxes and on taxes on production: indirect taxes levied on goods and services (and thus falling on all consumers regardless of their incomes) have been twice as important as such direct taxes as income tax. Personal income tax and corporate tax evasion has been widespread and notorious. Even in 1986, after democratic government reforms, fiscal receipts were still only 35 per cent of GDP compared with the OECD average of 45 per cent, and direct taxation still fell heaviest on wage-earners rather than on non-wage-earners such as landowners, businessmen, and self-employed professionals.

Spain's 'economic miracle' and its accompanying tides of migrants created enormous needs for housing, education, health facilities, welfare services, and infrastructures of all kinds; but the state's incapacity to raise finance, plus its inefficiency and venality, produced an increasing poverty of public services and amenities, and an off-loading of the burden of providing them onto private individuals and enterprises. In order to exercise their devolved powers, Spain's local authorities needed to be able to impose local taxes, augment their budgets from central government subsidies, and enjoy a greater share in the income raised by the central government: these reforms could not, however, be implemented until after 1975.

Manipulation of plans

The incoherence of urban growth in the Franco period was thus based upon the division of powers between many bodies, their lack of co-ordination, the weakness of local authority budgets, and a technical inability to tackle an often desperate situation. In these circumstances, town councils simply had to cede important functions to private initiative. The tide of immigrants arriving in the conurbations found neither infrastructures nor housing programmes to cater for its needs. The demand for dwellings was met by private improvisation, for example by the sale of parcels of land and by the construction of tower blocks. The guiding force behind such schemes was the desire for maximum returns in the minimum of time. The private sector was not only given state subsidies and a free hand; little effort was made to curb its illegal activities.

There were several loopholes in the 1956 Land and Urban Planning Act, the most serious being the absence of precise rules about the modification (through the medium of local plans) of already approved municipal and sub-regional plans (Montero 1972). Local authority impotence (but more often corruption) in the town councils, together with collusion with private economic interests, meant that this vehicle for reclassifying land could easily be exploited by speculators.

Residential and industrial classifications were extended at the expense of green zones and collective amenity areas (of the land envisaged for recreational purposes in the 1953 Barcelona plan comarcal, 1,290 hectares [46 per cent] of the total had been lost by the early 1970s). Local plans would alter sub-regional plan residential classifications to allow higher building densities. The majority of local plans, especially in the 1960s, were thus used to reclassify and revalue land which otherwise was not sufficiently profit-generating for the private entrepreneur.

Local authorities, meanwhile, would vacillate between applying controls or ignoring the issue; commonly, they validated illegalities a posteriori. Through their passivity, they failed to control the worst aspects of urban growth; for instance, they failed to fix land prices or to ensure that development companies installed basic infrastructures before constructing dwellings (as the law demanded). Frequently, local authorities only intervened and provided services after many years of pressure and complaint from citizens.

Private benefit and popular deprivation

Within the general climate of inconsistency created by constant changes in plans, the jungle of urban legislation at municipal level was manipulated - or ignored - by economically powerful or politically influential groups (Capel 1974). A weak official planning policy, coinciding with enormous population pressure (by 1970 some municipalities in the Barcelona region had already exceeded their population projections for the year 2000) and a corresponding demand for building land and housing, could not help but produce a speculative fever characterized by rocketing land prices, building anarchy, and galloping deficits in infrastructures and services.

The aggressive behaviour of landowners, development companies, building firms, banks, finance houses, and other private interests bore upon unelected town councils from the very beginning and accumulated wealth in favoured hands (Capel 1975). In the Barcelona sub-region, the real price of land increased by an average of 6.1 per cent per year from 1951 to 1978 (or 40 times over the period as a whole). Between 1950 and 1963, land values in Barcelona rose by over 900 per cent, compared to a general cost of living rise of 94 per cent. In the speculative building boom of 1954-72, landowners' incomes increased by 3,500 per cent (Roch and Guerra 1981). It was essentially the absence of democratic institutions and popular involvement which enabled private developers, in collusion with public authorities, to dominate and warp urban growth during the Franco period.

SPATIAL INEQUALITIES IN BARCELONA

In-migration and urban sprawl

Any map of Barcelona (see Fig. 2.2) reflects graphically the contrasts in urban morphology between the tortuous passage-ways of the medieval city, the grid pattern of the Ensanche (Cerda's nineteenth-century egalitarian vision) and the haphazard growth of the contemporary periphery (Bohigas 1963; Galera et al. 1973; Wynn 1979b). Under the Franco government, Barcelona was the second focus of attraction for migrants after Madrid. The 1970 census revealed that of the population of the Barcelona municipality, 53.7 per cent were born either in the city itself or in Barcelona province, while 43.8 per cent were first-generation immigrants from the rest of Spain, including the other three Catalan provinces. The influx of migrants, the exhaustion of building space, and the consequent prohibitive price of land, caused an outward spread of population and industry to a ring of adjacent municipalities (Fig. 2.3).

L'Hospitalet became Catalonia's second most populous municipality after Barcelona, and Badalona the fourth. Put another way, while Barcelona increased its population six times between 1860 and 1965, Sant Adria de Besos increased 68 times, L'Hospitalet 54 times and Santa Coloma de Gramanet 43 times (Carreras and Margalef 1977).

Under such pressure it was not possible to plan metropolitan growth properly. The ring of municipalities around Barcelona was ill-prepared to become the receptacle for immigrants and industries seeking cheap accommodation. Many became no more than working-class

Figure 2.2 Barcelona and adjacent municipalities

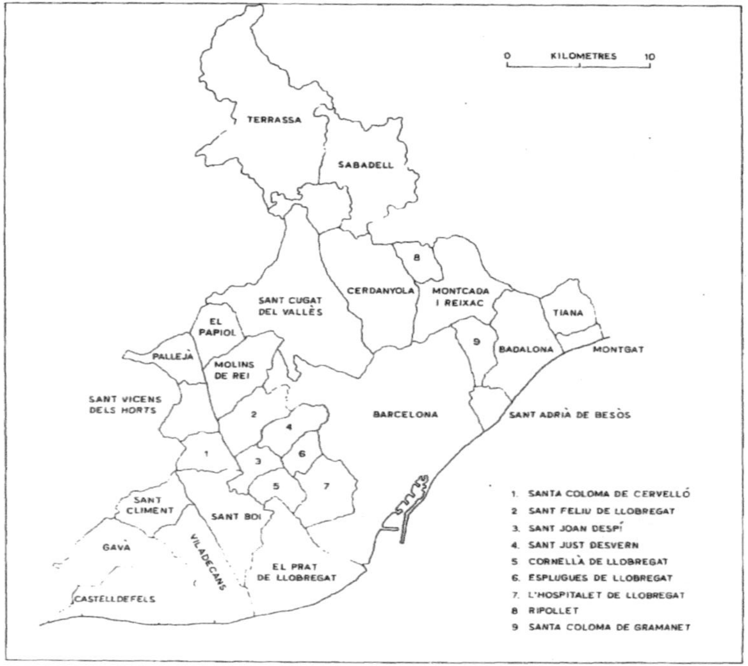

Figure 2.3 Barcelona: population growth in adjacent municipalities 1920-75

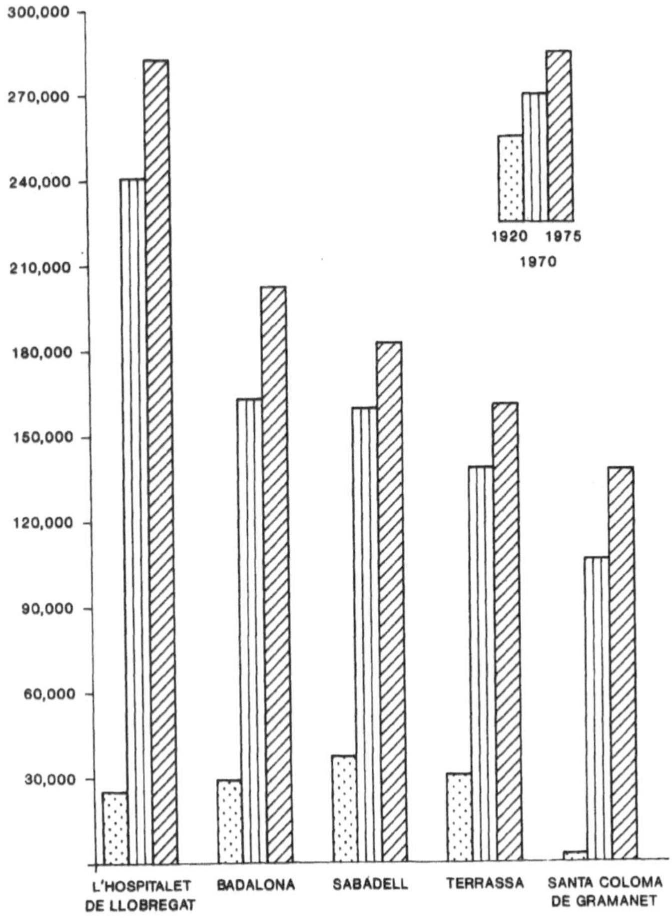

dormitories, social ghettoes, indiscriminate piles of housing and factories rather than planned components of a coherent city. Their proletarianization meant that their budgets declined in inverse proportion to their growth. Public penury and private speculation produced a degraded urban sprawl between the Besos and the Llobregat, spilling over into the Valles and threatening to spread to a radius of 50 kilometres around Barcelona (Wynn 1979a; Ferras 1977).

Spatial segregation

Under a political regime favouring private capital and renouncing any mediating role, access to urban resources, utilities, and amenities became a matter of affluent and powerful socio-economic groups satisfying their needs at the expense of the migrants and the poor (Lowder 1980). The lack of precision in local and partial plans, the connivance of planning authorities with business interests, the flouting of land-use zoning, the legalization of lucrative speculative developments (sometimes after payment of derisory fines), resulted in the bidding-up of central land values and the expulsion of working-class housing to land without infrastructures on the city margins.

The most accessible zones were developed for owner-occupiers - an inevitable product of a system in which developers operated on a basis of short-term loans and required substantial initial down-payments from prospective occupiers to finance construction. Short-term mortgage facilities also effectively excluded the lower-paid. Even groups benefiting from the 'economic miracle', such as skilled industrial workers, white-collar office employees, and civil servants, had to struggle to afford their 'key money' and subsequent payments: they had to be satisfied with dwellings with only minimum services (Serrahima and Marcos 1970). Quite moderately priced flats were often bought up by speculators for immediate subletting (Sala 1977).

While luxury apartments were going up in south-western districts of Barcelona such as Pedralbes, from which the highest profit margins could be obtained, working-class settlements were being built on the fringes of the city without what is generally accepted in Europe to be the minimum provision of infrastructure (Rodriguez and Sala 1976). In 1972 it was estimated that 225,000 families were seeking accommodation, yet the cheapest flats on the market were outside the reach of 64 per cent of Barcelona's households. While 40,000 luxury apartments remained unsold, living conditions on the periphery were deteriorating until they were almost beyond reform.

Squatter settlements and substandard housing

The most obvious nuclei of urban poverty, reminiscent of conditions in Third World cities, were the barracas. These occupied the wasteland and the slopes of hills, and lined the railway tracks and other public property, especially on the northern and eastern margins of Barcelona (COAB 1971). This phenomenon reached its peak in the 1950s and 1960s (in 1957 there were 52,377 people living in shanties on the slopes of the Montjuich hill alone); but the problem refused to go away entirely. The 4,041 barracas (containing some 22,000 people) which still existed in 1972 did not include those in the municipalities adjacent to Barcelona proper, nor the self-constructed, substandard dwellings which made up the greater part of housing in certain districts. Commonly, the barracas had neither electric light nor running water; an average unit of 30 square metres housed one or two families. It is a measure of the shortfall between housing demand and supply that shanty dwellers were often ordinary workers or self-employed people who either could not find, or could not afford, a better dwelling.

The housing problem of Barcelona and its satellites went beyond barraquismo and took the form of what Candel (1964) called 'barraques verticals'- substandard tenements and tower blocks in both the old centre of the city and in the new working-class districts. Overcrowding and squalor are still widely evident in the historic heart of Barcelona: the casco antiguo and the port area of the Barceloneta. The pre-Civil War Macia Plan had proposed a remodelling by partial demolition and the implanting of green spaces and communal services. However, improving conditions in the old city would have been a highly complex and costly operation in which neither the town council nor private enterprise wished to get involved during the Franco period, when there were so many other pressures and opportunities on the fringes of the city. The result was that a barely habitable 3 per cent of the city's dwellings housed 9 per cent of Barcelona's total population, a density of 531 persons per hectare.

The most lamentable feature of Barcelona's housing situation, however, was the anarchic peripheral growth produced by the race to house immigrants in the quickest and most profitable fashion. On the northern and eastern fringes of the city, especially, the order and coherence of the Ensanche (Cerda's nineteenth-century chequer-board) come to an end and are replaced by a confusion of high tenement blocks, self-built houses, shanties, and high-tension lines, erected without reference to any plan and often without licence (Wynn 1981). In the process of land parcelization from the 1950s onwards, each landowner decided what he would build on his land. A common way for development companies to acquire parcels of building land was by offering owners of smallholdings or old houses one or two apartments in the block to be erected: this explains the appearance of many high, narrow buildings with idiosyncratic shapes. Common problems were (and still are): irrational and truncated street patterns; lack of paving; absence of bus routes or journeys of as much as two kilometres to a bus stop (public transport services are most deficient in the peripheral districts where car ownership is lowest); no running water, so that entire estates depend on public hydrants or road tankers; irregular rubbish collection, or none at all, so that refuse accumulates on wasteland and building sites; inadequate sanitation services, clinics, hospitals, schools, recreational space, and even shops. Building densities too go well beyond acceptable limits; for example, in the district of Montbau there are 508 people per hectare; in Canaletes 527; in southwest Besos 612; and in La Mina 709 (Carreras 1980).

This chaotic process was repeated time and again, and not only by the private developer. Many public sector institutions did little more than fill the gaps left by private developers and had little effect on the property market or on building standards. At the end of 1970, the housing deficit in the Barcelona Metropolitan Area was still 136,053. The state bodies themselves contravened the planning regulations. Thus, in the housing estates of Verdun and Trinidad, built by the Syndical Housing Authority, building densities were respectively 3.7 and 3.9 times greater than those laid down in the sub-regional plan. There was, in fact, often a close correlation between the estates built by the Syndical Housing Authority (Obra Sindical del Hogar) and the worst instances of hurried construction, unskilled labour, bad workmanship, poor building materials, inadequate services, serious structural defects, and premature deterioration (Wynn 1980).

A notorious case is the Besos district, developed from 1960 onwards by the Patronato Municipal de la Vivienda de Barcelona to house people displaced from the shanties of Montjuich and elsewhere, and by public works schemes. Other developments have grown up around this nucleus - the La Pau estate built by the Syndical Housing Authority; tenement blocks built by the COBASA company and others along the Maresme highway - until the zone now houses some 83,000 people. The development might have been a showpiece for public housing policy; instead it was an arithmetic exercise in accommodating the maximum number of people in the smallest possible space. There was no provision for schools, hospitals, or recreational spaces; in the case of the La Pau estate, sixteen-storey blocks of flats were built without lifts.

In this unregulated urban market, better housing land was priced beyond the means of even the state housing institutions, which were themselves displaced to peripheral locations, often of very low environmental quality: good examples are again provided by the Syndical Housing Authority's subsidized working-class housing at La Mina (on reclaimed gravel workings) and Can Clos (adjacent to a municipal refuse tip). The peripheral location of working-class housing placed large populations of migrants on the margins of access to the urban amenities and employment which had lured them to the city in the first place. Workers faced long-distance travel to their factories, for example from San Adria de Besos in the north-east of Barcelona to the Free Port Zone in the south-west, while the most affluent citizens lived within walking distance of the central business district.

Privatization of public services

Logically, the immigrants required not only jobs and housing but also education, health provision, transport, and recreational opportunities; but with inadequate central and local government funds the unsatisfied demand for services had to be met by the private sector. But large segments of the population, inevitably the neediest, could not afford private services. The private sector, substituting for what would normally be public responsibilities, tends to mask the inadequacies of government. However, it is utopian to imagine that the free market system, of its own volition, is going to satisfy all popular needs. The private sector seeks profits and provides only those services which are easiest to introduce and which offer the greatest returns, such as schools and nurseries. Naturally, it shows no interest in other facilities which are no less desirable, such as parks and gardens, and it avoids the poorer districts or appears there in only a cheap and shoddy way. During Spain's 'economic miracle' the emphasis was on the rapid expansion of production, with a minimum of investment in nonproductive infrastructures. The inevitable result was serious shortfalls in social capital. Sometimes even basic services such as electricity supplies, water, drainage, road surfacing, and flood control had to be laid on and paid for by despairing residents themselves.

The lack of open space and even open air; the level of traffic pollution; the deplorable state of drainage, sewerage, and refuse disposal; the lack of supervision of food markets - all called for adequate health and sanitary provision. Yet Barcelona in 1970 had a deficit of 100,000 hospital beds; in some working-class districts there was only one dispensary for over 20,000 people; and in the whole Metropolitan Area there were only two welfare hostels. The World Health Organization considers an adequate hospital provision to be 10 beds per 1,000 inhabitants; in the 1970s the index in Barcelona was 5, and in the satellite towns of Badalona and L'Hospitalet only 1.4 and 0.3 respectively. The high rate of private medical provision (63 per cent of hospital beds) did not make up the deficiency.

Even if viewed only from the economic standpoint, the failure of the authorities to build enough schools, and the large numbers of children dropping out of the education system, boded ill for the future competitive position of Barcelona and Catalonia, face-to-face with new technologies and commercial rivalries within the European Community. In 1970 there was a deficit of 170,964 school places in the city; in District II (the Free Port area) only 40 per cent of children in the obligatory school attendance age group 6 to 13 actually had school places. Seventy per cent of education in Barcelona was in private establishments, often deficient in both equipment and quality of staff; in L'Hospitalet, an immigrant receptor municipality par excellence, the private sector provided 82.8 per cent of schools and accommodated 76.1 per cent of pupils.

In 1859, Cerda included 380 hectares of parks in his plan for the Ensanche; of these, the 1953 sub-regional plan retained 129 hectares; by the 1970s, 2 hectares of Cerda's parks had actually been created. Barcelona city has a theoretical ratio of 13.5 square metres of open space per inhabitant, considerably less than the majority of West European cities, and this limited area is very unevenly distributed. The highest ratio of open space per person, in line with other indices of urban well-being, is to be found in the West Residential District, whereas the working-class area of Poble Nou has only 0.3 square metres of open space per inhabitant. The same can be said of the satellite municipalities such as L'Hospitalet, Badalona, and Santa Coloma de Gramanet; part of the cost of their rapid growth has been the impoverishment of the natural environment and the irreparable loss of recreational space. The Tres Turons local plan illustrates the process at work (Tarrago 1971). The 1953 sub-regional plan extended and combined into one large urban park the two small existing Guell and Guinardo parks on the slopes of the Tres Turons hills. However, not only did the hills become an area of shanty settlement in the 1950s and 1960s, but the Barcelona Council's Tres Turons local plan of 1967 allowed landowners to build seven- to eleven-storey residential blocks on their south-facing slopes. The new Pelada Urban Park, opened in 1978, accordingly contained 15,000 people in dwellings ranging from high-rise flats to squatters' huts.

Industrialization at random

Industrial location is the key to urban planning and population decentralization, but Greater Barcelona did not possess a general plan for industrial estates during the Franco period. The Ministry of Housing, responsible for many of Spain's industrial estates, was conspicuous by its absence in the nation's most important industrial region. Until 1970, only one industrial estate, a mere 17 hectares in size, had been built by the Gerencia de Urbanizacion in the whole Metropolitan Area - a region in which 30 per cent of Spain's total population growth had taken place in the previous decade. Industrial location was reduced to a proliferation of estates promoted by individual municipalities; these often remained half-occupied because the communications network continued to focus upon Barcelona, and thus constituted burdens rather than assets to the town councils and local communities. Industrialists were left to their own devices to improvise roads, drainage, sewerage, and water supply, and they in turn have left the local authorities to struggle with the legacy. In the early 1970s, according to the Comision de Urbanismo, 50 per cent of the land zoned for industry in the 1953 sub-region plan had no services laid on; in the rest of Barcelona province the proportion rose to 80 per cent.

Neighbourhood Associations: a popular reaction to authoritarian mismanagement

The contradictions, distortions, and failures of the 1950s, 1960s, and 1970s produced a widespread scepticism about the ability of the government or the planners to exert any effective control over the growth of cities like Barcelona. Public campaigns against specific developments and against the general state of the metropolises were almost certain to be ignored if not physically repressed. The very gigantism of Barcelona and Madrid made it difficult to cultivate a collective spirit of protest, and there were no community organizations to speak for the working class and the immigrants. In the 1960s, however, the stresses generated by the disorderly growth of Spain's rapidly growing capitalist cities provoked the mobilization of the working class and the new petite bourgeoisie into one of the most active urban social movements in Europe - the Asociaciones de Vecinos (AAVV) (Neighbourhood or Residents' Associations) (Centre d'Estudis d'Urbanisme 1976). Especially strong in Barcelona (Puig 1974) and Madrid, this type of urban struggle focused upon such issues as the poor quality of house construction; unfair tenancy agreements and rents; and inadequate or non-existent infrastructures, basic equipment, collective services, and other amenities (Borja 1977a).

Although the campaigns and confrontations of the Asociaciones de Vecinos were primarily directed against the local authorities, state housing agencies, and development companies, they also played an important national political role in bringing about the crisis of Francoism in the first half of the 1970s. In the absence of legal opposition political parties and trade unions, they were often the only forcible democratic voice to challenge authoritarian, non-representative government (clandestine left-wing parties were frequently well represented in the AAVV) (Tarrago 1976). By exposing institutional inadequacies, deficiencies in planning and services, speculation, and corruption, they were able to flex and exercise working-class collective strength and reveal the inability of the rigid political system to recognize and satisfy legitimate aspirations.

The Asociaciones de Vecinos especially gained strength and effectiveness immediately after Franco's death in 1975, intervening successfully in the planning and decision-making processes and putting pressure on officialdom and private interests to improve urban living conditions nearer to the needs and desires of the majority (Borja 1977b). With the advent of true democracy in Spain in the late 1970s, this social movement went into decline, superseded by free trade unions and political parties, by the coming to power of socialist governments more responsive to social demands at central, regional, and municipal levels, and by the decentralization of decision making. But the AAVV, born out of the pangs of Spain's emergence from the semi-periphery, was a significant force for reform and played its part in speeding the country's transition to the Western European democratic core (Bariatua 1977).

CONCLUSIONS

Growth versus development

The expansion of Barcelona and its planning experiences illustrate the choice to be made by newly industrializing countries between economic growth (signifying an increase in production considerably greater than the increase in population, but without taking into account the distribution of that production and its benefits, nor how the social costs of rapid national economic growth are borne by different sectors of the population) and socio-economic development (in which the increase in production is accompanied by social, economic, and cultural changes which improve living conditions for the population as a whole).

Spatial planning and growth takes place in the context of conflicts between political groups, social classes, and vested interests (Blowers et al. 1982). From the particular Spanish political climate of 1939-75 stemmed several conflicts which enormously increased the social costs ordinarily to be expected from metropolitan growth: conflicts between technocrats and planners on the one hand, and political and other interest groups on the other; between a highly centralized government and local preoccupations; between a non-democratic regime and popular aspirations; between powerful associationist combinations (landowners, property developers, building companies, banks, finance houses, local politicians) and an urban proletariat trying to defend and improve its position - in brief, between those who gained from and those who paid the price of growth. Barcelona is an example of what can happen when urban growth is left to the free or unregulated play of market forces, with an authoritarian regime pursuing purely economic targets and private entrepreneurs amassing profits without regard to social effects. In the absence of a central government interested in the well-being of all its people and all its regions, the capitalist concentration of investment, labour, and production in areas 'useful' to big business produced an integration of space but a disintegration of society (Dear and Scott 1981).

The promise of democratization

Even before the death of General Franco in 1975, the Neighbourhood Associations and the Circulo de Economia of Barcelona (a private group of academics, economists, and industrialists) had campaigned against the urban chaos of the city and its periphery. Immediately following the end of the dictatorship, the municipal authorities themselves, especially the Metropolitan Corporation of Barcelona, rebelled against the burdens imposed by uncontrolled growth and central government constraints. As early as 1976, a new general metropolitan development plan revising (yet again) the 1953 sub-regional plan set about controlling peripheral sprawl and making private developers adhere to planning procedures.

The combination of Catalan autonomy and democratic municipal elections has provided local and regional public authorities with the will and the instruments effectively to direct urban growth, and has allowed the populace and its elected representatives to become involved in the preparation and implementation of plans. In 1976, new state housing policies introduced fresh administrative and technical regulations, and a Land Law Reform Act (revising the 1956 Act) has provided a new legal, administrative, and planning framework, including 'special plans' for remodelling existing built-up areas through decongestion schemes, new urban and community services, and improved traffic circulation and environmental conditions. Under this legislation, the central authorities (Ministry of Housing, Ministry of Public Works and Urbanism), the local authorities, and the Neighbourhood Associations have co-operated to remodel and re-equip housing estates and shanty areas. The active participation of residents themselves in the planning process has shown that there is no lack of interest, energy, or skills on their part. Not surprisingly, post-Franco Catalan thinking sees the future development of the whole of Catalonia occurring through a regional framework of the Catalans' own devising, rather than within the old provincial division favoured by the central government. The reconstitution in 1977 of the Generalitat (Catalonia's traditional form of autonomous government), and the restoration of autonomy in 1979 have devolved substantial powers to the Catalans. They are now responsible for their own economic development (including town and country planning), the regulation of agriculture and industry, transport, energy, and public works. It has thus become realistic to think once more of planning Catalonia as a unit - a return to the idea of 'Catalunya Ciutat' envisaged by Pau Vila in the 1920s, with Barcelona as the potentializing centre of a network of cities based on the resources of the different sub-regions. This concept finds favour among the new middle class of administrative, industrial, and service sector cadres, and also among students - especially among the ever more numerous students from the peripheral areas of Catalonia. Such elements do not admit the division of Catalonia into privileged and neglected regions; this is seen as a threat to the newly acknowledged unity of the region. The Statute of Autonomy empowers the Catalans to administer their own social services, education, social security, and cultural activities, and above all, to collect and spend their own taxes and to control local savings banks: none of the devolved powers would amount to much in practice unless they were accompanied by financial decentralization.

Recession and the end of growth?

To the well-intentioned student of Spanish affairs, there is much that is gratifying about the pace and achievements of the country's post-1975 transition to democracy, especially in political and social life. Nevertheless, several problems and questions remain as legacies of the boom period of unbridled urban growth. Even under authoritarian and grossly inequitable conditions, rapid economic growth did produce improvements in living standards for Spain's urban poor; the real problems have begun to emerge now that global economic activity has slowed and faltered, bringing unemployment and falling incomes, public as well as private, to the industrialized Western Europe which Spain has only recently managed to join. The costs of Spanish incorporation into the capitalist world economy are now being paid by those who participated least in the fruits of economic expansion. The very nature of Spanish industrialization up to the early 1970s, the spill-over into the Southern European semi-periphery of already mature smoke-stack industries, has made cities like Barcelona, Madrid, and Bilbao especially susceptible to unemployment. These erstwhile growth centres now exhibit the highest jobless figures in a country with the largest proportion of out-of-work in Europe. A particular reason for their predicament is the importance of the building industry (notoriously sensitive to cyclical fluctuations) as a major urban employer - precisely the product of the phenomena described in this study. At the same time, the economic recession and cuts in private and public investment and fiscal income have drastically reduced the municipalities' budgets for dealing with accumulated urban problems, to which have been added since 1975, rapidly rising crime and drug waves (Gonzalez-Berenquer 1977).

The present recession has also produced a standstill in the out-migration from Spain's rural areas to the cities (indeed, the populations of many rural areas are now so small and aged that they have no more people to contribute). Spanish scholars are already speculating on a process of 'polarization reversal' (Richardson 1980) of economic activity and job creation in favour of smaller cities, such as provincial capitals, which have proved more able to absorb fluctuations in economic fortunes, and which may offer more attractive environments for investment in the future (Fielding 1982). Many people in Barcelona, Madrid, and Bilbao are first-generation immigrants who retain strong links with their home regions and pueblos, and who might be tempted to return there, were they not imprisoned in their new homes by their mortgages and other debts. It is ironic that the sustained economic recession and high metropolitan unemployment may achieve more for the Spanish countryside and small towns than all the years and programmes of Francoist regional planning (Perry at al. 1986).

Has the phenomenon of the Spanish urban explosion ceased as quickly as it began? Or is Spain still at an emergent stage, so that when the current recession is over the forces of metropolitan dominance and agglomeration will once more reassert themselves, accumulating further amenities and services, bureaucratic activities, health and education provision, science and art in cities like Barcelona? Can industry be persuaded to decentralize over a territory of such size and with such difficulties of terrain and communications as Spain? Now Spain is in the European Community, will not the roles of Barcelona and Madrid be reinforced? In 1987, 33 per cent of all foreign investment in Spain went to Catalonia and a further 32 per cent to Madrid.

One concluding comment can be made unequivocally. Thirty-six years of undemocratic rule, coinciding with Spain's emergence from the semi-periphery and socio-economic transformation, caused an unprecedented metamorphosis of the urban fabric of the country and have left a legacy of irreparable damage to the physiognomy of Spain's cities.

Chapter Three

Urbanization at the periphery: reflections on the changing dynamics of housing and employment in Latin American cities

Alan Gilbert

INTRODUCTION

The process and pattern of urbanization in Latin America has been strongly influenced by international forces for more than four centuries. Today, urban living conditions are highly sensitive to the state of the world economy. Since most Latin American economies are heavily dependent on trade with advanced countries, on financial and capital flows from commercial banks and multinational aid agencies, and receive major investments from transnational companies, there is no way in which cities, or indeed much of the countryside, can be isolated from fluctuations in the world economy. Levels of employment in rural and urban areas are strongly influenced by changes in the external demand for Latin America's raw materials and manufactured exports. In addition, the external trade position affects internal levels of demand, which in turn influence living conditions in urban and rural areas.

Currently, much of the literature on urbanization in the periphery is emphasizing the integral links between the world system, national and regional economies, and the urban process (Armstrong and McGee 1985; Timberlake 1985; Drakakis-Smith 1986; Walton 1985a). Seemingly, the so-called 'New International Division of Labour' has transformed production processes in most Third World countries. Agribusiness has cut a swathe through traditional farming systems. Manufacturing activity has been increasingly tied to the international market. Capital flows into and out of the region have been facilitated by ever-increasing links with companies in the developed world. In addition, throughout Latin America, consumer tastes have been manipulated to resemble those in industrialized countries; Coca-Cola consumption, television viewing, and a car-based society are increasingly dominant.

It is the aim of this chapter to examine the links between the new international division of labour and the process of urbanization. I start from the assumption that, like all shifts in academic perspective, recent writing has exaggerated the changes that are occurring in the world system or at least the extent of those changes in particular parts of the globe. While profound transformations are affecting the urban areas of Western Europe and Eastern Asia, it is debatable whether recent processes of economic internationalization have brought similarly dramatic change to Latin America's cities.

In referring to the new international division of labour, I mean the internationalization of production that is often labelled the third new international division of labour. This chapter will explore this proposition by examining Latin American urban experience over the past thirty years, both before and after the new world pattern really began to emerge, and will attempt to answer the following questions. In what respects has the process of urbanization been modified by changes in the international economy? In what respects has urban morphology increasingly approximated to international patterns? To what extent have cities in different countries, or even in the same country, developed in distinctive ways? For example, are there clear similarities in the form of housing provision, in servicing systems, in employment patterns, and in urban living conditions? If there are variations, to what extent are these due to different modes of integration into the world economy or to different national cultural and social conditions?

I will also discuss the effects of the latest crisis affecting Latin America, specifically the recession that has affected most countries during the 1980s. I am less than certain that this recession should be conflated with the new international division of labour because I believe the effects of the two processes within Latin America are very different. This is certainly true of the social effects, with the 1980s crisis arguably undoing two decades of improvement for most groups within Latin American urban society. In order to discuss this process, I will examine trends in urban working and living standards over recent years. It will also be necessary to investigate certain conventional wisdoms about the nature of the 'Latin American crisis'; for example, the contention that urban living conditions have worsened through time. With respect to this contention, it will be necessary to consider whether poor servicing and residential crowding is more widespread, whether the relative expansion of self-help housing necessarily signifies worsening conditions for the poor, and whether there is any temporal trend towards higher levels of unemployment and towards worse immiserization through the 'informal sector'.

A final question relates to whether internationalization has led to Latin Americans losing control over their cities. Are the key decisions on employment, housing, and servicing now made directly or indirectly in Washington, London, Tokyo, and Frankfurt? What, for example, have been the consequences of IMF and World Bank conditionality on internal economic and social policy? Is planning now determined by the wishes of international institutions, catering for the demands of transnational enterprises, or do Latin American regimes maintain control over the critical determinants of urban growth? To what extent do local political considerations determine policy, especially now that the authoritarian blanket has been pulled back in so many countries? And, if Latin American governments truly lack urban autonomy, is their situation any different from that of earlier decades?

Clearly, many of the questions posed in this chapter are far too ambitious to be satisfactorily answered here. What I aim to do is raise, sometimes contentiously, questions that seem to be inadequately considered in the existing literature. I hope this effort will stimulate others to investigate further some of the points that I make.

THE NEW INTERNATIONAL DIVISION OF LABOUR

Before discussing the urban impact of the new international division of labour (NIDL), it is necessary to define with some care what is meant by this ubiquitous and intoxicating phrase. For most writers it signifies a major change in the spatial organization of world production (Palloix 1975; Jenkins 1984; Thrift 1986; Dicken 1985; Johnston and Taylor 1986). While views differ as to precisely when this reorganization began, there is consensus that it is characterized by an internationalization of the production process. In particular, it has been marked by increasing volumes of world trade, growing control over world production and trade by transnational corporations, consequent limits on national autonomy, a shift of production from core countries to parts of the periphery, the internationalization and greater mobility of capital investment, and the internationalization of currencies and producer services. What distinguishes NIDL from what occurred before varies from author to author. To Walton it means that 'footloose international capital and world-wide labour reserves, themselves capable of longer migrations, are now parts of a single system' (Walton 1985a: 4). This means that international capital has been strengthened at the expense of national labour and capital. As Singer puts it, 'not only is an increasing share of the output of all or most countries being exchanged through the world market, but ...more and more productive activities, taking place in several countries, are coming under the control of multinational ...firms' (Singer 1985: 27).

To other writers, it is the speed and unpredictability of change that is so important in NIDL. As Dicken points out: 'Reorganization, rationalization, and spatial change in TNCs are endemic in a competitive global market system' (Dicken 1985: 219). Indeed, Kaplinsky (1984) argues that almost as soon as it has developed, further changes in production technology and import policy are threatening the whole process of export expansion in NICs. Current trends towards protectionism in the developed countries, for example, are making the prospects of export manufacturers ever more difficult.

It is also clear that NIDL has involved much more than a shift in the location of industry (Thrift 1986: 28). To function effectively, decentralizing manufacturing industry has demanded synchronous changes in the world economy. A basic ingredient has been the internationalization of producer services such as accountancy, banking, advertising, hotels, and transport. It has also been necessary to develop an international currency with the concomitant internationalization of domestic currencies. As Schmitz notes:

> A buoyant transnational banking market developed over the 1960s and 1970s, specializing in borrowing and lending of currencies outside the country of issue, commonly known as the 'Eurodollar' market. In the 1960s, the currency supply was fuelled mainly by US balance of payments deficits (caused in particular by massive military and related expenditure abroad during

> the Vietnam War), and in the 1970s by the surpluses of the oil-exporting countries: the private transnational banks became the main conduit for recycling these 'petro dollars'.
>
> (Schmitz 1984: 11)

The result has been floating exchange rates, the creation of psuedo-currencies such as Eurodollars, and the internationalization of banking and capital markets.

A corollary of the above change is that deeper links are developing between national and international capital. Armstrong and McGee (1985) demonstrate how in the regional periphery of Ecuador, one family's economic interests have gradually diversified and spread from agriculture into urban activities and now beyond the region altogether. This example highlights how national and international capitals are becoming more interconnected. It is also clear that national companies are increasingly becoming international. A principal consequence of these changes in world organization is that the state has had to modify its role considerably. It has to negotiate and control the conflicts between national and international elites. As Walton puts it: 'The state is the focal point of this conflict, endeavouring through its own divided constituents to simultaneously abet the process and ameliorate its consequences' (Walton 1985a: 10).

As I have already suggested, the term NIDL is hardly new and several writers have agreed that the current transformation is probably the third major reorganization of the world economy. Paraphrasing Walton, the first reorganization was the incorporation of colonial regions as new material suppliers and grudging consumers for the economies of industrializing European states. The second consisted of industrialization within the underdeveloped periphery at the hands of a national bourgeoisie, or foreign enterprise, or both together. The third is the current process of globally integrated production and the multinational firm. As already intimated, I broadly accept this threefold division and will label the overlapping phases NIDL1, NIDL2, and NIDL3 respectively. Moreover, while accepting that something major has been afoot in the world economy, numerous qualifications and riders need to be added.

First, it is quite clear that certain parts of the world are more fully involved in the new international division of labour than are others. Indeed, most writers accept that

NIDL3's spatial impact has been highly uneven. Corbridge notes that 'the export of capital is still predominantly between the countries of the core, rather than from the high-wage countries to the low-wage countries' (Corbridge 1986: 62). Of course, this has been true of previous NIDLs, but so far at least, it seems probable that NIDL3 is far less embracing than either NIDL1 or NIDL2; indeed NIDL3 arguably excludes most of Africa and large parts of Asia and Latin America. The limitation has been that opportunities for involvement have been created but there is not room for every country to participate. Participation, according to Schmitz, is determined

> partly by location and geo-political significance; partly by the existence of a strong (repressive) internationally reliable regime; and partly by the existence of a technological infrastructure resulting from earlier import-substitution policies. Finally, state control over industrial development is held to be extensive and decisive in bringing about the dynamic growth.
>
> (Schmitz 1984:9)

Second, NIDL3 is still limited in the depth of its penetration and may continue to be limited. As Thrift notes:

> The greater level of integration of the world economy should not be overemphasized. The rise of a new world-economic order does not give overwhelming power to the multinational corporation, it does not give the banks total control, and it does not mean the end of national sovereignty. The internationalization of capital in its broadest sense, is a very messy business.
>
> (Thrift 1986: 62)

This suggests that autonomy for individual Latin American countries is not excluded, indeed the more successful Third World participants in NIDL3 may well be those that enter on their own terms. There are also strong reasons for believing that many countries, particularly those in tropical Africa, will never participate in NIDL3. Kaplinsky (1984), for example, argues that increasing difficulty of market access, radical technical change undermining the feasibility of competition from less-developed countries, the effects of IMF-type conditionality, and the impact on industrial growth of increasing military expenditure, will all limit the level of integration. Specifically he notes that:

> problems of market entry are forcing producers to locate closer to final markets. The costs of doing so are minimized by the introduction of radical electronics-based automation technologies. This bodes ill for export-oriented industrialization strategies in the Third World, in part because the technology is diffusing unevenly across sectors and economic space, and in part because it has the effect of increasing economies of agglomeration.
>
> (Kaplinsky 1984: 84)

Third, there is less than complete agreement about the causes of NIDL3. Some blame the falling rate of profit in the industrialized countries: according to Marxian analysis, an inevitable tendency in capitalism. Others attribute it to rising labour costs and trade union activity in the developed countries, some to changing technology and better communications, some suggest that the OPEC oil surpluses and the development of the Eurocurrency market accelerated 'off-shore' investment, while some blame greater governmental controls on pollution, safety, and working conditions in the First World as major contributory factors. The suspicion has to be that none of these factors has been as important individually as all of them together.

Fourth, there is some confusion among writers about whether the post-1973 world recession is independent of, or an integral cause of, or merely a component of, NIDL3. Certainly, the precise link between the world recession and NIDL3 varies from writer to writer and is usually less than adequately sketched. Often the two are conflated, somewhat artifically, into a single process. Since NIDL3 arose through multiple causes and cannot be attributed to any single stimulant, it would have occurred without the OPEC price rises which stimulated the first and second phases of world recession in 1973 and 1979. It is obvious that NIDL3 has been accelerated by the development of the Eurocurrency market, but the internationalization of production was under way before the world recession had commenced. It is even possible that NIDL3 would have developed faster without the recession. Controls on imports from NICs, for example, might well have been less inclusive had the industrialized countries not experienced the severe unemployment caused by the effects of the world recession.

It is more the world recession than capital restructuring that has brought higher unemployment to most industrialized countries.

NIDL3 AND ITS IMPACT ON LATIN AMERICA

To what extent is the internationalization of production changing Latin American economies and societies? In order to answer that question it is necessary to attempt some quantification of the supposed effects of internationalization. Presumably, whatever else is involved, a rising share of national production dedicated to the international economy is mandatory. This ought to be quantifiable in terms of the share of exports as a proportion of GNP. Table 3.1 shows that over the past two decades there has been a general increase in the proportion of GNP devoted to exports. This is, however, by no means a uniquely Latin American phenomenon, for a similar process has characterized most industrialized countries as well as most middle-income nations. Within Latin America there have obviously been variations in the degree to which foreign trade has increased. Brazil, Chile, and Mexico have increased their shares dramatically, the first two because of the export-oriented strategies, the last principally due to the recent discovery and exploitation of oil.

This increase seems to be a world-wide phenomenon, at least among most high-income and middle-income countries; it is only the low-income African and Asian countries that seem not to have increased their trade over the past twenty years. Certainly, the effect of the increase in world trade on Latin American economies should not be exaggerated. There seems little reason to believe that most Latin American economies have become overly dependent on exports. The majority of Latin American countries are less dependent on exports than (say) France or Britain, and few have expanded their exports to account for more than 25 per cent of GNP. Indeed, around the world it seems only to be the rather small economies that have very high levels of export dependence, for example Singapore, Hong Kong, Belgium, and Norway, together with the wholly oil-dependent, such as Saudi Arabia and Bahrain.

In so far as export dependence in Latin America has increased, the effects have been both welcome and unwelcome. When world demand (and consequently prices)

Table 3.1 Latin America: export contribution to Gross Domestic Product 1965–83

	GDP pc (US$)	Exports (%)		Manufactured exports share of total export goods (%)	
	1983	1965	1983	1965	1982
Argentina	2,070	8	13	6	24
Bolivia	510	17	19	4	n.a.
Brazil	1,880	8	8	8	39
Chile	1,870	14	24	4	8
Colombia	1,430	11	10	7	25
Ecuador	1,420	16	25	2	3
Mexico	2,240	9	20	16	12
Paraguay	1,410	15	8	8	n.a.
Peru	1,040	16	21	1	14
Uruguay	2,490	19	24	5	33
Venezuela	3,840	31	26	2	3
Low-income countries		6	9	24	50
Middle-income oil exporters		19	25	6	9
Middle-income oil importers		18	23	24	60
Industrial market economies		12	18	70	74

Source: World Bank (1985)

are rising, it is undoubtedly welcome; when world demand falls, the benefits of integration into the world economy are less obvious. What is clear, and constitutes an important point in the later discussion, is the degree to which export dependence encourages domestic instability.

It is important, however, to remember that export expansion is not only an effect of NIDL3. Indeed, the experience of Latin American countries is that each has been incorporated in a very different way: Chile has exported mainly primary products, Brazil has expanded manufactured exports, and Mexico has increased exports of primary products as well as manufactures. The confused picture is not simply a result of NIDL3, indeed most of the expansion can be attributed to the continuation of NIDL1 and NIDL2. Table 3.1 shows that only a few Latin American countries have managed to increase their manufactured exports in a major way; relatively few Latin American countries can be included among the NICs or the manufacture exporters. Perhaps this is not surprising, in that the major expansion in manufacturing exports from less-developed countries originated in the 'Gang of Four' (Hong Kong, Singapore, Korea, and Taiwan). As Dicken notes:

> In this global reorganization of manufacturing trade the increased importance of Asia as an exporter of manufactures is unique in its magnitude. ...By the end of the 1970s Asian countries generated three-quarters of all Third World manufactured exports compared with around one-third in the 1950s.
>
> (Dicken 1985: 40)

Add the export manufacturers of Southern Europe to the Gang of Four, and it is evident that Latin America has few members in the first division of the export league. Dicken's table of the world's leading twenty-five manufacturing exporters, for example includes only Brazil from Latin America. Admittedly his list of eleven proto-NICs includes Chile, Colombia, and Peru, but the point is that the effects of NIDL3 in Latin America can easily be exaggerated.

This is not to deny, of course, the penetration into domestic economies by transnational corporations, nor to reject the notion that Latin America has become more closely linked into the world economy in a variety of ways.

But, greater integration has come more through the extension of NIDL1 and NIDL2 than through NIDL3, particularly by Latin American countries increasing their primary exports. Most countries are successfully continuing to export traditional mineral exports such as copper, iron ore, and lead, while several are developing new sources of income through oil and coal. Similarly, traditional agricultural exports such as coffee, wheat, sugar, and bananas are being complemented by a range of products, including strawberries, soya beans, carnations, and marihuana. NIDL1 is alive and well and still recognizable in Latin America.

If NIDL1 is continuing, so too is NIDL2. Indeed, we can only agree with Singer when he asserts that 'multinational expansion in peripheral countries is effected mainly through import substitution' (Singer 1985: 32). In Latin America, it has long been clear that transnational corporations have participated actively in the region's industrial development. They have gained by supplying capital and intermediate goods to domestic industries making consumer goods. They have also invested increasingly in the more dynamic industrial sectors. Indeed, although the initial phases of import substituting industrialization 'may have been achieved by local enterprise, these countries' more dynamic industries had become progressively dominated by multinational corporations constituted in subsidiary national firms' (Connolly 1985). Since NIDL2 is continuing to develop, it is still providing a major route through which overseas investment is penetrating Latin American economies. This is demonstrated by the fact that although there are few important export manufacturers in Latin America, foreign investment has poured into certain parts of the continent. Thrift notes that: 'Of the total inflow of foreign direct investment into the developing countries, just six (Argentina, Brazil, Hong Kong, Malaysia, Mexico, and Singapore) accounted for between one-half and two-thirds of the total' (Thrift 1986: 27).

In so far as the internationalization of production is affecting Latin America, Brazil and Mexico are participating rather more than, say, Bolivia and Peru. But, as so much of the increase in foreign investment and other links with the world economy is developing through NIDL1 and NIDL2, it might be argued that Latin America has hardly changed. For the main beneficiaries of NIDL3 are the same economic giants that have long been recipients of overseas investment. That process has enabled them to develop first their primary export base, and then their industrial base. In turn, it has now created the potential (still not fully developed) to become export producers (Morawetz 1981). In short, Brazil and Mexico have long been more integrated into world production than Bolivia and Peru, and NIDL3 has not changed that pattern.

This pattern of world integration has been basically the consequence of different allocations of resources. To a considerable extent, therefore, the ability of different countries to participate in NIDL1 and NIDL2 was determined for them. Internal policies might affect a country's ability to take advantage of particular circumstances, but in general NIDL1 and NIDL2 depended greatly upon luck. Did the country have resources in demand on the world market, and did those resources stimulate internal economic expansion (Furtado 1970)? Similarly, the ability to participate in import-substituting industrialization depended greatly upon the size of the domestic market. It is no coincidence that Latin America's largest countries are also the most industrialized.

But, as has been well established in the literature, the world economy, while creating opportunities and establishing major limitations, does not determine outcomes. If it did, South Korea, with its lack of natural resources, would never have become a major exporter. Success in the world economy depends greatly upon national reaction. In Latin America, over the past fifteen years, those national reactions have varied dramatically between countries. Indeed, it is almost impossible to attribute the current situations in different Latin American countries wholly to the international situation. Opportunities have arisen and disappeared; the interesting question is why certain countries grasped the opportunities of the 1970s better than others and avoided the worst consequences of the 1980s. If the changes in the world economy created opportunities and potential problems, the actual outcome was a result of how the state in each country responded. Essentially, each state has a degree of constrained autonomy within which it can operate. It is the use made of this constrained autonomy that is critical.

During the period of boom from 1945-73 (and arguably later), different Latin American countries responded in various ways to the opportunities and constraints created by changes in the world economy. Nevertheless, it would be erroneous to ignore the similarities, especially up to 1973. During this period of world economic expansion, most Latin American economies followed similar approaches. All followed import-substitution policies and then gradually began to move towards export orientation during the 1970s. All became thoroughly urban, consumer societies. None, bar Cuba and Nicaragua, made any real effort to redistribute income. Most experienced a period of authoritarian government during the 1970s, and all but four nations were ruled by military regimes for part of that decade.

But too much should not be made of the similarities, especially since 1973; the internal responses to international change have varied considerably among the Latin American governments. While most took advantage of the easy credit available, Colombia was among the very few that did not go deeply into debt. While several opened their economies a little, Chile went overboard with its espousal of market forces. Some economies managed to grow rapidly while others scarcely prospered, even during the long post-war boom. There was also a tremendous difference among regimes with respect to the role of the public sector. In Brazil, Mexico, and Venezuela, the state sector was an essential element in economic expansion; in Chile, the Pinochet government privatized most of its functions. The point is that although there is a current crisis which has given rise to similar problems throughout Latin America, individual countries reached this point through very different routes and experiences.

It may also be argued that several Latin American economies have had their chances to achieve a deeper and more widespread process of development. For example, the Latin American oil economies have been in a much more favourable situation than the non-oil economies, even if currently it does not look that way. It is also clear that all of the larger countries have been favoured by access to a Eurodollar market which allowed them to sustain domestic investment and growth by maintaining levels of imports. Again, it does not look like that today.

THE WORLD RECESSION AND LATIN AMERICA[1]

If the effect of NIDL3 on Latin America has been partial and has affected mainly Brazil and Mexico, the effect of the recession has been far more widespread and dramatic. In this respect there are clear similarities between countries.

If the problems of different countries vary in severity, in part because certain regimes were unluckier than others, in part because of the wisdom, or the lack of it, in economic policy, there is little doubt that the autonomy of Latin American governments has been heavily constrained by the recession. Indeed, these constraints are demonstrated by the similar problems facing each of the Latin American economies.

Since 1980, most Latin American economies have been under threat and none has managed to maintain positive rates of growth in gross domestic product during the 1980-5 period. In some countries the situation has been little short of disastrous, with cumulative declines of between 15 and 30 per cent in Bolivia, El Salvador, Guatemala, Peru, Uruguay, and Venezuela. Throughout the region, the external debt has risen inexorably and the ratio of total interest payments to exports of goods and services has risen to over 30 per cent in Bolivia, Mexico, Peru, Argentina, Brazil, Chile, and Uruguay. Rates of inflation have risen generally, and particularly sharply in Argentina, Bolivia, Brazil, Costa Rica, Ecuador, Mexico, Peru, and Venezuela. As a result of the combined forces of inflation and limited wage rises, real incomes in the urban areas have fallen in numerous countries, most notoriously in Peru, where wages of manual workers fell by 40 per cent between 1980-5, and in the Mexican manufacturing sector, where they declined by almost one-third during the same period. Added to this severe cut has been a dramatic rise in unemployment. Current rates of urban unemployment have seldom been higher in Argentina, Colombia, Chile, Panama, Paraguay, Peru, Uruguay, and Venezuela.

But if there are common problems and tendencies afflicting the region, the interesting question is to what extent different economic and social policies might have avoided these outcomes? Prima facie, there was certainly a major difference between the processes that led to debt accumulation in Argentina and Brazil compared with those in Mexico and Venezuela.

Among the non-oil producing countries, the shock of the 1973 and 1979 price rises is a convincing explanation of why levels of external debt rose. Such economies were faced with the choice of contracting overseas debt, to cover increased oil-import bills, or of slowing domestic expansion. Most, and particularly Brazil, chose the former path and are now reaping the consequences. While internal policies, and sometimes corruption and poor investment, contributed to the crisis, national governments were placed in an appallingly difficult situation. Nevertheless, individual countries came out of ostensibly similar situations very differently. Colombia, admittedly partially self-sufficient in oil, responded with more adequate economic programmes than, say, Brazil. Chile, with its apparent initial success in opening up its economy, experienced a catastrophic fall as a result of adopting a mistaken exchange-rate policy in 1982 (Gwynne 1985). Apparently similar situations produced very different cumulative falls in GDP per capita between 1980 and 1985, ranging from 0.5 per cent in Colombia to 18.5 per cent in Guatemala.

The post-1973 situation was distinctly different in the oil-producing countries. There the oil price increases were greeted as a bonanza which would allow a major increase in domestic investment. In Ecuador and Mexico, high rates of investment produced high rates of economic growth throughout the 1970s. In Peru and Venezuela, however, the oil bonanza did not produce rapid growth, even though the latter invested hugely in heavy industrial and infrastructural development. In these countries, a mixture of expansionary government policy, the 'Dutch Disease',[2] overseas borrowing against future export revenues, and, in hindsight, a misguided belief in continued rises in petroleum prices and in the effectiveness of large-scale industrial development, combined to produce problems as serious as those in the non-petroleum economies (Ewell 1984; Cornelius 1985). Usually these problems arrived much more suddenly and spectacularly than in the non-petroleum economies and, as a result, 1983 was a particularly bad year for all the major producers in Latin America. Sudden falls in petroleum prices, together with external adjustment programmes, brought in 1983 major declines in per capita domestic product of 11 per cent in Bolivia, 5 per cent in Ecuador, 8 per cent in Mexico, 14 per cent in Peru, and 8 per cent in Venezuela.

There can be little doubt that the shocks to all of these economies were externally induced. Falling prices for exports, rising interest rates, and declining demand for export products were hardly conducive to internal economic stability. Similarly, given the rapid rise in external debt, the policies recommended by the IMF were clearly inappropriate in many cases. But the same might be said of most economies in the world. The question is whether different economic policies could have produced less traumatic internal effects. The provisional judgement in countries as diverse as Argentina, Chile, Mexico, Peru and Venezuela must be in the affirmative. In the oil economies, expansion should, in hindsight, have been much more controlled. Of course, banks flushed with Eurodollars contributed to the later crisis by the unwise expansion of credit facilities, but Latin American countries were hardly slow to take advantage of the easy terms. By borrowing privately, Latin American governments were able to 'avoid the influence of IMF conditionality on economic policy. Most importantly, they were able to sustain levels of imports above those they could have afforded if such loans had not been available' (Schmitz 1984: 11). That they failed to use this additional flexibility to put their own houses in order is unfortunate and has, in the long term, made their situation still worse (Cornelius 1985). Certainly, in the oil economies and in other countries such as Chile, domestic governments can be blamed for their inappropriate policies at least as much as they in turn can attribute the crisis to the effects of the external shocks.

Of course, now that they all face debt problems and suffer from economic recession, it is easy to show that there are major constraints on governmental autonomy over economic and social policy. There can be little doubt that this external control is highly damaging socially. Indeed, there are constant complaints that the US government is forcing political instability on the region through its insistence on monetary orthodoxy.

The main agency ordering change in government policy is, of course, the International Monetary Fund. Because the debt is so great, most Latin American governments have had to approach the IMF for help and this agency's prescription of financial stringency is affecting practically every country. But even for those countries anxious to determine their own economic policy and ready to bypass the IMF, it is difficult to avoid conditionality. This is because private lenders are increasingly reliant on the opinions of the IMF in determining their own policies to debtor countries. As Kaplinsky argues, 'this interlinkage of IMF, World Bank, and private bank conditionality is important, so that without the IMF "seal of approval" it is increasingly unlikely that significant resources will be made available to debtor LDCs' (Kaplinsky 1984: 85). In so far as they are trapped into following the conditions laid down by the IMF, most Latin

American governments are arguably following policies ill-suited to their economic and social needs. For the IMF has adopted a general policy for all countries which seems to owe as much to economic dogma as to enlightenment. The overall effect is certainly deflationary and is favourable to the opening up of the domestic economies to greater competition from abroad. The danger is that 'the form in which conditionality has actually been implemented is, despite some protestations to the contrary, blunt, and that the general effect is to enforce de-industrialization on the Third World' (Kaplinsky 1984: 87). Similarly, the IMF 'not only enforces restrictions on money issues and public expenses, but forces its clients to apply free-trade policies and to fix exchange rates at parity levels, so as to provide capital for all possibilities to maximize profitability, even at the expense of national interests' (Singer 1985: 38).

THE URBAN IMPACT OF EXPANSION AND DECLINE

During the economic expansion of the post-war period, the pattern of urban change throughout Latin America was fairly consistent. There were major variations in terms of the starting point, with countries such as Argentina, Chile, and Uruguay having begun to industrialize and urbanize quite early in the twentieth century, whereas Bolivia, Ecuador, and most of Central America did not begin to 'modernize' extensively until the post-war period. This difference in the pace of development should not, however, be allowed to disguise major similarities in the urbanization process. Very crudely, the pattern of urban development can be described as follows.

First, export production realized profits which were channelled directly and indirectly through the urban system. Most consumer imports were directed to urban centres and particularly to the largest, normally the capital, cities. These administrative and commercial centres also often acted as ports and transportation centres. Second, import-substituting industries came to be concentrated in the major urban centres. The combination of access to markets, availability of infrastructure, and the advantages of ready contact with government gave these cities a major advantage (Gilbert 1974). Third, rapid population growth, allied to a failure to bring about meaningful land redistribution, a broad tendency to commercialize agricultural production, and a drop in prices for domestic food products, all produced an impoverished countryside. Expanding opportunities in the cities, together with rural poverty, encouraged those with marketable skills to move to the cities.

Fourth, urban employment created limited opportunities for work in the 'formal' sector. These were rarely well paid and, as a result, many households were forced into 'informal' activities at low rates of pay. Households survived by having several members working. In turn, this source of cheap labour provided certain advantages for more affluent households in the city (particularly cheap services) and for the formal production sector generally (Bromley and Gerry 1979). An employment market developed with a continuum between formal and informal sectors with low pay characteristic of the low-skilled in both sectors.

Fifth, expanding office employment, as government, commerce, and industry grew, helped to create a gradually increasing middle class, especially in the fast-growing economies of the region. In parts of the southern cone, arguably half of the urban population became middle class. A consumer-based society developed on the technology, advertising, and products of companies originating in Europe and North America. Latin American cities formed part of the world market for cars, computers, and Coca-Cola.

Sixth, the built environment was created through a mixture of 'formal and informal' processes. The middle- and upper-income classes lived in residential areas similar to those found in North America; areas that had all the amenities of modern life, from infrastructure and services to cars and refrigerators. A commercialized and increasingly large-scale building and urban real-estate industry developed to supply this housing. Meanwhile, the poor were relegated either to rental accommodation in decaying parts of the city or increasingly to living in self-help accommodation in highly segregated parts of the city. Self-help settlements received irregular servicing but did provide opportunities for those with regular and adequate incomes to establish a foothold in the urban property market. Despite the informality of the systems developed to provide housing for the poor, self-help houses were regularly sold and formed part of the commercialized land and housing market.

Seventh, as cities grew in size and complexity, the state came under increasing pressure both from the residential population and particularly from industrial and commercial groups, to provide adequate services and infrastructure, such as access roads, drainage, water, electricity, education, and health facilities. Improvements occurred most rapidly in those areas which were essential for industrial and commercial development, not least because these sectors could afford to pay for new services. The poor residential areas benefited only partially from the new services, but most city residents were better supplied than the rural populace.

Eighth, governments developed an ability to perform certain functions for the population. As demands for services and infrastructure developed, there was a recurrent fiscal crisis, but not one that was any worse than that which had always afflicted Latin American local government. Indeed, the finances of most large cities were much healthier than those of small cities, and these in turn were healthier than those of most rural areas.

Needless to say, this brief summary over-simplifies reality, but since the existing literature contains numerous descriptions of each of these processes there seems little reason to dwell longer on them here (Morse 1971; Portes and Walton 1976; Butterworth and Chance 1981). It is more interesting to relate the process of urban development to the process of boom and arguably bust that has characterized Latin America since 1970. More particularly, how did the boom and especially the subsequent recession affect the autonomy of urban areas in Latin America? Even within this more limited theme, I shall pay special emphasis to aspects which may have been somewhat erroneously treated in the literature.

It is extraordinarily difficult to provide firm links between the changing economy and the process of urban development. Part of the difficulty rests in separating, or even identifying, the urban dimension (Castells 1977). One method of separation is to confine discussion to the construction of the built environment, but such containment eliminates many aspects of change critical to the welfare of the poor. It also removes from any analysis many of the key ingredients determining welfare levels. For example, changes in the built environment are determined to a considerable extent by the rate of economic growth, the distribution of income, and the levels of unemployment and underemployment. Another problem is that the urban population constitutes a majority in Latin America, which means that any discussion linking the world economy with urban areas is covering a vast range of issues. Finally, there is the problem that since there is a great deal of overlap between the three phases of the new international division of labour, and indeed with the recent recession, it is often difficult to distinguish the effects of the different phases on the urban process. My approach therefore is to examine a selected, but hopefully important, range of urban issues over time: what has occurred over the past thirty years, and how has NIDL3 and the recent recession actually changed or accelerated the processes involved?

Metropolitan expansion

An important point, but one which is well described already, is the vast increase in the proportion of the population of every Latin American country living in urban areas (Table 3.2). The vast movement of people from the countryside stimulated rapid growth and in later years raised the rate of natural increase in the cities as the young migrants bore their children in their new environment. While migration continues to be an important element in urban growth, the bulk of that growth now comes from natural increase (Fox 1975).

Table 3.2 also shows that urban development has been characterized by a process of concentration in the larger metropolitan areas. In most countries during the past thirty years, the largest cities have grown the most rapidly. Recently, however, three factors have started to reduce the pace of metropolitan expansion. These retarding forces have no doubt come as welcome relief to the many Latin American governments which have consistently worried about the rate of metropolitan growth. The first is that birth rates have begun to drop quite dramatically in the region's urban areas, thereby heralding a slowing of metropolitan growth. Second, the pool of potential migrants in the rural areas has begun to decline throughout the continent (Fox 1975; Williams and Griffin 1978). Indeed, in Argentina, Uruguay, Southern Brazil, and even parts of Andean America, the pace of out-migration over the years has been so rapid that the population of many rural areas is now declining and is in a minority throughout the region (Table 3.2). Third, there are some signs of a process of spontaneous deconcentration occurring around the major cities of the continent. This process of 'polarization reversal' has been clearly apparent in the state of Sao Paulo

Table 3.2 Urban and metropolitan growth in selected Latin American countries

	Population (%) living in settlements with over 20,000 inhabitants		Urban population (%) living in cities with over 500,000 inhabitants	
	1960	1975	1960	1980
Argentina	59	70	46	45
Bolivia	23	29	0	44
Brazil	28	45	35	52
Chile	37	66	38	44
Colombia	51	48	28	51
Ecuador	28	36	0	51
Mexico	29	39	36	48
Paraguay	16	23	0	44
Peru	29	45	38	44
Uruguay	61	65	56	52
Venezuela	47	64	26	44

Source: CEPAL Statistical Summary of Latin America 1960-80, World Bank (1985)

during the 1970s, where smaller cities began to outpace the growth of the metropolitan centre (Townroe and Keen 1984). While there is controversy over the source of this change (Storper 1984; Townroe and Hamer 1984), it seems to have little to do with government efforts at employment decentralization. Arguably, the tendency is more clearly linked to the failure of the urban authorities in Sao Paulo to resolve the urban diseconomies that are all too obvious in that city (Kowarick and Campanario 1986). Whatever the causes, it is not a surprising tendency. Similar patterns of deconcentration within the immediate hinterland of a major city have been observed throughout North America and Western Europe. An incipient process of 'polarization reversal' has also been apparent in numerous other metropolitan regions in Latin America for a number of years (Rofman 1974; Gilbert 1974). This trend can be expected to accelerate in the near future, especially in the vicinity of the largest cities such as Mexico City and Buenos Aires.

What is difficult to establish, however, is the precise link between this process and NIDL3. It may be significant that Sao Paulo is the first city to experience a major process of urban deconcentration given its more obvious links with NIDL3. Nevertheless, I do not think the link is strong; the tendency has been incipient for a number of years.

A more obvious link between NIDL3 and changes in the national settlement system involves the growth of regions clearly concentrating on manufacturing exports. However, as I have argued, there is relatively little export manufacturing in Latin America, and certainly very few plants which are devoted entirely to export production. Consequently, there are few examples of urban regions which are booming as a result of links with NIDL3. The obvious exception to this statement is northern Mexico, where the growth of maquiladora plants is one of the sources of rapid urban growth in the border cities. Even in Mexico, however, the impact of NIDL3 should not be exaggerated, for the maquiladoras of the border zone account for only 2.3 per cent of Mexican industrial production (Calcagno and Jakobowicz 1981). In the rest of Latin America, apart from a small number of free trade zones in Brazil, Colombia, and the Dominican Republic, there has been very little export-oriented industrial development. Admittedly, without the world recession there might have been more export manufacturing. Two oil-producing countries, for example, had both invested heavily in steel plants with an eye to export markets (the Mexicans in Ciudad Lazaro Cardenas and the Venezuelans in Ciudad Guayana). But the world recession and consequent world-wide excess of steel production has rendered this initiative unviable.

In general, export-oriented manufacturing has made relatively little impact in Latin America and therefore can have had little effect on the urban system. Perhaps a more significant, if possibly temporary, link is that between urban growth and the world recession. Even if it can be supported only by anecdotal evidence, it seems clear that where urban areas have been suffering from acute problems, migration has slowed and some workers have even begun to move out of these cities. One example of this process is Santiago, Chile, where extraordinarily high levels of unemployment seem to have stimulated a movement of workers to the countryside in search of work and sustenance. A parallel phenomenon, although with a different spatial impact, is occurring in Mexico. It is regularly reported that the pace of illegal migration into the United States is currently accelerating. As long as the fortunes of the US and Mexico economies move in opposite directions, and the peso continues to weaken against the dollar, Mexicans will try to move into the United States. The spatial effect in Mexico is both to reduce metropolitan expansion and to boost the population of the border cities. The latter would seem destined, in current circumstances, to increase their role as staging posts for migrants and day-commuters to the American labour market.

However, the precise effect of the recession on the urban settlement system in Latin America is difficult to evaluate because, despite its severity, the depression has been short in duration. Nevertheless, it is possible to hypothesize that since urban expansion has traditionally been linked to economic development, economic decline should help to slow urban growth. Needless to say, certain caveats need to be made to such a crude argument.

First, the effects of the recession on rural areas will clearly differ. For example, some subsistence agricultural regions may well retain potential migrants during a period of high unemployment in urban areas, whereas export agriculture areas shedding labour because of the recession might produce additional cityward migrants. Since the particular effects of recession will touch each rural area differently, the need to migrate will vary from region to region. Second, since the recession is affecting cities with different levels of severity, growth in some may slow more dramatically than in others. It is likely, for example, that the dramatic effects of the Mexican recession on Monterrey will slow migration to that city more than the more subdued effects on Guadalajara will slow migration to that city.

Despite these caveats, it seems likely that a sustained recession will lead to a slackening of metropolitan growth in many parts of Latin America. Given the problems facing the largest cities, that might even be considered by some to be a desirable process. It is certainly an intriguing possibility that a brief burst of recession may achieve more in terms of urban deconcentration than years of regional planning! However, most observers will not greet this development with pleasure, for the costs of recession are likely to cause far greater hardship than any gain from slower metropolitan expansion. In so far as the longer term prospects for some Latin American economies seem distinctly worse than for others, we may well see different trends in urban and metropolitan expansion. The immediate economic future for Peru and Chile, for example, is distinctly more gloomy than that for Brazil or Colombia; it will be interesting to see what effect this has on the pace of urban growth.

The differentiation of urban production

It seems clear that during the twentieth century, the basis of urban production has tended to diversify. Of course, there was some diversity even during the colonial period: compare mining centres, such as Ouro Preto or Potosi, with religious centres, such as Popayan or Puebla, or with vice-regal capitals, such as Lima and Mexico City (Morse 1971). Nevertheless, in the past most cities had similar administrative, religious, commercial, and manufacturing roles. During the twentieth century, the development of manufacturing and the emergence of large state bureaucracies have led to a greater diversity.

Growing differentiation was caused initially by different links with the export economy. Cities at the centre of agricultural export zones played different roles to centres in subsistence agricultural regions. In turn, these variations created different opportunities for industrial development. Some cities developed industrial activity, whereas others remained essentially as commercial and retail centres. The manufacturing growth of Sao Paulo, Monterrey, and Medellin was highly distinctive in their respective countries. This pattern of specialization continued to develop during NIDL2 as import substitution increased the manufacturing base. Indeed, manufacturing through import substitution brought about important changes in the national urban system. In general, such manufacturing activity was located in the capital cities and entrepreneurial centres. It led to the expansion of a few centres and the relative decline of most others. In particular, large-scale manufacturing production often displaced that of small towns and even rural areas, thereby accentuating specialization (Gilbert 1974). In addition, government action sometimes created new specialist centres, sometimes industrial cities such as Ciudad Guayana and Ciudad Lazaro Cardenas, and sometimes administrative centres such as Brasilia. The overall trend during the NIDL2 phase was towards diversification and the accentuation of incipient tendencies towards primacy in the Latin American system.

In its limited development in Latin America, NIDL3 is encouraging further diversification. In so far as the new world economy is indeed a 'mosaic of unevenness in a constant state of flux', NIDL3 is bound to have a differential impact on cities throughout the region. This difference is most clearly seen in Mexico, with the border cities playing a decidedly different role in the Mexican, and indeed, wider North American, economy than Guadalajara or Puebla (Arias and Roberts 1985).

But, as I have argued throughout this chapter, the effects of NIDL3 have been very limited in most parts of the region. The result is that the employment structure in most cities looks rather similar. There are some industrial and some port cities, but most are dominated by tertiary activities. The majority of employed people work in retail, commercial, and office jobs. The proportions of the population working in the 'informal' sector do vary, but more from country to country than from city to city (see p. 104). There are cities where the labour force is dominated by women, but beyond the Mexican-United States border, rather few. I would, therefore, stress rather less firmly than Armstrong and McGee the 'growing divergence within Third World societies between their primate cities and their regions and the rural areas' (Armstrong and McGee 1985: 49). This diversification was apparent in the sixteenth century and has continued to emerge slowly ever since. It is not, I believe, any more a feature of NIDL3 than it was of NIDL1 or NIDL2.

Undoubtedly, there are important differences between cities, and the economic and social consequences of these differences are being demonstrated very clearly by the current recession. For given that cities play different economic and social roles in every national urban system, the effects of sudden economic change will affect each differently. Cities with major administrative functions may be less affected by economic recession than are commercial centres; mining towns and ports are more likely to experience wide fluctuations in economic activity as a result of world trade cycles. By contrast, a national capital with a mixture of administrative, commercial, and domestic industrial roles, may suffer less gravely from fluctuations in the wider economic environment.

What is clear is that the world recession, and national responses to it, are affecting industrial capacity and employment in different cities in very different ways. Just as in the United Kingdom, manufacturing in the north of the country has been very badly affected, so cities in Latin America have suffered differentially as a result of the current recession. In Colombia, for example, the recession has produced very different unemployment rates. As in the past, the employment structures of certain cities seem to absorb fluctuations in national economic fortunes better than others. Since 1975, unemployment in the port city of Barranquilla has almost always been higher than that of Bogota. It has also demonstrated wider fluctuations, owing to its greater reliance on international trade. Similarly, the industrial city of Medellin seems also to have consistently higher levels of unemployment than Bogota. In 1984, at the height of the recession, Bogota had an unemployment rate of 12 per cent compared to 15 per cent in Barranquilla and Medellin. In Bucaramanga, it was only 8 per cent. Similar differences in unemployment levels during the recession have also been apparent in Mexico (compare, for example, Monterrey and Guadalajara) and in Chile during the 1970s (for example, Concepcion and Antofagasta) (Gwynne 1985).

In so far as there are differences in the productive base of cities, such variations in employment and unemployment are not surprising. But, in general, these differences have been caused more by NIDL1 and NIDL2 than by NIDL3. The current recession is emphasizing the importance of these differences, but they have been apparent for some time.

Infrastructural development

Industrialization requires the construction of a complicated system of infrastructure. Roads, electricity, water, sewerage, telephones, and the like are required not only to produce manufactured goods, but also to support the labour force. Compared to previous periods in Latin American history, both the need for and the provision of infrastructure are much greater. Of course, access to the vastly increased pool of services and infrastructure does not benefit all Latin Americans equally. As a general rule, however, infrastructure and service provision have improved markedly in most cities, and especially rapidly in those places where industrialization has been sustained over a number of years.

In the past, infrastructure has sometimes been provided by individual manufacturing companies, but in most larger cities the main responsibility for organization has rested with the state. This responsibility has gradually brought about major changes in the administrative procedures and structures of the state. Although the organization of service provision has varied greatly between countries and cities, there has been a widespread tendency for the state to rationalize its bureaucratic machinery. The needs of today demand the creation of modern, efficient enterprises to provide the main forms of infrastructure and services. Without this improvement, the continuing growth of the modern sector would be impaired as indeed would the very working of the complex structure of the modern city. As a result, there have been clear moves in most cities towards an increasingly bureaucratic and technical method of operation. The result has been to create agencies which have managed to supply the cities with increasing quantities of power, water, and transport. This move towards efficient state corporations has been manifest most clearly in those activities which are of direct concern to manufacturing and commercial activity (Gilbert 1981; Gilbert and Ward 1985).

In Bogota, which is by no means unique, there have evolved modern, technical agencies which supply electricity, water, and telephones. These, however, have not developed in functional areas less critical to capital accumulation. These technical agencies began to develop at a time when industrial companies were demanding better services to support their expansion. Larger-scale provision required more technical procedures and large loans. In Bogota, the local banks demanded that changes be made in the form of agency operation before loans would be forthcoming. They asked that partisan political involvement in agency affairs be reduced and that a commercial method of operation be introduced which would guarantee the repayment of the loans. The 'modernization' and technification of public companies, therefore, was not externally induced; it was essentially stimulated by domestic pressures. It was only later, when major loans were contracted from abroad, that the major multilateral agencies became centrally involved. Since the 1960s, the World Bank has been a major provider of credit for public utilities in Latin America and its loan conditions have required a more commercial, technical, and efficient form of agency operation. The criterion of a return on capital has been an essential ingredient in setting tariffs. Without this return, external loans could not be repaid and the future expanison of capacity would be threatened. In this sense, service provision has undoubtedly become an area where local control over decision making has been severely compromised.

This process occurred quite widely during NIDL2, although local responses to the need for technification varied considerably. External pressure was not felt strongly, for example, in the few countries which could finance their own infrastructural development. In Venezuela, public utilities expanded with limited signs of increased efficiency. High deficits and low tariffs were often combined with poor services (Gilbert and Healey 1985). Elsewhere, countries with slow rates of industrial and commercial expansion did not manage to improve their public utilities. Without a large industrial sector to consume and finance the investment, the key ingredient for expansion was missing. The improvement of infrastructure and services was, therefore, highly variable across the continent. Nevertheless, the expansion was still widespread and involved both domestic and external investment on a vast scale.

In this respect, Connolly is correct when she argues that foreign investment in NICs requires a great deal of infrastructural development, the provision of which generally falls on the national state: 'If this is the case, then we are forced to consider the possibility that the NIDL also implies a global reassignment of state functions regarding investment in fixed capital and the consumption fund' (Connolly 1985). It is not a phenomenon of NIDL3 alone, however, for in Latin America it has long been part and parcel of the process of import-substituting industrialization. It is also a process that worked reasonably well under conditions of continuous economic expansion. In those cities which were industrializing rapidly, the expansion of electricity, water, and telephone supplies grew apace. Where commercialized agencies had evolved, future expansion was planned and the costs were largely paid for by industrial and commercial customers. In so far as residential use became important, there was sometimes a measure of cross-subsidization. It is arguable, therefore, that during the post-war boom, increasing external influence over local decision making in the field of service provision was no bad thing, as it led to vastly improved capacity, and it gave local governments new sources of influence and control which compensated for their loss of complete autonomy over the service agencies. For example, if local political parties were forced to limit their direct political involvement in the day-to-day activities of the main public utilities, the increased expansion in servicing capacity allowed them to offer more services to low-income settlements and thereby to build larger political constituencies. In general, the local and national states seem to have been prepared to compromise to international pressure over the way in which services were provided. They recognized that without major improvements in services and infrastructure the economic future would be threatened. Local autonomy was both limited and increased by international lending and influence in the public utility sector.

With the recession of the 1980s, however, a major problem has arisen. The major loans required to guarantee future expansions of capacity have now become very expensive to service. The combination of devaluation and rising interest rates has put pressure on many local authorities and public utilities. In many cities, this has led to sudden deficits in the balance sheet or to increases in tariffs. In certain cities, the problem has reached crisis proportions. For example, the local authority in the Federal District of Mexico is very heavily in debt to foreign banks; in 1983, 93 per cent of its current income was scheduled to go on debt servicing (Connolly 1985: 4.10). Other cities, such as Bogota, have lesser problems but the same tendency is still apparent.

The situation in the 1980s, therefore, is that a relatively good system of infrastructure and services has been developed in the larger cities, but the cost of these services has suddenly risen. Given the currently limited financial capacity of local governments in Latin America, this has increased pressure to raise tariffs to the consumer. In this sense, the Latin American consumer is paying the direct cost of rising interest rates in the United States. Today the benefits of increasing capacity are proving very expensive.

The problem has been magnified in many cities by the failure of previous administrations, despite external pressure over the years, to charge appropriate tariffs. Mexico City is currently suffering badly, in part at least, because it has long maintained large subsidies through its oil revenues, a source of income that has now been vastly reduced. The cost of operating the metro, for example, hugely exceeds the income received from fares. Currently, internal financial pressure from the IMF, during the seemingly constant renegotiation of the Mexican debt, is forcing the authorities to face the political repercussions of raising the cost of transport: this is always a politically sensitive issue in any Latin American country. Of course, the nature of these repercussions is highly variable from place to place, but few governments are in a position to reduce the services provided. As a result, most Latin American urban administrations are faced with mounting financial deficits and are being forced to raise tariffs and/or to increase sources of taxation.

This pressure on local authorities in Latin America is currently widespread and is undoubtedly leading to higher prices for services. It is also leading to an extension of the tax base, and even to illegal settlement areas. It cannot be wholly coincidental that 1986 saw major initiatives in both Bogota and Caracas to legalize extensive areas of self-help settlement. Although there is a certain level of popular demand for such legalization, the increased property tax yield may well be the most prominent motive for the recent change. In any event, in the near future there is likely to be a major shift away from subsidies in many Latin American cities. Such a shift is not wholly detrimental to the poor, because there is a sound argument that general subsidies are both inefficient and inequitable. As Linn (1983) has argued, subsidies usually bring benefits to the rich as well as to the poor. And in so far as many of the poor may not gain access to subsidized services, the subsidy is regressive. Consider, for example, subsidies for water in a city where many of the poor do not receive a domestic supply. Not only are they excluded from water provision, but their taxes are being used to subsidize higher-income groups. The current difficulty is that local administrations are being forced to rethink the question of subsidies while, at the same time, general tariffs are rising because of the external debt situation. In some cities, service costs are escalating rapidly. The consequent political effects are no doubt exercising many government minds in Latin America today.

Changes in the employment structure of Latin American cities

Throughout the post-war boom, the formal employment sector expanded rapidly in every Latin American city. In certain cities, such as Sao Paulo and Mexico City, manufacturing growth contributed strongly to this expansion. Elsewhere governmental, financial, and commercial activities were the principal generators of formal sector employment. At the same time, there was little sign of any diminution in the relative importance of informal sector employment. Indeed, figures compiled by the Programa Regional del Empleo para America Latina y el Caribe (PREALC) and by the Economic Commission for Latin America, suggest that the urban informal sector has maintained its relative size at around 30 per cent throughout the 1950-80 period. Table 3.3 shows that the urban informal sector has even grown in a few countries. While an accurate definition of the formal and the informal sector is very difficult, and efforts at quantification of each sector often prove spurious, these figures do allow us to speculate about the recent relationship between formal and informal sector growth.

Such speculation is important because a number of conflicting ideas have been put forward about the nature of employment change in Latin America. These can be crudely categorized into three different prognoses for the future relationship of growth between the formal and informal sectors. First, the modernization approach would suggest that as the formal sector expands, the size of the informal sector declines (Lewis 1954). This was the classic pattern in Western Europe and the United States, but not one that seems generally to have been replicated in Latin America in recent years. At the same time, it is correct in so far as Table 3.3 shows that the more affluent Latin American countries have smaller proportions of informal sector workers than do the less affluent. The ratio between formal and informal sector employment in urban Argentina, Costa Rica, Uruguay, and Venezuela is between 3:1 and 4:1 whereas in Bolivia and Ecuador it is less than one. In addition, the modernization view is correct to the extent that the formal sector has increased more rapidly than the informal in a number of countries, although by no means in the majority.

Second, initial versions of the marginality thesis supposed that, because formal sector manufacturing employment was increasing more slowly than urban population growth, the informal sector was bound to expand more rapidly than the formal sector. Modern technology meant that few workers were required by large-scale industry and, therefore, much of the labour force was not even useful as an industrial reserve army to keep down the

Table 3.3 Latin America: Economically active population in urban formal and informal sectors 1950-80

		Economically active population in:		
		formal sector	informal sector	urban areas
	1950	75.9	24.3	72.0
Argentina	1970	80.9	19.1	81.6
	1980	77.0	23.0	84.4
	1950	72.7	27.3	39.2
Brazil	1970	72.1	27.9	53.5
	1980	72.8	27.2	62.1
	1950	64.9	35.1	62.9
Chile	1970	76.1	23.9	69.8
	1980	72.9	27.1	74.2
	1950	61.0	39.0	39.2
Colombia	1970	68.6	31.4	56.4
	1980	65.6	34.4	64.9
	1950	64.8	35.2	33.2
Ecuador	1970	42.1	57.9	40.9
	1980	47.2	52.8	48.1
	1950	62.2	37.4	34.5
Mexico	1970	65.1	34.9	52.1
	1980	64.2	35.8	61.5
	1950	53.1	46.9	36.0
Peru	1970	59.0	41.0	50.5
	1980	59.5	40.5	58.8
	1950	81.4	18.6	77.8
Uruguay	1970	79.3	20.7	81.0
	1980	76.9	23.1	82.3
	1950	67.9	32.1	51.1
Venezuela	1970	68.6	31.4	71.3
	1980	79.2	20.8	79.0

Source: Wilkie and Perkal (1983)

wage rate (Nun 1969; Quijano 1974). According to this interpretation, there was a surplus population which was marginal to the production process and which was forced to resort to employment in the informal sector. Given the lack of a social welfare system, the poor would survive through employment involution; people would subdivide jobs between themselves (McGee 1976). Impoverishment would be an automatic corollary of a growing informal sector. According to this argument, the proportion of informal to formal sector jobs would tend to increase through time, especially in those cities where economic growth was very slow. In fact, the data in Table 3.3 suggest that there is no consistent pattern. In the three countries experiencing a rate of per capita GDP growth of 1.0 per cent or less between 1970 and 1980, the formal/informal sector ratio increased in Venezuela, remained fairly constant in Peru, and fell in Argentina.

Third, the CEBRAP (Centro Brasileiro de Analise e Planejamento) type of approach provides different explanations of the link between the growth of the two sectors (Oliveira 1972; Moser 1978; Roberts 1978). According to this school of thought, companies operating under conditions of peripheral industrialization, subcontract many less profitable or uncertain activities to informal sector workers because this strategy minimizes risks to the company. In such circumstancs, the informal sector grows not because of a lack of formal sector growth, but as a direct result of that expansion. Certainly, this idea has found much support in the literature, which has demonstrated beyond doubt that there are innumerable links between the activities of the formal and informal sectors (Bromley and Gerry 1979). It is also compatible with the fact that informal sector jobs have grown very rapidly in cities where rates of economic growth have been high. Clearly, rapid expansion of large-scale industrial and formal sector retail and commercial activities should also generate increased demand for domestic servants, street vendors, and small-scale repair shops. Certainly, this kind of process seems to be operating increasingly in what Friedmann (1986) has called global or world cities. In Los Angeles there has been a clear pattern of informal sector employment and of sweated labour linked with the booming formal sector in that city (Morales 1984). Similarly, Sassen-Koob suggests that in New York:

> the existence of major growth sectors, notably the producer services, generates low-wage jobs directly, through the structure of the high-income life-styles of those therein employed and through the consumption needs of the low-wage workforce. Even a technically advanced service industry such as finance generates a significant share of low-wage jobs with few skill or language proficiency requirements. High-income residential and commercial gentrification is labour-intensive. And so is the massive array of low-cost service and goods-producing firms selling to the low-wage workforce.
>
> (Sassen-Koob 1986: 99)

I would argue that similar processes have been occurring throughout Latin America over the past thirty years. The reappearance of an informal sector and low-wage jobs is new in New York and Los Angeles, but it is a phenomenon that has never disappeared in Latin America.

It is my impression that all three of these hypotheses have a certain validity and apply to varying extents in different parts of Latin America. Indeed, it would be surprising if there were any real consistency in the pattern of employment change throughout the region. Since the economic structures of different cities vary so much, and because the economic records of different countries are highly diverse, it is not surprising that patterns of formal/informal sector employment are likely to vary through time and space. This possibility seems to be increasingly recognized, even by those who have argued from a broadly Marxian perspective that the informal ('petty commodity') sector is bound to be incorporated by capitalist production. The tendency to dissolution predicted by some writers seems to have been constantly accelerated and reversed (Armstrong and McGee 1985). The recent regeneration of informal and black economies in the cities of developed countries (Mars 1982; Pahl 1984) underlines the flexibility of the capitalist system. Whether formal/informal sector jobs will increase or decrease, can be predicted with little confidence. The actual outcome depends upon many factors including technology, the income distribution, the pace of economic growth, and position within the world and national economies. In short, it is an unpredictable consequence of a combination of international, national, and local circumstances.

Most work on the formal/informal sector relationship has been conducted in the context of economies experiencing positive, albeit sometimes slow, rates of economic growth. In so far as that condition has changed during the 1980s it is interesting to comment on what effect negative growth rates may have had on Latin American cities. One scenario is that unemployment has risen so rapidly in most cities because the absorptive capacity of the informal sector has been limited by the recession in the formal sector. Of course, this may be a temporary phenomenon as people adjust to the new circumstances. Certainly, if the slump were to continue, then 'urban involution' or 'marginalization' would be a likely response to the direct threat of hunger. Indeed, the case for a marginalization process similar to that argued by Quijano (1974) seems much more convincing under conditions of recession than in an economic boom.

Certainly, Ginatempo (1985), Sandbrook (1982), Singer (1985), and Booth (1985) have all hinted that the informal sector is only functional to the needs of capital in specific circumstances. A particular instance is where the capitalist sector is producing for export. Then the availability of cheap, informal labour to which risky business can be subcontracted is undeniably useful. In addition, the informal sector is useful in reducing the cost of reproduction of the labour force. On the other hand, the informal sector is less functional when the capitalist sector is producing for a domestic market. It can be argued that it is even dysfunctional when the formal sector is attempting to produce goods for a mass market. In such circumstances, low pay in the informal sector actually reduces the market for, say, the textile and clothing industries. While the costs of the clothing industry are certainly subsidized by the informal sector, the very poverty of the people employed in that sector prevents expansion of the textile industry. There are circumstances, therefore, when the benefits of the informal sector to the capitalist sector are at best dubious. In such circumstances, the informal sector may not be needed by capital either as a labour force or as an industrial reserve army. 'The inevitable fate of this labour force, within the existing system of production, is to take refuge more and more permanently in the "marginal pole" of the economy' (Quijano 1974: 399). Under conditions of recession, the marginal pole may increase in size. This is particularly likely in certain declining regions of a country. Such a model of 'overpopulation' has been suggested by Ginatempo for southern Italy. In this region, 'the low-income population is facing a worsening of its already poor living conditions. This population, in order to survive, accumulates income from precarious activities and from welfare assistance and has limited access to the consumption market and services provided by the state' (Ginatempo 1985: 106-7). Remove the reference to welfare assistance, and the analogy with many parts of urban Latin America appears close.

It seems that there are at least two processes occurring simultaneously in most cities. First, economic growth in the formal sector creates all kinds of opportunities for informal work. Second, those who cannot obtain work in either the formal or the 'dynamic' informal sector either remain unemployed or seek work in the 'marginal' sector; for many the latter is their only chance of survival. What occurs in any given city depends upon the dynamism of the formal sector but also on the rate of growth of the pool of potentially employed workers. Clearly, these different processes are hidden in most of the debates about the informal sector. Indeed, perhaps it is this very concept, together with those masquerading under other names, that is confusing our understanding. For, as has been widely recognized, the sector is very difficult to define and is really a catch-all for a highly heterogeneous group of activities. Included within it is everything that does not fit elsewhere. In order to understand what is happening in Latin American cities we need more detailed breakdowns of the employment sector. Approaches such as that of Sassen-Koob could helpfully be employed in Latin American cities.

Nevertheless, it seems likely that a temporary increase of workers eking out an existence through some kind of marginal employment is a likely outcome of the current recession. Begging, crime, street vending, fire-eating, and other forms of 'marginal' employment may well be a survival strategy to face out a recession. Fortunately, given the general growth characteristics of Latin America, most economies are slowly recovering from the current economic setback and, therefore, many of these 'marginal' activities will contract. Nevertheless, in some countries, it seems possible that some portion of the population will be rendered permanently 'marginal'. The degree of expansion of this sector will depend upon whether there is a continuing supply of labour requiring work which cannot be absorbed by the formal or the dynamic informal sectors. So long as there is a large migratory flow from the countryside, or an influx of city-born young people into the urban labour force, the informal sector will continue to be needed as a source of low-income jobs. I hope I am wrong, but forecasts of the dissolution of the informal sector seem hopelessly optimistic in these days of robotic technology, and when many economies face increasing difficulties in integrating themselves more fully into the world division of labour. Indeed, the problem for many Latin American countries is not the effect of NIDL3 on most parts of the region, but its absence.

Urban consumption

Consumer tastes in Latin American cities have been dominated by foreign influences ever since the Spanish and Portuguese arrived. During the twentieth century the source and breadth of that influence have changed, emanating more now from North America and Northwestern Europe, and reaching down through the whole hierarchy of Latin American society. The car, urban architectural design and planning, and the consumer society have come to dominate Latin American cities. Topography and climate apart, the region's cities today look very much alike. As Felix notes:

> The 'international demonstration effect' is not a recent creation of Hollywood and Madison Avenue, but has been a behavioural trait of the Latin American affluent classes since the Conquest that...has probably been unmatched in its intensity by the behaviour of the affluent of other LDC regions.
>
> (Felix 1983: 7)

There are several elements of this process that repay further comment. First, during the twentieth century, developing patterns of urban consumption have had a profound effect on the built environment. In particular, the growth in car ownership and the imported preference for suburban-style housing has encouraged low-density residential patterns in elite and middle-income areas. In turn, this has been expensive in terms of infrastructural investment, particularly that devoted to roads, water, and sewerage. Road construction in particular has become a major industry in most large Latin American cities. Most large Latin American cities have good roads, particularly in the higher income residential areas. Bogota with its series of transport construction plans over the years, Caracas with its motorways, and Mexico City with its ejes viales all fall into this category. The generalization fits less well in the lower-income parts of the city, but there other services are arguably even worse.

Second, the globalization of the culture of Latin American societies has helped incorporate increasing numbers of workers into the urban labour force. If taxes were a principal way in which the British incorporated the labour force into paid labour in their African colonies, the lure of consumer durables has played a similar role in Latin America. 'It is now possible to be aware of distant places, of life-styles, of consumer goods through the vicarious experience of the electronic media. ... These media ... perform a vital role in creating a global shopping centre, (Dicken 1985: 114). In Latin America, parts of this shopping centre have been open to the urban poor. While noting important variations between countries, many consumer durables have diffused down the social scale (Warren 1980; Wells 1983; Felix 1983). The frequent, and often overly critical, observation that most slum dwellers have televisions is borne out in Table 3.4. Most poor homes in Buenos Aires, Caracas, Mexico City, and Sao Paulo do have televisions, and almost all have radios. It is probable that it has been the allure of consumer goods that has attracted many to the higher incomes of the cities (Armstrong and McGee 1985). In Malaysia, they argue, it is the desire to participate in modern consumer society that has been the principal force driving women from the countryside to the towns. Since women have long migrated proportionately in greater numbers than men to the cities of Latin America, this argument is possibly even more true in these countries. Without espousing a 'bright lights' rationale of migration, with the implication that the migrants are somehow misguided, the globalization of consumer wants has clearly played a major role in urban expansion.

Third, Felix (1983) argues that consumer tastes have sucked in foreign imports and have helped destroy rural artisan industry very quickly, compared with what happened, for example, in Japan. Such tastes have also led to a distortion in the production process. For example, Brazil produces a very high ratio of cars to trucks, compared to what was produced in Japan when the latter was at a similar stage of development. In short, consumer tastes have helped

Table 3.4 Latin America: persons per television

	1960	1980
Argentina	45.8	5.3
Bolivia	-	16.4
Brazil	59.6	8.2
Colombia	105.0	11.5
Chile	-	9.1
Mexico	56.0	9.3
Peru	307.9	20.7
Uruguay	101.2	8.1
Venezuela	30.5	9.1

Source: Wilkie and Perkal (1985)

distort the development path and even accentuate income inequalities. This argument is a clear variation on the, in hindsight less convincing, arguments of Furtado (1972/3) who argued that the Brazilian model of development favoured income concentration in order to re-create markets for constantly changing consumer durables. Sufficient numbers of new models of cars and televisions could only be sold to high-income groups if the incomes of these groups were increasing rapidly. It seems more likely that income concentration was a more influential factor on consumer demand than vice versa, although in Brazil it is possible that both processes were occurring.

The demand for consumer goods was clearly a stimulus to, and a consequence of, the process of import-substituting industrialization. Over the long term, however, the rampant development of consumerism in Latin America has damaged the prospects for sustained economic growth by putting heavy pressure on the balance of payments. As Felix argues,

> excessive preference of the Latin American middle and upper classes for jam today ... has been a major reason why periods of fast economic growth and rising inequality have tended to terminate rather quickly in balance of payments crises, leading to crises of governance. Capitalist regimes in Latin America that impose income-concentrating policies may induce fast economic growth, at least when external conditions are

> favourable; but unless they also constrain consumer behaviour, they cannot sustain that fast growth for long.
>
> (Felix 1983: 29)

Adapting this argument, it is easy to argue that the current effects of the world recession were accentuated in Latin America by rampant consumerism. This is most obvious in those places which most flourished during the boom of the 1970s: Caracas, Mexico City, and Sao Paulo. In each of these cities, consumerism was notable for its exaggerated presence. Oil revenues in two cities, and the benefits of Latin America's most sophisticated and largest industrial machine in the last, helped produce consumer and real-estate booms that far exceeded the long-term sustainability of the economic machine. Just as individual families bought consumer durables and new homes on credit, so national and local governments contracted new loans on the prospect of a continuation of the world boom. Increases in government expenditure in Mexico and Venezuela went up in direct proportion to increases in oil prices (Cornelius 1985). Loans were contracted on the basis of future production. Like their societies, these governments wanted tomorrow's jam today. The fall in oil prices has now left jam on the face.

The impact of the world recession has been accentuated, therefore, by Latin America's elastic demand for consumer goods. As a consequence of excessive spending during the boom, the domestic slump has been very great. The slump has had a double impact on domestic demand through increasing prices and falling incomes. It has raised the price of new consumer goods due to high rates of inflation in most countries. It has raised the prices of essential products, thereby dramatically cutting real incomes and the surplus available for purchasing consumer durables. In some cases, too, incomes have fallen through short-time working or unemployment. In those countries where consumer durables are provided predominantly by domestic industry, the economic impact of this process has been considerable. On the social effects, it is more difficult to comment. Clearly, consumer wants remain unchanged even if purchasing power falls. I suspect that little will change as a result of the recession and that when economic growth accelerates, as it surely will in most of the region, the hunger for consumer goods will be maintained. And, in so far as consumer durables have been bought on credit, the majority of the population are still paying off what was contracted during the boom.

Residential development

The construction and development of residential and commercial building has long been an important source of profit for indigenous elites in Latin American cities. Profits have been generated through the production of building materials, the construction process, and perhaps most notably, through land speculation. Indeed, during the early phases of urbanization it was natural for rural landowners to begin to invest in real estate in the urban areas; land was a business with which they were thoroughly acquainted. The combination of economic and urban growth was bound to bring profits and it therefore attracted much indigenous capital. The profits could be increased still further by making suitable contacts with private business and with the state. The latter was necessary in terms of planning permission and to obtain public works contracts. Throughout Latin America the state was a major investor in infrastructure, housing, and roads - and political links were an important source of capital accumulation. Indeed, in the oil economies, where the state has controlled the majority of investment funds, it has determined which groups will participate in, and which will be deprived of, a share of the oil wealth (Gilbert and Healey 1985).

Perhaps surprisingly, the Latin American construction industry and real estate business seem to be mainly in local hands. This is certainly true in Mexico (Connolly 1985), Colombia (Gilbert 1981), and Venezuela (Gilbert and Healey 1985). Unlike the Middle East and certain other parts of the Third World, the transnational engineering and construction industries seem to be mainly involved in building ports, mass transit railways, and industrial plant. Residential and commercial development seems to have remained mainly in local hands. In Venezuela, it is controlled by large national conglomerates, in Colombia a handful of national companies handles the most profitable real estate developments. Indeed, in many provincial cities there has almost been a division of profits between international, national, and local capital. In Valencia (Venezuela) for example, the first has been prominent in manufacturing and construction, while local groups have been prominent in manipulating the urban land market and residential development.

With respect to residential development, the large construction and real estate companies have been concerned only with elite, middle-income, and government housing projects. They have been little involved with housing for the poor. This poor majority has been unable to afford homes in serviced, legalized commercial settlements, whether those homes were built by state or private initiative. The poor's response has been to build their own homes and the number of families living in self-help settlements has increased as a proportion of the total urban population throughout the last thirty years or so (Table 3.5). Poor families have been forced to obtain land through a variety of illegal processes. The precise mechanism has varied according to what has been tolerated by the local and national state. Certain governments have permitted political supporters to invade land; elsewhere large-scale invasions have been prohibited by the authorities (Collier 1976; Gilbert and Ward 1985). In the latter case, the poor have been required to purchase land illegally, in ejidal areas, or on private land without planning permission, or on land beyond the service lines (Ward 1982; Gilbert et al. 1982). These settlements have almost invariably occupied the cheapest and environmentally poorest land in the city. Distant hillsides, areas with difficult drainage, and land close to industrial areas, have been the typical preserve of the poor in Latin American cities.

Self-help housing has played an increasingly important role in accommodating Latin American urban populations. Fifty years ago, the urban poor were mainly housed in rooms in the central city area. While the vecindades, corticos, inquilinatos, and conventillos still exist in the central areas of most cities, today there is a much greater possibility of the poor constructing their own self-help accommodation. The state throughout Latin America has been forced to accept that the vast increase in the urban labour force could not be accommodated through large-scale rental housing. The response has been to tolerate the expansion of quasi-legal housing alternatives that allow the poor to construct their own dwellings. Even if the state in most places has only recently begun to accept self-help as a philosophy (Turner 1967; Lloyd 1979), they have long accepted self-help in practice. Without such a solution the state would have been forced to provide more state housing or be threatened by a major housing crisis. As it is, despite the obviously low

Table 3.5 Relative growth of irregular settlements in selected cities

City	Year	City population	Population in irregular settlements	Percentage
Caracas	1961	1,330	280	21
	1964	1,590	556	35
	1971	2,220	867[i]	39[i]
	1985	*	*	61[ii]
Lima	1961	1,846[iii]	317[iii]	17[iii]
	1972	3,303[iii]	805[iii]	24[iii]
	1981	4,601[iv]	1,150[iv]	25[iv]
	1981	4,601[v]	1,460[v]	33[v]
Buenos Aires	1956	6,054	109[vi]	2[vii]
	1970	8,353	434[vi]	5[vii]
	1980	9,766	957[vi]	10[vii]
Rio de Janeiro	1947	2,050	400	20
	1957	2,940	650	22
	1961	3,326	900	27
	1970	4,252	1,276	30

Source: unless otherwise noted, United Nations (1979)

Notes:

- * not available
- i OMPU (1972)
- ii FUNDACOMUN (1985)
- iii Official census figures cited in Sanchez-Leon et al. (1979)
- iv Official census figures cited in Dietz (1987); these figures are probable underestimates
- v Matos Mar (1985)
- vi Estimates based on census data for total population multiplied by shares of villa miserias cited in column three
- vii Proportion of housing units in Gran Buenos Aires classified as villas miserias; Abba et al. (1986)

standards of construction and servicing in many cities, there can be little doubt that much of this housing is adequate, judged by the traditional accommodation standards of the poor. Over time, many families have managed to construct thoroughly adequate houses, obtain services and infrastructure, and as a result establish a strong foothold in the commercial housing market in the city. In addition, the consolidated self-help areas now form the main source of rental accommodation for the recently arrived, the young, and the poor of Latin America's cities (Edwards 1982; Gilbert 1983). Despite the unfavourable press that self-help housing still receives in most Latin American newspapers, the self-help home is usually more spacious than, and has several other advantages over a rented room. For this reason, most Latin American families would choose to occupy a self-help house if they could. Indeed, government support for self-help housing is increasingly becoming the norm in Latin American cities and most illegal settlements soon become legalized.

NIDL3 has had little effect on the housing process. On the other hand, the world recession and its impact on Latin American economies has been substantial. The construction industry is always cyclical; indeed, its fluctuations are normally far more marked than those of other sectors. The current recession has had a major impact on the construction sector and this is one reason why urban unemployment rates are at present so high in Latin America's cities. Without government contracts and with even middle-class incomes declining, demand for building has been limited; in the oil-rich economies, government construction budgets have plummeted in recent years.

If it has become less profitable to build, it has also become more expensive (even for middle-class families) to buy a house or apartment. Even for the poor, the rising real cost of building materials and, where invasions are prohibited, of land, have made even self-help an option open only to the regularly employed. The outcome has varied from city to city. In places where land invasions are strongly discouraged, the rising cost of self-help seems to be leading to construction on smaller plots and in time is likely to lead to fewer families building their own homes (Gilbert and Ward 1985). This decline in self-help seems very noticeable in cities where recession has bitten deeply, and where authoritarian regimes have prevented land invasions taking place. The number of allegados (guests) in Santiago has reputedly increased dramatically as a result of the recession. Similarly, many young adults who might otherwise have constructed their own accommodation are likely to be renting rooms. It is only where political democracy is highly competitive or where, as in Brazil, the unsavoury fist of authoritarian government is receding, that land invasions seem to have increased in frequency (Sachs 1983).

The current recession, therefore, is affecting the housing situation in several ways. First, it is reducing employment in the construction sector. Second, it has led to increased crowding in the housing stock. Third, it has cut budgets for servicing the low-income settlements. Fourth, it has reduced the possibilities for capital accumulation open to local elites. Fifth, it has reduced incomes throughout society and therefore limited the effective demand for housing. In this sense, the current recession has brought further complications to an already difficult situation.

Standards of living

A constant theme in the literature on Latin American cities concerns the declining standard of living of the poor. Many writers insist that there has been consistent immiseration of the Latin American urban poor. While there has been a sad deterioration in living standards during the recent crisis, even for many among the middle class, there seems to be little doubt that the long post-war boom improved living standards in most cities of the region. Felix notes that the post-war period brought many more benefits for the urban poor, certainly in comparison with what he calls the immiserating, export-led growth of the pre-1930s. He argues this point partly on the basis of trends in a Physical Quality of Life Index (PQLI) composed of three variables: infant mortality, life expectancy at age one, and adult literacy.

> In 1950 the PQLI ranged from 36 for Guatemala to 77 for Argentina, while in the mid-1970s, it ranged from 43 for Bolivia to 90 for Puerto Rico. All seventeen Latin American countries for which the index has been computed raised their PQLI between the two dates, although in varying degree.
>
> (Felix 1983: 10)

Although the improvement of national living standards has been all too limited, the general pattern is clear. If this is the case across Latin American nations as a whole, it is clearly true of the major cities which have benefited most from the process of economic growth.

Even if urban conditions in Latin American cities obviously vary, the impression so often given by the literature - that the living conditions of Latin America's urban poor have been constantly falling - is highly erroneous. Not only have most cities managed to absorb their rapidly increasing populations, but most have actually improved the levels of infrastructure and service provision over time. Table 3.6 shows that the proportion of households with access to piped water has increasing in most cities. Today, electricity is to be found in practically all the low-income settlements. Transport is poor, but underground railways and crowded buses get just about everyone to work. In addition, the quality of housing, as measured by indices of overcrowding, space per person, and service provision, shows few signs of deteriorating. Between 1970-80, for example, Mexico City's population grew by around 5 million people, and yet the proportion of households with access to electricity, water, and sanitation increased (Gilbert and Ward 1985). Similarly, service provision in most Colombian, Mexican, and Venezuelan cities improved during the 1960s and 1970s. Although increasing proportions of Latin American urban populations are living in self-help housing, the general consolidation that tends to occur over time in the structure of the housing unit, suggests that this process has brought gradual improvements to the housing stock rather than any general deterioration. Certainly, living space would not seem to have become more restricted in most cities, the proportions of owner-occupiers has increased and service provision has generally improved.

A similar conclusion has been reached in a detailed study of living conditions in a single city, Sao Paulo (Wells 1983). Based on a wide array of survey and census information, it shows that living standards in that city improved gradually during the period 1930-75, even for the poor. This conclusion is very different from that of most Brazilian writers, who depict the nature of the Brazilian development process as one of constant immiseration. Such an argument is clearly apparent in the analysis of Da Camargo et al., who state that:

Table 3.6 Urban water supply for selected Latin American countries

	Population with water supply to, or close to, house (%)		Population with sewerage (%)		Dwellings with electric light (%)	
	1959–61	1976–79	1960–61	1967–77	1960–63	1970–76
Argentina	65.3	69.3	42.1	33.4	84.7	-
Bolivia	55.8	81.9	40.4*	42.4	76.4	76.2
Brazil	54.7	74.6	54.5*	34.3	71.5	75.6
Colombia	78.8	86.9	61.4	67.8	83.4	87.5
Cuba	56.6	91.7	32.4*	46.6	-	98.6+
Chile	73.6	92.9	59.5	62.4	86.3	-
Ecuador	58.2	89.3	53.2	71.7	78.5	84.3
Mexico	67.5	73.2	70.4*	41.1	-	80.7
Peru	59.5	79.0	51.3	54.9	50.7	58.3+
Uruguay	73.7	93.2	50.8	61.7	87.8	89.2
Venezuela	54.5	94.1	30.2	52.3	-	-

* 1964 + 1981

Source: Wilkie and Perkal (1985)

> if it is true that malnutrition, unsatisfactory health conditions and inadequate housing - to cite only a few of the symptoms of poverty in the city - are graver in other regions, it is equally clear, that these 'problems of underdevelopment' ... are not disappearing in Sao Paulo, rather, in many cases they are tending to increase.
>
> (Da Camargo et al. 1975: 21)

More recently, it has been suggested that the recent history of Sao Paulo has brought further deterioration:

> economic progress has in fact been marked by the growing pauperization of significant sectors of the Paulista working class. ... Throughout the 1970s, economic growth guaranteed job creation incorporating large numbers of urban workers. This at least cushioned the social consequences of wage deterioration. The present decade, however, has drastically reversed this process, as the number of jobs is dwindling and unemployment has sharply risen.
>
> (Kowarick and Campanario 1986: 163, 167)

It seems to me that these analyses were overly pessimistic. Despite the severe holding down of industrial wage rates and the pay of low-income groups, the pace of economic growth in Brazil was so fast, that all social groups benefited (Brundenius 1981). Da Camargo et al. are correct in arguing that the benefits of growth were distributed regressively; the rich certainly gained more from the economic growth process. Nevertheless, Wells' analysis clearly demonstrates that up to 1975, workers in Sao Paulo were also gaining from the process of incorporation.

Equally clearly, the recession of the 1980s has had profoundly negative effects on the welfare of the poor in Sao Paulo and in most other cities. It is not yet clear whether the income gains of the past fifty years were erased during the recent plummeting of real incomes, although this is conceivably the case. What has certainly not been erased is the general improvement in sanitary standards and participation in the consumer market, particularly with regard to consumer goods. If the recession were to continue, then the poor would clearly lose many of the benefits that they have gained.

The lesson seems to be that even under authoritarian and inequitable conditions, rapid economic growth does produce benefits for the neglected majority. The real problem comes when there is slow economic growth, or worse, when growth rates become negative. In such circumstances, unemployment rates rise, real incomes fall, and not only the poor but even the middle class begin to suffer. This is obvious in the extent to which real incomes have declined recently in Latin American cities. The consequences of such falls in income have been demonstrated in many cities over the past couple of years. But, it is equally clear that economic growth, if it can be recaptured, does raise incomes even for the poor. This was shown in Brundenius' (1983) comparison of Brazilian and Peruvian living standards; whereas the bottom quintile of the Brazilian income distribution has increased its real income, the same quintile in Peru has seen both an absolute and a relative decline in real income.

CONCLUSION

NIDL3 has had relatively little impact on the organization of production in most Latin American countries. Most of the changes that have occurred since 1960 have been related to an increase in the export of primary products (NIDL1) and to continued growth of industry through import substitution (NIDL2). Nevertheless, the integration of Latin American economies into the world system has increased, as it has for most other parts of the world.

This increased participation has both provided opportunities and created major problems. During the post-war boom, the opportunities were in the ascendant, and most Latin American economies prospered as a consequence. As urban areas were major beneficiaries of this expansion, they grew and prospered. Metropolitan cities in Latin America saw an expansion in the size of the middle class (Portes 1985b) and most saw an improvement in general living standards. Needless to say, the local impact of the expansion varied greatly between countries and between cities. Some oil economies grew dramatically, other economies sometimes grew slowly; in places the urban poor participated, albeit less fully than the rich.

Most of the benefits of economic expansion accrued to the metropolitan and larger cities. The built environment was transformed by the demands of industrial, commercial, and transnational capital. Manufacturing activity and related producer services demanded better infrastructure and services. Much of the cost of infrastructure was built on the basis of loans from outside the region. While economic growth was continuing during the 1970s, the debt posed little real problem. Some autonomy over decision making was relinquished but the advantages gained were clear. Electricity provision and water services became available to the mass of the population. If autonomy in the form of local patterns of clientism had to be forsaken, then most local elites determined that it was a price worth paying.

Another cost of higher living standards was that Latin American cities looked and felt increasingly like those of North America. This was a mixed blessing, but one to which many Latin Americans willingly acceded. They wanted cars, suburban housing, supermarkets, and other features of the American dream. It was unfortunate that local cultural values were weakened and that environmental pollution increased. It was a pity that cities became larger and larger and more difficult to control. But, in general, this integration into world consumerism was what most Latin Americans demanded. If there was a loss of autonomy, it was a price they deemed worth paying and it was, after all, a precedent that had been well established under NIDL1.

The impact of economic growth on the urban areas, therefore, was generally positive. It led to prosperity for many and to limited participation by the majority in the fruits of economic expansion. Of course, the development of Latin America's cities occurred in a thoroughly inequitable way. As a consequence, the poor suffered immensely from bad working and living conditions. Even if they had wanted to protest, the opportunities to do so were severely limited in many countries by authoritarian and repressive military regimes. But, on the whole, while economic growth continued there were some gains even for them; most poor people in Latin America seem to have accepted that fact phlegmatically.

The main problem has come now, with the end of the boom. For the current recession has brought major falls in salary levels, has increased the fiscal crisis of the state, has cut both private and public investment, and has forced the state into making choices that it had hoped to avoid. In places, the impact of the recession has been truly dramatic. In part, this has been an unfortunate and unforseeable consequence of a downturn in the world economy. All countries have been affected and Latin America has shared in the general gloom. Nevertheless, it is quite clear that most Latin American countries have suffered to a greater extent than most developed countries, and the recession has severely limited local autonomy over urban change. The cut in government budgets - imposed largely from outside - has brought problems. Without economic growth, the ability to invest in the provision of essential services is curtailed. But the situation in Latin American countries has been accentuated by a triple problem: first, they have contracted debts that need to be repaid; second, in order to renegotiate their debts they are being told by the lending agencies to implement policies of governmental financial stringency; and third, they have often postponed reforms which should have been made under growth conditions. It is for these reasons that local and national governments are being forced into making very difficult political choices.

The fact that Latin American governments are being faced increasingly with harder choices about government expenditure and investment would be tolerable in a period of economic prosperity. But the situation is that they are affecting certain reforms at the same time as the population at large is being severely affected by falling real incomes. In addition, they are being pressed from outside to improve their domestic housekeeping, often in ways that are arguably very misguided. Of course, many would recognize that some of the changes are both necessary and should have been made during the good years. For example, most Mexicans realize that metro fares have long been too cheap; similarly in Venezuela, the price of petrol internally has been kept too low. But, to make such politically sensitive reforms at a time of grave economic recession, and at the behest of outside lenders, is a triple burden. The unavoidable consequence will be greater hardship for the majority of urban dwellers and particularly for the poor. There are also likely to be some intriguing political realignments as a result.

NOTES

1. All the figures in this section are taken from CEPAL 1985.

2. The Dutch Disease affects some countries facing a

resource boom. The boom affects the structure of the economy in two ways. As described by Neary and van Wijnbergen (1986):

> the first is the 'spending effect': higher domestic incomes as a result of the boom lead to extra expenditure on both traded and non-traded goods. The resulting increase in the price of non-traded (relative to traded) goods causes the output of the traded-goods sector to decline. A second effect emerges if the booming sector shares domestic factors of production with other sectors, so that the boom bids up the price of these factors. The resulting 'resource-movement effect' reinforces the tendency towards appreciation of the real exchange rate (defined as the relative price of non-traded to traded goods). The result is a squeeze on the tradable-goods sector. These predictions are common to most models of the Dutch Disease and have been observed in a variety of countries.

They have clearly been apparent in Latin America in both Mexico and Venezuela.

Chapter Four

Industrialization and household response: a case study of Penang

Mei Ling Young and Kamal Salih

INTRODUCTION

Historically, the advent of industrialization, whether autonomous or dependent, has always been accomplished by transformation of the organization of production, labour forms, and territoriality. It has involved a decline of subsistence production in the household, an increase in wage labour, and massive movements of people between territories. It has also been accompanied by urbanization produced by the spatial concentration of population and productive activity, of social infrastructure, and of the means of collective consumption, and has been characterized by urban forms that reflect different degrees of rationalization of production structures, formal or informal. Each of these transformations has had social effects in the technological, territorial, and sexual division of labour, in labour force formation, and in the levels of welfare of households. The state has usually been the prime mover, bringing about increasingly complex forms of state activity. On a world scale, industrialization has required the incorporation of newer territories both as sources of raw materials and as markets, and the accumulation of capital through the internationalization of productive activity.

The process of industrialization on a world scale has been uneven, both territorially and temporally. The cycles and rhythms in the development of the capitalist world economy result in a differentiation between zones of capital accumulation and appropriation and zones of consumption and reproduction of the labour force. Historically, this has led to a specialization between territories through changing divisions of labour during the various epochs of expansion and contractions. Recognizable, though shifting, cores of the capitalist world economy have taken their turn in exercising hegemony over the system as a whole during periods of expansion, while new zones of accumulation have emerged during the contraction phases as the older centres of accumulation declined. The downturns in the development of the capitalist world economy represent periods of crisis, during which such a restructuring of the previously dominating model of world accumulation have taken place. This restructuring has affected not only the international division of labour, but also the internal production structures, labour force formation, and the household economy. The nature and extent of these transformations reflect local, cultural, socio-political, and economic specificities.

The secular transformation of societies, central or peripheral, under the impetus of the self-expansion of world capitalism, masks the cyclical patterns in the development of the world economy. The present world conjuncture, which began to take shape in the late 1960s and the early 1970s, suggests a period of reorganization of the post-industrial economies as a result of technological change, increased competition from newer centres of world production, and stagnation. In the transitional economies of the periphery, some graduating to semi-peripheral status in the 1970s, rapid industrialization meant constant adjustment pressures in economic and social organization. It is true, however, that in peripheral formations the industrial transformation is incomplete and is limited to certain branches of production by the constraints and possibilities of the new international division of labour.

The impact of industrialization cycles on the household economy has been the subject of considerable examination in recent years. The focus of historical demographers, social historians, and historically oriented sociologists and economists concerns the theoretical and empirical issues involved in the household economy. These issues include the role of the household as a production and consumption unit, the sexual division of labour, the relationships between patriarchy and the capitalist mode of production, the domestic labour debate, and household adaptive strategies.

In post-industrial societies, the problems of structural unemployment, that is an increase in surplus population and the diffusion of informal activities caused by stagnation of the world economy in the present conjuncture, have renewed the interest of western researchers in the household economy. For the industrializing societies of the periphery, a surplus population has always been an endemic issue, just as the informal sector, in response, has been a persistent structural condition of dependent industrialization. In the 1970s, a number of these peripheral economies achieved high growth rates through rapid industrialization under an emerging new international division of labour. This was based on the redeployment of capital to specially created export-processing zones in the periphery, where cheap labour was used to assemble components into consumer products or other intermediate products for export to the industrialized countries. This cheap labour is usually drawn from surrounding rural areas or urban surplus population. The incorporation of this new labour force constitutes a process of social transformation of major importance in recent times. The question is whether this implies a genuine proletarianization process.

The purpose of this chapter is to consider some issues related to labour force formation and the role of the household economy under the above conditions of industrialization, and its conjunctures, in a peripheral formation. The particular case study reported here is part of a larger project examining the impact of recent rapid industrialization on social change, using the household economy as the immediate site of the unfolding macro-processes.[1]

The section which follows discusses the theoretical and empirical issues concerning the relationships between industrialization and household processes, drawing on experiences in the post-industrial societies. The Penang case study is a culturally specific experience of the impact of macro-social processes on the household economy in the periphery. These issues are elaborated upon later in the chapter, when two techniques, namely narrative individual case studies and family life course, are used to highlight (and empirically to reflect) some of the issues on household and individual responses to industrialization. It is clear from these case studies that household practices, including adaptive strategies, differ according to household circumstances. This variety of household practices suggests not only the cultural and local specificity of the household economy in the production process of the whole economy, but also the different ways it articulates with the broader processes.

INDUSTRIALIZATION AND THE HOUSEHOLD ECONOMY: A THEORETICAL EXCURSION

The argument suggested here is reflected in Figure 4.1, which relates the household economy in the periphery (or semi-periphery) to the industrial cycles in the centre. As argued in the introduction, the present world conjuncture results from changes in the international division of labour which led to the emergence of new centres of accumulation in the periphery and the decline or stagnation of old centres of accumulation. In specific zones of rapid industrialization, urbanization took place which contributed to significant changes in the local economy, including labour force formation. One significant development was the creation of a new female industrial workforce in these zones, drawn largely from the rural areas and, therefore, out of domestic production. During a conjunctural downturn, this local labour may be retrenched and is reabsorbed into the household economy. At the same time, due to the nature of export-platform industrialization, the incorporation of labour into the new zones was insufficient to absorb the pre-existing surplus population. This produces a situation where the informal sector in the periphery functions as the site of reproduction of the labour force during cyclical downturns (local conjunctures), as well as representing a persistent structural phenomenon of dependent industrialization. Stagnation or decline in the central economies, however, produces local conjunctures which heighten the process of 'informalization' and household practices such as survival strategies for labour affected by the recession (see Redclift and Mingione 1985; Pahl and Wallace 1985). While such post-industrial societies achieved high levels of industrialization during the pre-existing division of labour during earlier periods, the present world conjuncture has seen a decline in employment in the centre generating a surplus population that could not be supported by the state owing to its own fiscal crisis. As a result, a resurgence of self-provision of services occurs which recaptures the important role of the household economy and domestic labour.

The concern with the relationship between industrialization and the household economy under capitalist development revolves conceptually around the changing forms of the family during the capitalist transition (Levine 1977). As succinctly stated by Harris:

Figure 4.1 Industrialization: conjunctures and the household economy

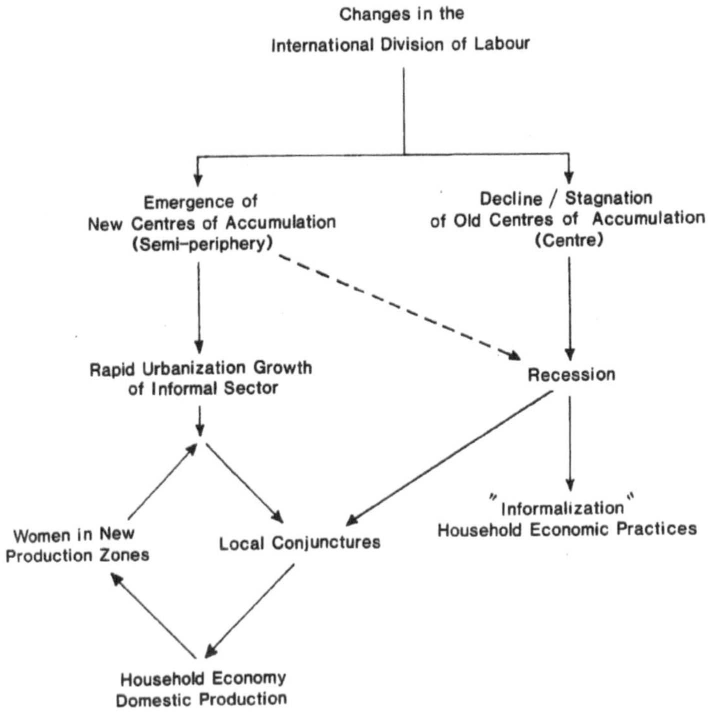

> The family form which emerged from industrialization is to be understood ... as the result of the interaction of the development of a capitalist mode of production with pre-existent family forms.
>
> (Harris 1983: 185)

This involves a transformation from an agrarian economy in which the household is a production unit to one in which the household is linked to the productive process only through the wage labour of the male. In this transformation, three transitions within the household economy take place:

(i) from domestic labour to wage labour;
(ii) from a co-operative unit between husband and wife in the domestic mode of production to a unit which excludes women from the productive process; and
(iii) from a productive unit to a unit of consumption.

Thus it can be seen that:

> The development of the factory system of production brought into being a new form of family whose functions were the reproduction of labour power and consumption. Generational reproduction includes biological reproduction, sexual regulation, and socialization, whilst daily reproduction involves domestic labour. The family is also the agency for the reproduction of class and gender relations, that is the transmission of property and the inculcation of patriarchal ideology on both a daily and a generational basis. As a unit of consumption, it is essential to the process of commodity circulation.
>
> (Beechy, cited in Harris 1983: 196)

The domestic domain over the last decade has become a major focus of theoretical attention. Redclift has identified three issues in this debate. The first traces the development of the argument concerning the persistence of pre-capitalist subsistence production based on the household in less industrialized countries. This is epitomized by the work of Meillasoux, 'who sees the continuing exploitation of the domestic economy as fundamental to capitalist development itself, and the family under capitalism as the direct successor of the domestic agrarian community' (Redclift 1985: 93-5).

A second strand of thought concerns the informal economy under advanced industrialization, since restructuring and the decline of full employment 'have also revealed the importance of economic processes that are not encompassed by commodity relationships or by the accepted definitions of the capitalist mode of production' (ibid).

The third source of theoretical concern on the domestic domain is the feminist critique 'which has emphasized the limitations of the definitions of the economic system which exclude the reproduction of use values and the reproduction of labour power, processes that lie at the core of the conception of the household and domestic sphere' (ibid).

This last strand emphasizes the question of the sexual division of labour and the role of patriarchy in explaining the subordination of women in the capitalist mode of production. These various, sometimes conflicting, elements have focused attention on the changing relationships between wage work and 'other' forms of work not directly organized by capital. Redclift concludes that,

> despite competing accounts of its nature and status, the household and the domestic domain have emerged not merely as the incidental location of the workers' natural instinct for self-preservation ... but as fundamental to the analysis of the economic system.
> (ibid.)

Based on the above arguments, the household is a significant unit for the study of societal change. Not only is it the most common and basic socio-economic unit but, as Laslett notes, 'the interplay of its size and structure with economic and demographic development, make up an intricate adaptive mechanism' (Laslett 1969: 11). According to Netting (cited in Netting et al. 1984: xiii), in a peasant agrarian economy the household is perhaps the most flexible and responsive social grouping, sensitive to minor, short-term fluctuations and a prime means by which individuals adapt to the subtle shifts in opportunities and constraints that confront them.

How do we define the family and the household? Without going into the complicated cultural contexts and debates (see Netting et al. 1984), suffice it to note that the major difference between these two closely related concepts is that the household is a task-oriented residential unit while the family is a kinship grouping, which may not be localized. However, it is through the commitment to the concept of family that people form households to enter into relations of production, consumption, accumulation, transmission, and socialization. Thus households are

> a primary arena for the expression of age and sex roles, kinship, socialisation, and economic cooperation where the very stuff of culture is mediated and transformed into action. Here, individual motives and activities must be coordinated and rendered mutually intelligible .... Decisions emerge from households through negotiation, disagreement, conflict, and bargaining. The

> decision to marry, to build a house, to take in a relative, to hire a maid, or to migrate are seldom made or acted on by isolated individuals, because such decisions necessarily affect household morphology and activities.
>
> (Netting et al. 1984: xxii)

In this study, we extend the household concept to that of the household economy, incorporating family members who do not co-reside (in a household) if they remit money and/or services regularly to the household. These are usually out-migrant working children who may or may not be married. This concept is associated with income pooling as a mode of household organization (Wallerstein 1982). The notion of income pooling represents a distillation of household practices that have come to be known as household adaptive strategies, and which emphasize the empirical dimensions of the study of the household.

Historical demographic studies on the family and household in Europe and in the eastern states of North America, based on family reconstitutions derived from censuses and parish records, have shown how family structures have changed due to the interplay of the processes of mortality, nuptiality, timing of marriage, marital fertility, and migration (see Laslett and Wall 1972; Levine 1977; Lee 1977; Hareven 1978a). These basically quantitative studies have been complemented by the research of social historians in France (Aries 1962; Flandrin 1979) and England (Stone 1979), who have shown how the system of values and norms within families, as represented in the roles of parents, strength of kinship and family ties, and the prolonging of childhood leading to the importance of the role of the family in education, have changed as the authority of the patriarchy, church, and state has shifted. Stone concluded that family change is the ebb and flow of the battle between competing interests and values represented by various levels of social organization, from the individual to the national (Stone 1979: 424).

Another group of research has shown how families, within the constraints of the family cycle and the life course, adapted to a rapidly changing economic environment. Hareven and Langenbach (1978), in their compilation of oral histories in Amoskeag, New Hampshire, and Hareven (1982), in her study of the New England families responding to industrialization, showed how these families functioned as crucial intermediaries in labour recruitment, and acted as active agents in an interaction with the factory system. There was evidence of how families allocated labour, investment, and resources to cope with the vicissitudes of changing industrial needs. Depending on the stage of the family life cycles, decisions were made as to whether members of the family were to be sent out to work, remain in school, or be forced to stop school to look after younger siblings at home so that the mother might be released for factory work. As more children entered the labour force and could contribute to the pool of family income, older siblings were allowed to marry and form their own households.

The argument for a household approach is a clear indication of the disenchantment with the individual approach. The household approach asserts that decisions, and consequently actions, are not isolated events but are part of family decisions where different members are part of a collective unit. To take this concept one step further, to argue for a life-course approach, defined as 'the changing pattern of interdependence and synchronization among the life histories of family members' (Elder 1978: 17) is to accept the view that the family is not a static unit but a changing entity over the life course of its members (Hareven 1978b: 1). Rather than view the family life cycle as cross-sectional stages in the development of the family, the life course examines both the individual and collective family development as a meshing together of roles and decisions. Consequently, rather than identifying stages, the life-course approach examines transitions and individuals' movements through different family configurations, and analyses the determinants of the timing of such movements. Thus, the life course examines the interactions between four types of time (see Hareven 1977; 1978b for an elucidation of these concepts).

'Individual' time concerns the life history of an individual, from birth to death, as he or she passes through the major biological and cultural events of life, assuming different roles, responsibilities, and so on. 'Family' time is closely associated with the family development cycle. Family and individual decisions affect the timing of transitions such as leaving home, entering the labour force, marrying, setting up a new household, child bearing, launching children from home, and widowhood. Crucial to this concept is the process of family adaptation and change over time; the timing, arrangement, and duration of events in the life course. How the family responds and adapts to changing circumstances is constrained to some extent by the family life cycle. For example, parents may be at their most solvent when their children have grown up and are contributing to the household income. Conversely, they may be at their poorest when their children are too young to work or when they themselves are too old to work and have no children remitting money to them. 'Industrial' time refers to the trade and demand cycles of an industry - for example, in this study, the swings of the electronics industry. Finally, 'historical' time is synonymous with chronological time, which concerns the forces of history and society on the household. An understanding of how these 'times' interlock - how individual time is enmeshed within family time, how during industrialization both are affected by industrial time, and how all of these are encompassed by historical time representing the major economic, political, and social processes - provides insights into the dynamics of social change.

Tilly and Scott (1978) approach the issue of family change in France and England through the analysis of women's work as part of the changing family functions which are identified in this progression: subsistence family economy (where members of the household produce solely to maintain themselves, such roles being defined by sex and age), family wage economy (where members of the household are engaged in wage employment), and finally the family consumer economy (where the role of the married woman as mother and housewife is paramount and husbands and unmarried children are the main family wage-earners).

What these diverse studies show is that the evolution of the family faithfully reflects the broader economic, political, and social conditions of a country. Thus, the household, as a unit of production, consumption, accumulation, and socialization is a decision-making unit which mediates between the individuals collectively and the forces outside. As such, the household in its process of change becomes an adaptive unit, a microcosm of the broader structures at the village or community level, from the regional to the national and international levels. But the relationship between the family and the outside world is a dialectic one. While the households change as a result of socio-economic transformations, so too do households themselves contribute to the macro-forces and themselves are evidence of the transformation.

Demographic processes operating within the context of the family may also be seen as household adaptive strategies. On a macro-level, population pressure sets in train a number of responses in agricultural households (see Boserup 1970). On a micro-level, in Japan during the eighteenth and nineteenth centuries declining mortality within the household changed the practice of nuptiality, caused delayed marriages, increased celibacy, and led to declining marital fertility (T.C. Smith 1977). Also in Japan, parents used infanticide as a means of regulating child spacing and the ratio of male to female children in the family in response to decreasing land sizes (T.C. Smith 1977; Hanley 1977). These processes were reversed when there was land availability resulting in high fertility rates (McInnes 1977). Similarly, different inheritance systems, especially that of partible and impartible land in Germany in the seventeenth and eighteenth centuries, caused peasant households to respond differently (Berkner 1977). Thus, where the inheritance system is partible, the number of households increases, as does population. Where impartible inheritance systems exist, there is increased out-migration and celibacy practised by non-heirs. Families adapt inheritance systems to their own needs, as evidenced in the divergence between inheritance customs and practice.

> The inheritance system sets limits, creates problems and opportunities and evokes certain types of behaviour which conform to it or avoid its consequences. Inheritance laws and customs are things that the peasant must deal with in planning a strategy which will reach his goals; they do not determine the goals or the strategy.
>
> (Berkner and Mendels, cited in Hermalin and van de Walle 1977: 110)

This review of the theoretical and empirical sources of the links between industrialization and the household economy brings out an interesting fact: that of the convergence of interest on household practices, to use Mingione's (1985) terminology, within developed industrial countries and that found in the transitional industrial economies of the periphery. We might conclude here with Pahl and Wallace's contention that their work on detailed empirical studies on the household and household work strategies on the Isle of Sheppey

> sharply undermines the more optimistic scenarios of those ... who see people being able to liberate themselves from the tyrannies of markets, exchange values and the capitalist relations of production to something more productive and satisfying in the cocoon of the so-called 'domestic economy'. ... The whole burden of our argument rests on the reality that there is only one economy and that a household's position in that is fundamental in determining its positions in other economic spheres.
>
> (Pahl and Wallace 1985: 224)

The thrust of the project being partially reported here is an empirical study which takes the household as its basic unit of analysis in order to examine the fundamental processes in the transformation of the household in development.

INDUSTRIALIZATION AND HOUSEHOLD CONJUNCTURES: THE PROBLEMATIC OF LABOUR FORCE FORMATION IN PENANG

Since 1972, Penang had seen tremendous changes in industrialization and urbanization, centred around the main conurbation of George Town and Butterworth, and spreading outwards towards the new townships along the eastern coast of the island (Fig. 4.2). From an area of net out-migration, due to a stagnating economy before the 1970s, the industrialization which followed transformed Penang into the second most important metropolitan centre in Malaysia after the Klang Valley, where the capital, Kuala Lumpur, is located. This industrialization is based on the creation of free trade zones (FTZs) and other industrial estates, which had drawn migrants from the surrounding region. Some 50,000 had been added to the local industrial workforce, the bulk of which consisted of single female school-leavers in their first job (see Salih et al. 1985; Armstrong and McGee 1985). In the process, the relationships between traditional activities, namely, small-scale industries and agriculture, and those associated with the entrepot trade, were changed (Chan 1980; Salih and Young 1987).

The focus on processes of export-platform industrialization and the particular labour force formation it produces in Penang raises a number of issues:

Figure 4.2 Penang: industrial and urban centres

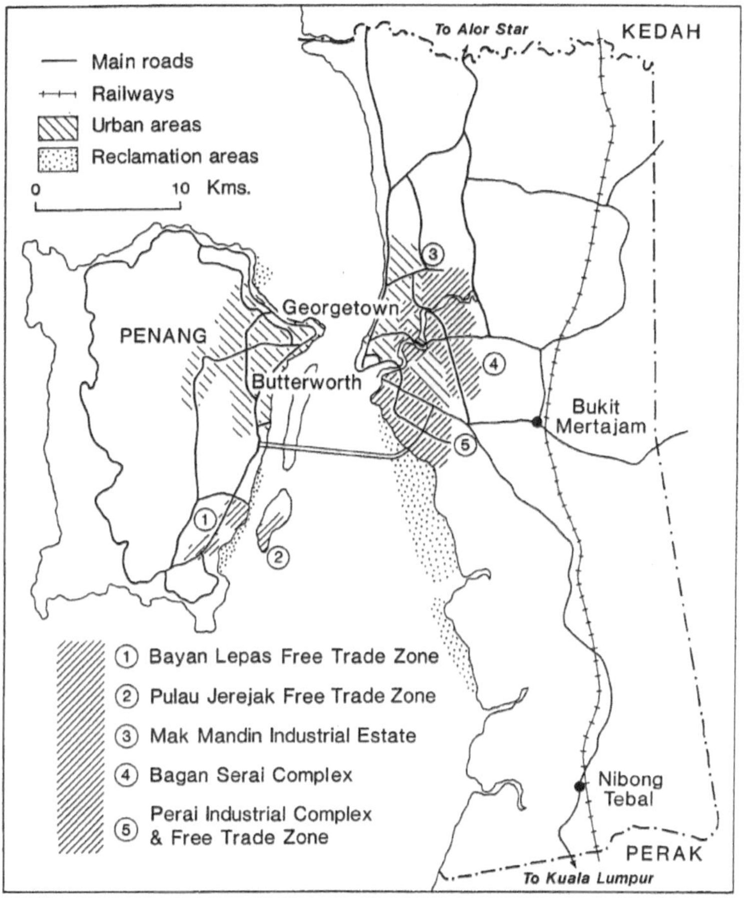

(i) the long-term role and impact of transnational corporations, particularly those in the FTZs, on the rest of the economy;

(ii) their labour commitment and the possibility for developing endogenous industrial capacity;

(iii) the process of proletarianization and development of labour relations;

(iv) the growth and stability of the labour force, and with the impact of conjunctural factors, the possibility of labour reabsorption into the urban economy; and

(v) the impact on the urban informal sector and rural areas from which the new workers are recruited, including the effects on family formation and emergence of new social organizational forms.

Among the processes in these so-called world market factories which have direct impact on labour experience and the processes of labour formation are: technological conversion and the practices of labour shedding (including lay-offs); increasing the number of temporary workers; and reducing indirect labour. These practices are most evident during downturns in the trade cycle (Fig. 4.3), but a process of long-term adjustment towards more capital-intensive technology appears to be occurring in the face of rising labour costs (Salih and Young 1985; Salih et al. forthcoming).

Another issue is the impact of rapid industrialization in Penang on the incorporation of migrant workers from rural households into the urban labour force. Given the particular process of labour force formation due to export-platform industrialization in Penang, a 'peculiar' process of proletarianization has occurred. The incorporation of the single worker into the urban labour force does not represent a complete proletarianization process in the classical sense, in view of her continued familial links to the peasant household. However, the new worker is subject to work relations and consumer behaviour patterns which constitute a broader process of proletarianization (McGee 1984) leading eventually to a severance of these familial ties. When eventually the single worker is completely absorbed into the urban economy, through processes of new family formation, through the subsequent migration of the original household, or other factors, this proletarianization process will be completed. If the proletarianization process is incomplete, owing to instability as a result of the lack of

Figure 4.3 Penang: monthly average employment in the electronics industry

Source: cited in Salih & Young (1985: 39)

labour commitment by both the multinationals and the workers, then several conservation processes will operate including circular mobility of labour, return migration, and the continuation of rural-urban links through household processes.

The single worker may be displaced through labour-shedding strategies. The concern is how these displaced workers cope with the situation. Are they reabsorbed, and if so, into what companies and industries? Do they return to the rural areas? What is the impact on the rural household? (see Young and Salih forthcoming).

This last question may be most fruitfully studied through household processes, since the impact would be determined by the nature and type of household, its stage in the family life cycle, the structural links of the household to the local economy, etc. Thus the processes and structure of particular types of households will determine the nature of labour participation and reproduction within the context of overall household adaptive strategies to changing social and economic situations.

The concern for these labour force processes, of their formation and reproduction, within the household economy relates to the issue of social costs of rapid industrialization, especially that engendered by export-platform corporations. By analysing the process of labour participation as part of the overall household adaptive strategy in response to its changing environment, it may be possible to determine who bears the cost of labour in these export-platform industries. This is to address the question of whether this form of industrialization and labour participation benefits the sending rural household (principally through remittances) or the informal sector household; or whether the rural household and the informal sector (through various household mechanisms, including the reverse wealth and obligation flows) subsidize the cost of labour incurred by the multinational. This is in addition to incentives, subsidies, and infrastructure already provided by the government to promote these off-shore investments.

Structural change and the Malay family

Before examining the cases illustrating the household processes in our study area, it is instructive to identify the major changes in Malaysian society in recent times that can be attributed as sources of the transformation of the Malay family. We may highlight the following forces as crucial:

(i) the increasing comprehensiveness of Malaysia's integration into the modern world system;

(ii) the dramatic surge towards industrialization in the 1970s (albeit narrowly based and distorted) produced by the latest phase in the development of the world economy associated with the redeployment of capital to the periphery;

(iii) the rapid but concentrated form of urbanization engendered by that particular industrialization, which has led to the creation of a more permanent Malay urban community;

(iv) the introduction of a new workforce characterized by a predominance of single female labour in the urban economy, and the increasing participation of women in the labour force;

(v) the rise and expansion of a Malay middle class as a result not only of the industrialization and rapid urbanization process, but also of the expansion of the role of the state in development;

(vi) the restructuring of society associated with state policy, for example, the New Economic Policy and the 70 Million Population Policy; and

(vii) the re-emergence of Islam as an important force in everyday life and public affairs.

On the first issue, an analysis of the Malaysian economy in the present conjuncture and for the future clearly requires an understanding of its historical articulation within the world system (Salih 1981). It also requires an understanding of the interplay of internal social forces produced by that articulation, which in turn determines its future mode of accommodation within the world system. Such an analysis can be broken down into three basic themes: first, the nature of the incorporation of the Malaysian economy into the world system, and the structural elements that this process has produced; second, the social forces and groups which are created by this process, which in turn determines the particular trajectory of the system as a whole; and third, the formation and role of the state in mediating this process of incorporation and the social basis of its power. It is the unfolding of social processes at this level of determination which creates the conditions and constraints for the transformation of the

Malay family in a historical sense. It remains for us to specify more precisely how these processes translate into a determinate structuring and reconstitution of the Malay family in the present time. The analysis also must comprehend the implications of the cycles and rhythms of the economic system as a whole which bear on the functioning and cycles in the family domain itself. An example of this is the impact on the family life cycle of the export of labour from the household economy during the expansive phase of development of the larger economy, and the reabsorption of that labour in times of crisis.

Within the framework of Malaysia's development within the world system, it is possible to analyse the recent surge in the country's industrialization as a combination of developments in the internal and external sphere. Further capitalist penetration of the Malaysian economy led initially to the differentiation of the Malay peasantry into a 'traditional' sector and a semi-feudal smallholder sector producing for the world market (Salih 1981). The traditional role of women in peasant households, as manifested in reproductive and productive activities, began to be transformed by the generalization of Malaysia's incorporation into other sectors, accelerated by the extension of schooling and the successful entry of some women into governmental, professional, and service sectors.

This produced a tension between the role of women in the domestic domain and in the public domain which may finally be resolved only by an ideological transformation of the gender relations and sexual division of labour within the Malay family (see Mohamad 1984). But this resolution is threatened by cultural and religious reassertions, such as those associated with the resurgence of Islamic fundamentalism in Malay society in recent years (see Narli 1984). Thus, the increased participation of women in the labour force may not really represent an emancipation, but rather a reconstitution of the role of women within the sexual division of labour. This phenomenon is currently creating a powerful conservative tendency within the Malay family, which may lead to a reassignment of roles. This could be reinforced further by the introduction of such policies as the 70 Million Population Policy, unless its implementation takes into account the dual role of women in modern society.

The reference to state policy above emphasizes the important role of the state in the transformation, or conservation, of the Malay family. The rapid incorporation of women into the industrial workforce, engendered by the outward-oriented industrialization pursued by the government in the 1970s, has given rise to severe contradictions in the overall position of women in the domestic domain, in the factory, in the new urban milieu, and in the larger society. In other spheres of economic and social development, the state directly affects the structural transformation in society (the maintenance or demise of the informal sector, the extension of the peasant family through promotion of settler families in FELDA schemes, or the conversion of smallholdings into mini-estates) which impel changes in family structure and processes. The New Economic Policy (Malaysia 1970), in seeking to eradicate poverty and restructure society with regard to ethnic imbalances, provides new opportunities for Malays in the educational and economic spheres. This will fundamentally affect the Malay family and lead to intergenerational differentiations in outlook, attitudes, modes of behaviour, and family relations.

ECONOMY UNDER LOCALIZED INDUSTRIALIZATION: CASE STUDIES OF HOUSEHOLD PRACTICES

We have chosen four case studies to illustrate the issues of household practices. They are divided into two types. One is based on individual life-history narration, the second comprises the reconstruction of individual life histories integrated into a life course. While these two methods attempt to investigate the same processes, the story-telling technique is more loosely structured and therefore less quantitative (in spite of probing) in comparison with the life-history matrix. However, the main criticism of the life-history matrix is its structured nature which cannot capture the atmosphere and subtleties derived from the story-telling techniques.

Narrative study of two women: Mak Teh and Yati

The first two case studies of Mak Teh and Yati were chosen because they represent quite typical examples of in-migrant production workers in electronics in the FTZs of Penang.[2] Data abound on the characteristics of these inmigrants to Penang (see Salih et al. 1985). We know they are female, mostly Malays aged 16-25 years, with at least six years of formal education. They migrate from rural areas where their parents are involved mainly in the padi and rubber sectors. They seek jobs commensurate with their education and which they cannot find in the villages; they are recruited in walk-in interviews and recruitment drives in the villages; they live in high-density rented rooms with a sprinkling in hostels, and with very few urban services available to them. Many maintain close links with their households by remitting income. Little is known, however, about first, the actual structures and processes within the households from which they come; second, their impact on these households in terms of their contribution and how it is used; third, what happens when they are retrenched; and finally, their future as permanent in-migrants to cities (the proletarianization question). All these questions fall within the ambit of the household economy and household strategies within the life course.

Mak Teh belongs to the earliest intake of workers in the FTZ. She has been working for twelve years and therefore has experienced some of the previous recessions in electronics. We are able to gain insights into her relatively long experience as a worker, and as a working daughter, with all its ramifications on the household she has physically left behind in the kampung, but of which she still feels so much a part. In contrast, Yati has had only five years' work experience as she was incorporated into the Penang workforce in the early 1980s. From these two individual case studies, we will draw out the issues pertaining to the life course, such as sex and sibling order; apsects of gender in the division of labour within the domestic domain; how and why these girls left their homes for wage work; the nature of their links to the household in terms of income, services, and obligations; their role and impact in the household economy; and finally, some of their hopes for the future and fears of retrenchment at the current time.

Mak Teh

Mak Teh, aged approximately 36, twelve years as a production worker in a semiconductor factory.

Aptly known as Mak Teh to everybody, she looks and behaves like a mother. She is a plump and quiet lady in her mid-thirties. Many people would think that being an unmarried woman at this age would cause an identity crisis in her life. Perhaps there is one, but her crisis is more mundane - it is that of keeping her aged parents and dependent siblings well fed and well looked after. Consequently, her real fear is that of retrenchment.

Her work

Mak Teh finished her primary schooling and left to 'jaga adik dan anak saudara' ('look after brothers and sisters and their children') before coming out to work. She joined Monolithic Memories Inc. (MMI) in 1974, two years after MMI was set up in Penang. She remembers those days when MMI had a fleet of only two factory buses compared to the present fleet of over forty. She started with a daily wage of about $2.70 and her take home pay per month amounted to abut $60-70.[3] Rent only cost about $3 in the beginning and through the years it has spiralled to $30 a month. 'It [wage] was enough then, since prices of things were lower - food, bus fare, and even things like sarongs and clothes. I remember how I used to wear the new sarongs across the ferry so that I won't be taxed.' She reminisces about those early years of working life when Penang was a free port. 'I was scared of the city then but now it's OK,' Mak Teh said on her adaptation from the kampung to the city.

She works at the wire-bonding section which consists of about thirteen girls who have been in service for about seven years or more. They are the experienced hands and have been in this section ever since they started work in the factory. There has not been much automation in this section. Basically their work is still the same as before, old machines were replaced by new ones. She is not sure if these are upgraded machines. 'The target level has not changed much, over 100 per day (can't remember the exact number) because we do the wire-bonding manually.'[4]

Because of her long work experience, she remembers other bad times. These were brief spells, lasting only two to three months in 1974-5, 1977, and 1980. This year is the worst she has ever experienced. Her present daily pay is slightly above $20 and so she can still cope with pay-cuts and forced holidays. 'We just have to spend less and cook more often instead of eating out.'

Mak Teh's family

Ayah and Mak (Parents) Both Ayah and Mak are about 70 years old but Mak looks older than Ayah. They were married at the age of 18 or 19 and have lived in the same kampung ever since (see Fig. 4.4).

Ayah is very thin, almost just bones and wrinkled skin, clad only in a sarong during the day. When it rains, his bones hurt him so much that it makes sleep impossible. He has just run out of his supply of medication which he purchases for a fee of $5 (tablets and services) from a friend who buys it in Sungai Petani, the nearest town. Despite his age and poor health, Ayah is mentally alert and remembers past events vividly. He was the fourth in a family of five boys. His parents were poor and involved in kerja kampung.[5] Ayah entered school at the age of 12 and stopped after three years to do all sorts of work available in the kampung. Later his father passed away and he went to Bujung to live with a relative ('that was how we met'). Ayah worked as a rubber tapper on an 'orang putih's'[6] estate. Then he married Mak and moved in to live in her mother's house.

His wage as a rubber tapper was then about 50 cents a week, amounting to about $18 or $20 a month. When the Japanese came, Ayah and Mak already had three children. Nobody could work privately for anybody then and like everyone else he had to work for the Japanese. Even his eldest son, who was about 10 years old, had to work. Two unfortunate events occurred during this time. His little daughter died. Later, he was rammed by a buffalo which caused his health to fail over the years. After the Japanese left, he again worked in a Chinese rubber estate as a mandor or overseer. He left the job because it was too much work and responsibility for the amount of money he was paid. He became a contract tapper in an orang putih's estate but had to stop working about fifteen years ago because of his poor health and advancing age.

Mak is short and has all the looks of a tok (grandmother) except that this tok has very inquisitive looks and expressions behind a pair of smudgy glasses. A very responsive and articulate woman, she speaks expressively and with openness. She has a catching laugh which probably helped her and the family through those difficult early years. As she puts it, 'cerita senang makcik tak tahu, cerita susah yang makcik tahu' ('stories of hardship Auntie knows, stories of prosperity Auntie doesn't know'). She is the second

Figure 4.4 Mak Teh's family map

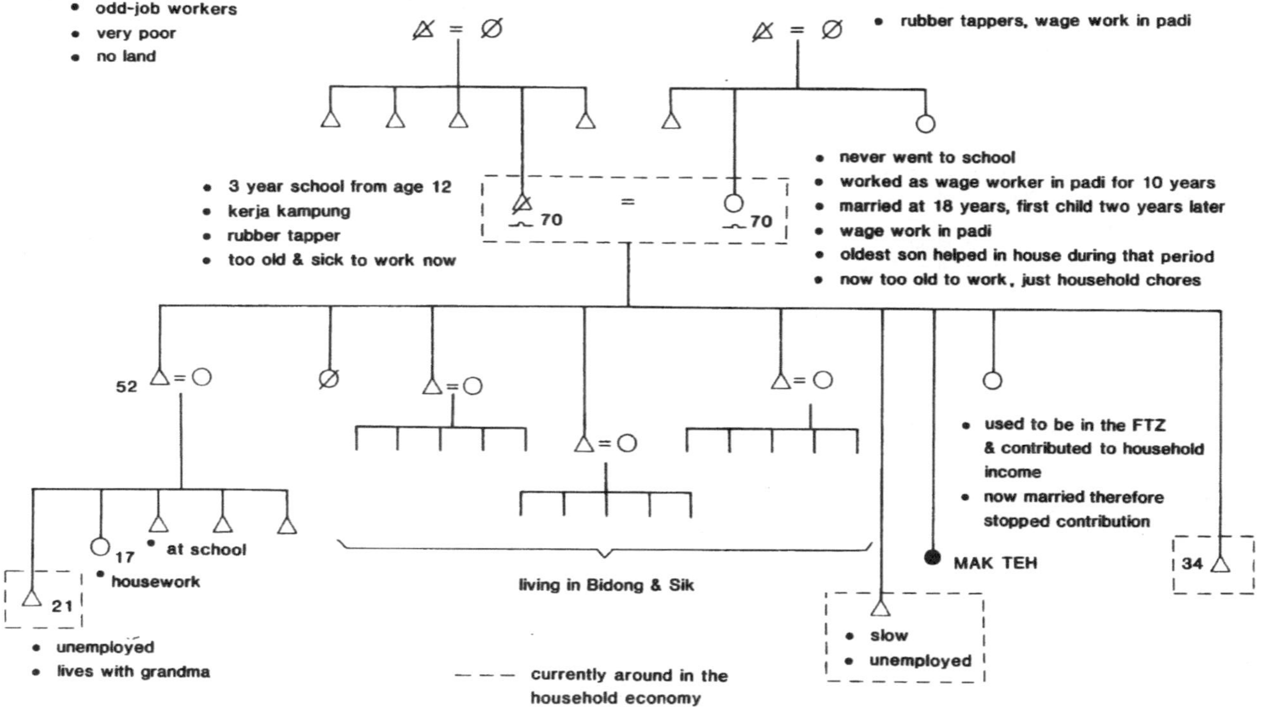

child in a family of three. Her parents worked as rubber tappers or ambil upah (waged workers) as padi planters. Mak says that she started to work at a very young age, pointing to her grandson who is about 11 years old. She was also ambil upah as a padi planter. Mak never went to school. She got married at the age of 18 and had her first child two years later. Marriage did not confine Mak to the house since it was just not possible to live on Ayah's wages alone. Mak continued working in the fields, usually leaving the house early in the morning, returning in the evening. The children were left under the care of the eldest son who bathed and fed the rest. Out of sheer necessity, all her sons are good cooks and housekeepers. The youngest son can even sew.

Mak has stopped working in the fields now. She does household chores, cooking and washing for the rest who are still living with her - Ayah, two sons, and a grandson. Mak Teh has bought a pack of Amway soap powder[7] for Mak so that she need not scrub the laundry, especially jeans which are the boys' favourites. 'These boys will soak their clothes for days and I wash them because I can't stand the sight and smell. I don't have the strength to wring them so I just hang them on the line. I only sweep the floor when it is dirty, maybe once in two days. If there were girls around, it would be different. These boys are not bothered,' Mak says without a trace of complaint.

Mak Teh's siblings Mak Teh's oldest brother and his family live in the same kampung. They are rubber tappers. His eldest son, who is 19, is unemployed and lives with his grandmother (to avoid his father's wrath after being involved in an accident). The next daughter is 17 years old and looks after the house when the parents are at work. She also sees to household chores and minds the younger ones. Mak thinks that she is so thin because she is too tired to eat properly. This family does not contribute to Mak's household, and is in fact quite dependent on her; the grandchildren often eat there. Mak paid for her grandson's treatment after the accident and also pays for other extraneous expenses.

The second child was a little girl who died during the Japanese occupation. The next three sons have their own families and are living in Bidong and Sik. The fifth son is not married. Although he eats and socializes at Mak's house, he lives in another house which is just opposite Mak's. He moved out because he was sick of washing dishes in the house. He appears to be rather effeminate, is a bit slow, but is said to be a good cook. He formerly sold bee-hoon (rice-noodles) in the kampung. At present, he is unemployed.

After Mak Teh comes a younger sister who also used to work in the FTZ and contribute to the family income. Since her marriage, she has stopped her contribution and lives in Seberang Perai with her husband.

The youngest is a male who is 34 years old and still single. He too went to primary school and stopped, preferring to work rather than stay in school. He has held all kinds of jobs, the last was at the biggest rubber-processing factory in Semiling. He was there for a few years but left when he started falling sick due to the gases produced in the rubber factory. He did register for a short course in agriculture but went late and could not catch up. Now, he just jaga kerbau (looks after buffaloes). They have some buffaloes. Every evening he cycles to locate their whereabouts because in the kampung there is no land for the buffaloes to wander about in, so they are let loose. He appears soft-spoken and shy. Mak Teh says that he is not demanding.

She has thought about purchasing a motor bike for him but at the same time she is afraid that he might get into an accident. 'This is something which I have to think hard about,' Mak Teh says, 'but he does not demand that he wants it. Well, if I don't buy him the bike, the money is enough to pay for his wedding expenses.' She seems so much more concerned about his future than her own.

Mak Teh's contribution Mak Teh's contribution to the family had been a joint one before the others got married. Of all the single children left, she is the only one with a regular job. What choice has she but to be the major supporter of the family? They have an account with the Pakcik[8] who comes into the kampung to sell vegetables, fish, and meat. Mak Teh settles this major expenditure at the end of the month. During the harvesting season, they sometimes receive rice from the neighbours.[9]

The house

The house is still in the process of renovation. The work was started about five years ago, being done bit by bit. Mak Teh says that so far it has cost them about $5,000-$6,000 and this was due to the fact that they coud not afford to purchase all the necessary materials at once. The latest stage is cementing the floor. Mak Teh has been the major contributor towards this renovation. They are not sure when the whole house will be completed. But when Mak Teh talked about it, she spoke positively about building the bathroom in the house and putting all the crockery into the kitchen. Building a proper bathroom in the house is of much concern for the safety of her ageing parents. It seems that Ayah had been affected by some spirits when he went out to the bathroom early one morning but luckily for his kuat semangat (spiritual strength) he was able to withstand the spirits and recovered fast.

They did not hire a tukang rumah or house builder for this job. It was done entirely by a relative who was learning the skills of constructing a house. There is electricity and piped water in this kampung, all done just before the last general election. This house has electricity but no piped water. Water comes from a well outside the house. The house is sparsely furnished. There is a set of cushion chairs, all pushed against the wall. Years ago, one of the brothers purchased a black-and-white TV or what they call TV dua warna (two-coloured). The floor is lined with linoleum and straw mats which Mak wove.

The kitchen is still the old part of the house since the renovation has not reached this portion. There is a firewood stove and two kerosene stoves. There is a lot of crockery, all purchased for the weddings of the sons and daughter. 'When the kitchen is ready, I can put all these into the cupboard,' Mak Teh said.

There are some chickens running around the house. Mak used to rear about fifty of them but there was an outbreak of disease and all but a few died. They could get as many as twenty eggs per day and some of them would be sold or given away.

Yati

Yati, aged 20, five years as a production worker in the semiconductor industry.

Yati is tall and pretty. Initially, she appears shy and reserved, but gradually she opens up. Although she is 20 years old, she has matured more quickly than others of her age through child-minding her siblings at a young age and four years of working life in Penang as a production operator. She was one of the daughters who was taken out of school to look after her younger brothers and sisters when Mak had to take on a full-time job in the estate.

She started working in the FTZ when she was 16 years old. She followed her cousins from the same kampung who were then working in the factories in Penang. She started with Intel as a temporary worker for three months, went to Litronix, and is now in Mostek. Intel was a 'good' company in terms of the working environment but she could not stay on because she was hired as a temporary worker. 'Hari Sabtu I stop kerja di Intel, hari Isnin saya dapat kerja di Litronix. Senang saja nak cari kerja di kilang' ('On Saturday, I stopped work at Intel, on Monday I had a job with Litronix. It is easy to find a job in the factories'). But she did not stay on for long in Litronix because 'kilang itu kecil dan pada masa itu ada personal problem dengan operator lain. Masa itu budak lagi. Sikit-sikit saya naik marah' ('the factory was small and I had some personal problems with other operators. At that time, I was still a kid and prone to get angry over minor things'). After a year she left for Mostek. Although she has worked for four years, her take-home pay is not the equivalent of what she would be receiving had she stayed on in one factory. Each time she started with a new factory, she started at the bottom rung. She has been with Mostek for two years and her pay is about $240 per month. She may be starting with yet another new rate of pay when Mostek hands over its operation to Thomson.[10]

Yati has no steady boyfriend although she has many admirers. She said that so far she has not been on a single date. She wants to work first, but if she does get married, she does not want to work any more. According to her, when her friends get married, they usually continue working until the first child comes along before they quit, that is if both husband and wife are working in Penang. If the man is from another state, naturally the girl will leave. One of her room-mates is getting married soon and she, too, will leave.

Although she has been in Penang for four years, she saw very little of it before she met Faridah. They are now the best of friends, doing a lot of things together, buying the same things and sometimes wearing the same beads and shoes. When she visited Faridah and saw the number of things she had bought for the house, Yati decided to do the same. She purchased a cupboard for the kitchen and a show-case cabinet for the house.[11] She also does a lot of crocheting.

Yati's Family

Ayah and Mak Ayah is tall and still youthful-looking at 53 years. His parents were early settlers who came and buka (opened up) this area. The children all inherited a huge piece of tanah pusaka (hereditary land) with brothers and sisters living nearby. His parents were rubber tappers. He cannot actually remember how many brothers and sisters he has - approximately eight or nine, he thinks. He is the fifth child. He went to school for a few years.

Upon reaching the age of 20 in 1952, he left for Singapore to work in the then Far East Military Army as a cook. He remained there for fifteen years and returned in 1967. It was in Singapore that he married Mak (in 1960, at the age of 28) and started a family. From the old photos, he wore the 'man about town' image - dark glasses, Elvis Presley hair-style, and trendy outfit - full of gaya (style). He as a 'British protected person' while in Singapore. Life was easy then with all basic needs (food, shelter, and health) taken care of by the army. By the time he returned in 1967, he already had six children.

He returned to the kampung. With a gratuity of $2,000 he built the present house. He also brought back from Singapore items like a refrigerator and furniture which were later sold off when tragedy struck. Although he went for a resettlement course as a barber before coming back, he never used this acquired skill at all. He worked at different jobs but was mainly tapping rubber in an estate for about ten years. In between, they had already started selling pieces of the furniture and later the refrigerator, since there was no electricity in the kampung and they needed the money. It was about eight years ago that he became very sick - according to Mak, orang buat (bewitched). He became mentally ill and what followed then was the family's loss of their bread-winner. By then, there were already nine children, seven of them of school-going age. From here onwards, Mak took over and became the main narrator and actor in this story. Ayah no longer worked, he remained in the house, devoting much of his time to planting flowers and rearing birds (which he still does today).

Mak is short and plump and looks like a Japanese woman. She believes in having more than enough food for the children. Because of the large number of children, she never buys kueh (cakes); she has to make them. Being the only child, her childhood days were spent playing. When she grew older, she helped her mother make kueh to sell. When she was 16, she went to Singapore to help look after her relative's children. It was in Singapore that she met Ayah and married him at the age of 19. The first child arrived a year later and from then on, she gave birth every year. She was a housewife in Singapore, busy with the six children who were born there. She says that during those days, life was comfortable. Yati adds that she had heard from others that her mother had bangles upon bangles of gold on her hands then.

Mak continued to give birth almost every year after they returned from Singapore. However, the last five babies were more widely spaced, probably for the following reasons: her age, financial strain, her husband's sickness, and her own entry into the labour market. She was no longer confined to the home as a housewife and producer of babies. Within this difficult period, two of her girls died. She attributed their deaths to sickness.

When father started falling ill, she helped him tap rubber and that was the first time she ever worked outside the house. And when Ayah's condition worsened, she became a full-time general worker in the rubber estate, earning $6 a day to feed the whole family. The remaining furniture and jewellery were sold. Mak had to appeal to the Welfare Department for help. They were given $20 a month, which was barely enough, but with further appeal this was raised to $90, which was a tremendous help. But things were still very tight financially and they could not afford to send all the children to school. At this time there were seven of them in school, and transport alone cost about $80 a month. Because the eldest son was already in form three, it was decided that he should complete it; the next son was in form two while the third was in form one. The second eldest daughter (the eldest daughter died at a young age) had just finished standard six, the next three were in lower primary classes. With mother working full-time, it was necessary for someone to look after the younger ones. The choice of who had to stop school was quite obvious - the eldest daughter - in spite of her being a good student and enjoying school. Yati cried when she left school.

Life went on with Mak trying to make ends meet in whatever ways she could. The children helped out with odd jobs here and there. Over the years, Mak's pay had increased by only $1, making it $7 per day today. She says that she felt the financial relief when Yati came out to work in

Penang about four years ago. The family is still receiving aid from the Welfare Department, but Mak thinks that will finish next year since most of her children are now working. She hopes that all her children can find regular (tetap) jobs soon and then she can enjoy better days.

Yati's siblings Yati comes from a big family. She has thirteen brothers and sisters but three of them died when they were young (see Fig. 4.5). She is the eldest of the girls but in terms of seniority, she is the fifth. The youngest sister is only four years old. The four brothers before her went to school as far as form five, the eldest is in the University of Malaya, while the other three are working in Kuala Lumpur. The sister after her has just sat for her form five exams (SPM). 'Adik saya pernah duduk di sini. Dia cuti. Saya tak bagi dia kerja di kilang. Ti rasa Ti seorang kerja di kilang cukuplah' ('My younger sister stays here. She is on holiday. I don't want her to work in a factory. I feel that my working in the factory is enough'). Her expressions reveal more than her words. She speaks with a rather cynical smile, as though she were saying 'I know what it is like to work in a factory and no sister of mine will be in my shoes, whatever the cost.'

Five of them were born in Singapore. When they returned to this kampung asal (village of origin) Nenek (grandma) was till alive in the village. It was therefore easy for her mother to raise the thirteen children in the kampung because she could leave them with Nenek or with cousins when she went to work.

Call them the equivalent of the Waltons, or a soccer team if you wish: there are altogether thirteen of them living in this one-bedroom house. So when it came to bedtime, all were lined up from one end of the house to the other.

The eldest son is 24 years old and a recent graduate from Universiti Malaya. He had always been a scholarship holder, ever since secondary school. Mak says that she hardly spent any money on this son's education. In fact, Mak had asked him to stop when he failed his form three (SRP) exams but he was determined to try again. In form six, he was awarded the Bank Negara scholarship. When he obtained a place in Universiti Malaya, he had to seek help from relatives and borrowed $600 for fees, etc. While at university, he was on a Public Services Department scholarship. For his additional expenses - for example, the

Figure 4.5 Yati's kinship map

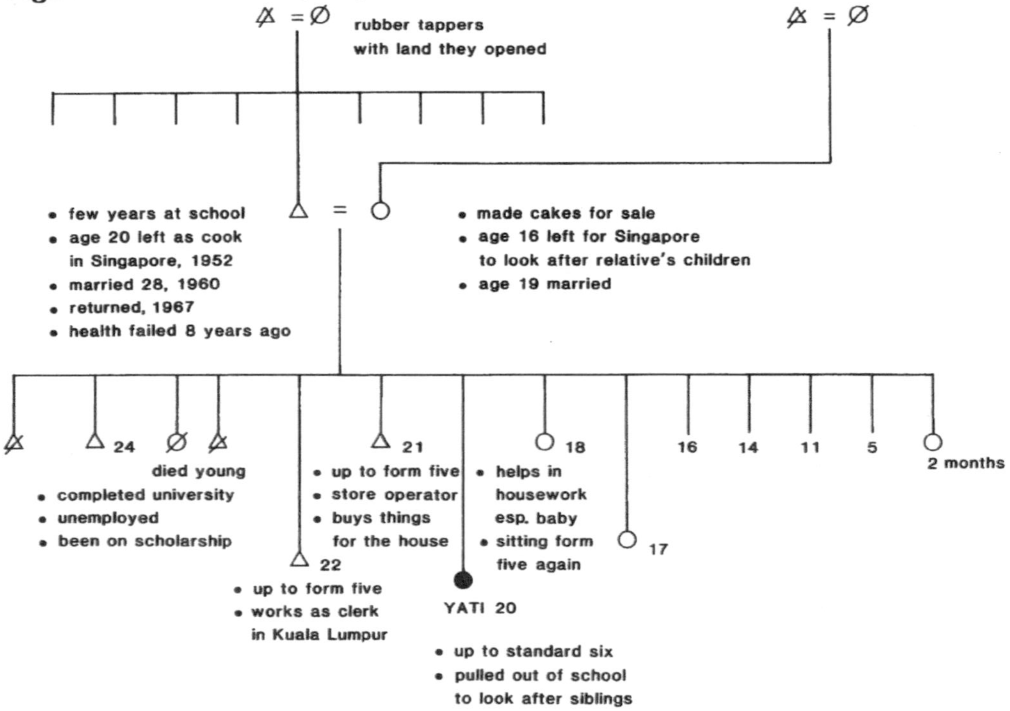

recent convocation - he asked for money from Yati. He graduated with a degree in public administration. With no job offer till today he works part-time for a relative who manufactures soya sauce in the kampung. He also nets fish from the nearby tin mine once in a while. He sells off the bigger fish and rears the small ones until they are of an edible size.

The second child was a girl who died young. Next comes another son, 22 years old. He went to school until form five. Working in Petaling Jaya at Dutch Baby Industries Berhad (making condensed milk), he comes home quite regularly. Mak says that he does not earn much and with such escalating costs of living in Kuala Lumpur, he can only maintain himself.

The next son is 21 years old. He, too, left school after form five. 'We were not that interested in going to school. Only my elder brother was hard-working. Furthermore, with the family situation and lack of perangsang (encouragement) from my parents, we were not good in class,' he says. He is working on a daily basis for a relative who runs a soya sauce factory. He used to work in a funfair group, operating the merry-go-round, but found it boring. Although he was paid $25 per night, it took him away from home and he did not like it. Working for the relative as a store operator, he goes out for deliveries, coming home daily. He makes $18 a day. Mak says that it is usually this son who buys things for the house. So far, he has bought a rice cooker, an electric kettle, and an electric fan. Mak says that when he sees other people buy items for their homes, he does the same.

Yati is close to this brother. While he takes interest in things like what kind of hair-style or clothes suit Yati, the other older brother chides her for not wearing the tudong (to cover the head).

After Yati comes 'Mah. She is taking her SPM again this year. She helps around the house, taking care of the little baby girl, helps Mak cook, wash the clothes, and clean. She is a great kueh-maker and has expressed to Yati how convenient it would be if they had that special machine to make a certain type of kueh. She intends to go for a typing course when the exams are over.

A'kak comes next. She is 17 years old. She takes on less responsibilities than 'Mah at home. Both 'Mah and A'kak make some extra money during the fruit season. They have some mangosteen trees in the compound. During the fruit season, they tie them up in bunches and sell them by the roadside or bagi dua[12] with a Chinese man from town. The last season they each made about $500. 'It's so easy. Just go and look for mangosteens. We don't need any capital at all,' 'Mah said. The rest of the brothers and sisters are 16, 14, 11, and 5 years old; while the youngest is two months.

The house

The house was built in 1967 when there were only six children. Now there are thirteen of them, plus Ayah and Mak, and the house has not undergone any major renovation or extension. Thus, space is being used to the maximum. There are two televisions in the sitting room. The smaller one, which is not working now, was given to them by a relative. On the opposite side of the hall stands a bigger TV which they bought second-hand for $124. It was bought about three months ago to prevent the kids from hanging around the relatives' houses late at night. Pushed against one side of the wall is a table, and standing in the middle portion of the wall opposite the table is the show-case cabinet which Yati bought. It is indeed an odd place for it, but this is the only available space. This cabinet is full of sets of crockery, cups, plates, etc. which Yati bought, to be used only on occasions like Hari Raya.[13] Right in the middle of the room, hanging from the beam, is baby's cradle, with just enough space for swinging. A fan stands next to the TV.

The kitchen is simple. Meals are served on the pangkin.[14] There are two old chairs with cushions, which Yati calls 'my mother's old-fashioned chairs', just next to the kitchen door. The three prominent items are the cupboard, rice cooker, and electric kettle. At night, the two bicycles are pushed into the kitchen.

Household income

The average monthly income derived from various sources and expenditure is as follows:

Mak's salary ($7 x 26)	$182
Welfare Department per month	$90
contributions from sons and daughters	$100
Total	$372

Expenditure:

marketing ($3 x 30) + rice + essentials	$150
kueh, biscuits, etc	$20
education (transport, pocket money, etc.)	$70
bills - water and electricity	$10
miscellaneous	$20
Total	$270

Additional and/or seasonal income depends on the time of year. For example, a major fruit crop earns them some extra money.

Fears of retrenchment

Yati's greatest worry now is the much talked-about retrenchment in Mostek. Although, officially, none of her friends has been retrenched, rumours have it that retrenchement will be on a 'last in, first out' basis, i.e. it will start with the operators whose number is 5,000 and below.

'Jika Yati kena buang, Yati nak pergi Kuala Lumpur. Abang Yati ada di Kuala Lumpur' ('If I am retrenched, I shall go to Kuala Lumpur. My brother is in Kuala Lumpur'). He was then in his final year at Universiti Malaya on a scholarship. Yati has no idea what course he is doing since they hardly meet and even if they do, 'dia tak cakap banyak dengan Yati' ('he does not speak much with me'). If she is retrenched, she plans to stay with her brother and try to get a job in one of the many factories there. Yati is certain that 'saya tak nak balik kampung. Tak ada kerja di kampung. Lagipun sudah biasa di sini. Ada freedom' ('I do not want to go back to the village. There's no work in the village. Furthermore, I am used to the life here. I have some freedom'). Or she will try to get a job in Ipoh. She could commute daily between Ipoh and her kampung, which will cost her $1.20 per trip. She hopes that the recession will be over in two years' time and she will come back to work in Penang. 'We don't have any land at home to work on. Even the rubber trees don't belong to us. My brother has not found a job yet. It will be easier for the family if he has. If I work I can at least buy my little brothers and sisters some new clothes at each Raya. I feel so sorry for them. My mother is, expecting another baby very soon. Entahlah [I don't know].'

And during all our conversations so far, the impression that she has given of her role in the family is modest. But having spoken to her family members, it is very clear that she is important and that everybody who is capable of working within and outside the house matters very much.

From the cases of Mak Teh and Yati, together with the impressions gathered from the rest of the sample of detailed individual and family histories, some important issues may be highlighted.

It is evident that the withdrawal of their female workers from the domestic economy represented a major transition for many households. This is particularly true when we compare this current generation with the females of their parents' and grandparents' generations. Their mothers generally had very little education, and were withdrawn from school in order to look after their siblings and do housework so that their own mother could be released to work in padi or rubber as unpaid family workers. The essential difference between that generation and these young girls is that the modern generation substituted wage work for family labour to sustain the household economy. This inter-generational difference reflects the changing circumstances and opportunities. While their mothers had little to no education, and if there were any, it was often in religious schools, this modern generation has been able to take advantage of widespread education in rural areas made available by the policies of the independent state in the 1960s. In the 1970s, this was reinforced by the universal education system, by more scholarships for the Malay bumiputras, and by the New Economic Policy which made employment restructuring, that is the incorporation of Malays into the modern wage sector, its major goal. As a consequence of the employment creation policy and the increasing number of youths entering the labour market, transnational corporations were encouraged to set up factories in newly created FTZs.

A second feature is the issue of the life course. It is clear, not only in the present young workers' generation but also in their parents' generation, that sibling order affects the opportunities open to children. The older ones are almost always the ones who were denied the chance to go to school, or to continue school; being forced to leave school early to assist with family economic activities, either in the fields or in the house. Thus, it is not uncommon to see a great disparity in both education and occupation between the older and the younger children; the difference ranging from a padi farmer with three years' schooling to a younger sibling who is in university.

An added dimension is the question of sex order. In Mak Teh's family, her older siblings (although boys) had to do housework because Mak Teh was sixteen years younger. But, by the time she was in primary six, she was withdrawn from school to look after her siblings. Similarly, Yati too was forced to leave school, although she was the fifth child, and was a good student and enjoyed school. In this sense she was the proverbial sacrificial lamb. It was her withdrawal from school, and subsequent entrance into the wage sector, that enabled her older siblings to continue their education to form five: the eldest successfully completing his university education. The future of her younger siblings at school is now assured because Yati is working and contributing to the family income.

In many ways, both families have become dependent on the working daughter (Salaff 1981). While both the girls initially left home to work in the factories, and to assist with the economically depressed family situation, later this sense of responsibility towards their families increased as their families became more dependent on their incomes for necessities as welll as for extras. At times, the working daughter's income is even taken for granted by family members simply because of its regularity, which is so different from the income from village work. In the same way, if any of the other members of the household needed extra money, for some extraordinary expenditure such as medical expenses, she is the one they turn to because of her assured income.

This sense of responsibility grows in the girls - as reflected in statements by both Yati and Mak Teh, and other girls. Yati wants her younger sister to continue schooling so that she does not have to work in the factories. Mak Teh, in a rather maternal way, toyed with the idea of buying her brother a bike, or else keeping the money for his marriage. It is this sense of responsibility that has made these girls delay marriage to well beyond the national average age for marriage. Similarly, this responsibility gives them an important role in decision-making and an authority in the household, as clearly evidenced in Mak Teh's case. Of course, this is related to the length of time the girl has been working and to the extent of the household's dependence on her income. What has emerged in our observations is that while these factory workers are often reticient and meek in their urban context, they are quite different in their rural setting, having a quiet sense of authority.

The monetary contribution to the household may be direct or indirect. Indirectly it enables their parents' income to be used for other things, to provide a bit of flexibility to an otherwise insecure and tight economic situation. More direct contributions are purchases for the household (such as consumer durables, furniture, food, and extra clothes for the family), paying for the siblings' tuition, and contributing to house maintenance. The amassing of crockery sets with brand names like Arcoroc and Tupperware; furniture such as beds, cushion sets, dining tables, cupboards, and even modern kitchen cabinets; and electrical appliances such as televisions, rice cookers, fans, and blenders is done slowly owing to their limited budget. The bigger items are usually bought on hire purchase. But the almost universal tendency of these girls to buy expensive brand names shows a certain circuit of direct buying which operates among the factory workers. It raises the issue of rising ostentatious consumerism amongst the factory workers. For example, a set of Arcoroc plates and cups costs more than a production operator's average monthly wage of $250. Of crucial importance here is the locking-up of money in consumer durables, many of which are seldom used, except for special occasions. Except for house renovation and some investment in jewellery, no money is being channelled into the buying of assets such as land or agricultural machinery. Is this because it has never been a tradition for females to invest in agricultural land, which they merely inherit rather than buy (see Young 1983 for a discussion on Malay land inheritance); or is it because their own incomes are far too small and insecure (unlike government wage work) for such big outlays? Yet this pattern is dissimilar to the findings in other Asian countries, especially Thailand, where female out-migrants remit money for investment in land in their villages (Young and Salih 1986). Finally, has the disinterest in agricultural work, especially padi and now rubber, displayed by the younger generation of rural Malays (as evidenced in vast tracts of abandoned smallholder padi and rubber land, and labour shortage in the estates) created a general atmosphere that promotes a complete break with agriculture in this generation? This is an interesting question because it has important implications for the Malaysian economy as a whole.

The stage of the life cycle of the family is also affected by the sibling and sex order of the children, as can be deduced from the earlier discussion. Often the dependence of the families on the working daughter is affected quite dramatically by personal tragedy, such as the death or serious illness of a father or the major bread-winner. Usually when this happens the burden falls very heavily on the mother, who is forced to explore all sorts of ways of making ends meet. Such strategies can clearly be seen in the case of Yati's mother, where as her husband's health failed, she was forced to work outside for the first time. As his health worsened she had to change from part-time to full-time work on the estate. In such circumstances, it is inevitable that the children will suffer, in some cases they are withdrawn from school (if there is no assistance from the Social Welfare Department and their schooling is not subsidized by scholarship) and forced to work in order to help with the family income. In Yati's case, she was withdrawn from school at precisely this time and started work in the factories at 16. It is also interesting to note that in both case studies, siblings died young during particularly sustained, economically depressed times.

It is also evident from our cases that female children are more responsible and caring about their families than male children. Thus female production workers will find ways and means to cut down on their expenditure in the urban context so that they can remit more money home. Male children often cannot find the extra money to send home, and give the excuse that the cost of urban living is very high. Instead, money is often remitted from the village to them.

The question arises as to what happens if the daughter has to stop work. Two processes usually spell the end of the working daughter's contribution to household income. The first is marriage and the starting of her own family. There is not enough data to show conclusively that the girls leave their jobs at the time of marriage. It appears to depend upon whom they marry and whether the husband lives in Penang. For example, those who marry civil servants stop working. Those who marry men from outside Penang naturally have to leave. But girls who marry fellow workers remain working until the birth of their first child. But even this is dependent on whether they have a relative such as a mother or aunt who can look after the baby. In fact, the practice of giving away children to grandparents to look after (often in the kampung) while the parents work in town is common, and a prevalent household strategy among Malays.

The second way in which the contribution of the working daughter to the household income pool may cease is as the outcome of retrenchment. At the time of the research the factories were retrenching workers due to the recession. Yati, Mak Teh, and other girls lived in fear of losing their jobs. Some felt shy about going back to the village and being stigmatized for having done something wrong in the factory. But the overwhelming concern was how their families would cope without their incomes.

The above questions and observations have been posed from the perspective of the working daughter in the household context. As in the discussion of the family life course, much depends on the situation of the household and the adaptive strategies adopted by each household to cope with each contingency and to adjust to the long-term macro-changes affecting the family. However, these strategies are constrained by the stage of the family life cycle and its life course: so the options that they choose have to be determined within these constraints. This is better seen through the analysis of individual life histories within the context of the family life cycle. This is the focus of the next two case studies.

Life course case studies

The two cases of household history as illustrated in Figures 4.6 and 4.7 summarize the life histories of household members (including those who have migrated but contribute to the household income, or are subsidized by the household and therefore participate in the income pooling). They examine with more accuracy the transitions of individuals within the life course. Unlike the individual case studies of Mak Teh and Yati, which have touched on family members in a more qualitative sense, the household case studies are able to depict the integration of the concepts of 'individual', 'family', and 'industrial' times, meshed together within a household in 'historical' time (see Hareven 1977; 1978a; 1982). Thus we are able to see in a clearer time perspective how personal tragedies such as the death of the main income-earner forces the household to adapt and cope in various ways, for instance by migrating, or by increasing the number of jobs for the other household members. Likewise, we are able to capture, albeit rather starkly, the different configuration of labour input within the domestic domain and productive work of family members through time. As in the individual case studies, the same issues of sibling order, sex order, labour input, and income pooling, etc. within the household economy and household strategies are examined.

The diagrams are arranged with historical and family time along the horizontal axis and the parents and family members arranged along the vertical axis, according to sibling order and sex. Symbols are used to indicate the major transitions in the individual life histories such as birth, schooling, migration, job changes, marriage, and their own family formation. In this way it is possible to see the interrelationships between a family member's transition with the rest of the family and the significance of that transition at particular junctures of the family life cycle when read vertically.

Figure 4.6 consists of a female-headed household with five children, the eldest son of which is already married with five children. It is a poor household, belonging to the lower echelons of economic status in the village. It has just the basic essentials, possessing only a bicycle by way of transport and only recently acquired a black-and-white television set. The life history of the head of the household was affected by the death of her father, which resulted in her migration to stay with an uncle. After marriage, when all her children were still at school and her youngest daughter only 9 years old, the most traumatic event affected her family, that is the death of her husband. From being the rather carefree full-time housewife of a policeman, with time even for cooking classes, and a regular income and barracks provided, she was suddenly left with five school-going children, no income, and no housing. She brought her family back to her village of origin where her brother provided her with a small house. She was now forced to survive by taking on multiple jobs such as sewing, tapping rubber, working in the padi fields earning about $100 per month during the planting and harvesting season, and selling cakes in the school and during Hari Raya. She also took on all types of kerja kampung.[5] She received only a $190 widow's pension per month for the first nine months, before settling at $250 per month. This meant that she struggled to

Figure 4.6 Household I

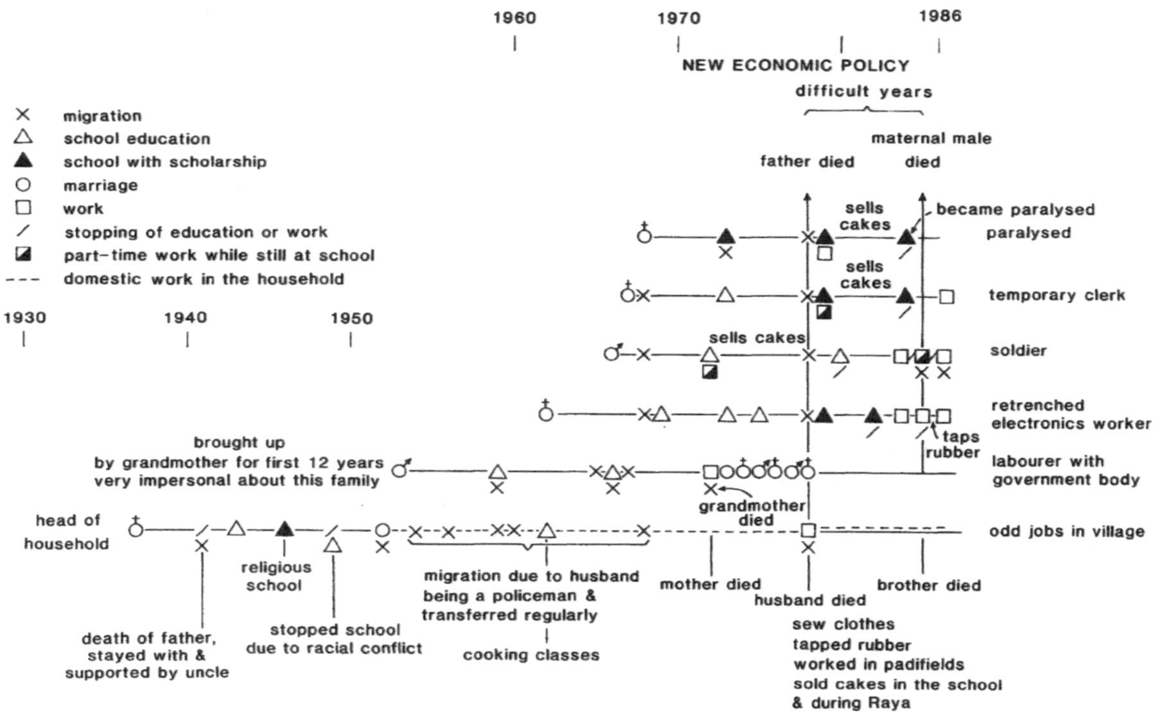

keep her children at school. Fortunately all received scholarships and were able to remain in school, supplementing their incomes by selling cakes for their mother.

It was during this economically depressed period that the eldest daughter, while still in the sixth form, left to work in the factories. She became a major contributor to the family income, contributing all of her wages ($230) to the family, but suffered much stress when she was retrenched this year. To augment the family income she has tapped rubber since her retrenchment, while applying for all types of work. A few months later the fourth child, a daughter, got a temporary job as a clerk for $200 per month (she gave her mother most of it), which gave the family some reprieve, although they worry about what will happen when her job ends.

When the daughter started to work in the electronics factory, the household income situation appeared to ease somewhat, but the situation worsened when the eldest daughter was retrenched, followed by the sudden paralysis of the youngest who had just reached form five. Her medication is expensive. The boys in the family do not contribute. In fact, the eldest is hardly perceived as part of the family, having been brought up entirely by the grandmother. He was left with his grandmother because his father, as a policeman, was transferred regularly which would have interrupted his son's schooling. After the death of the grandmother he lived only one year with his parents and siblings before marrying. After his marriage, his mother had to give him $300 to set up house (this was before his father's death). As for the other son, he could hardly make ends meet. His mother gives him $20-$30 whenever she can, in spite of the fact that he is unmarried and is earning as a soldier. In this case, the family's economic situation is balanced rather precariously, depending very much on the two girls who are able to work.

The second household (Fig. 4.7) consists of husband and wife and six children, with the oldest two girls already married. In contrast to the first case, at least three children have worked in the electronics factory and two are still actively contributing to the household income. With the capable entrepreneurship of their mother as a trader, this is a household well above the average economic status in the village, unlike the first household. The house is completely renovated and full of consumer durables, furniture, refrigerator, cabinets, colour television, carpets, and the like.

Figure 4.7 Household 2

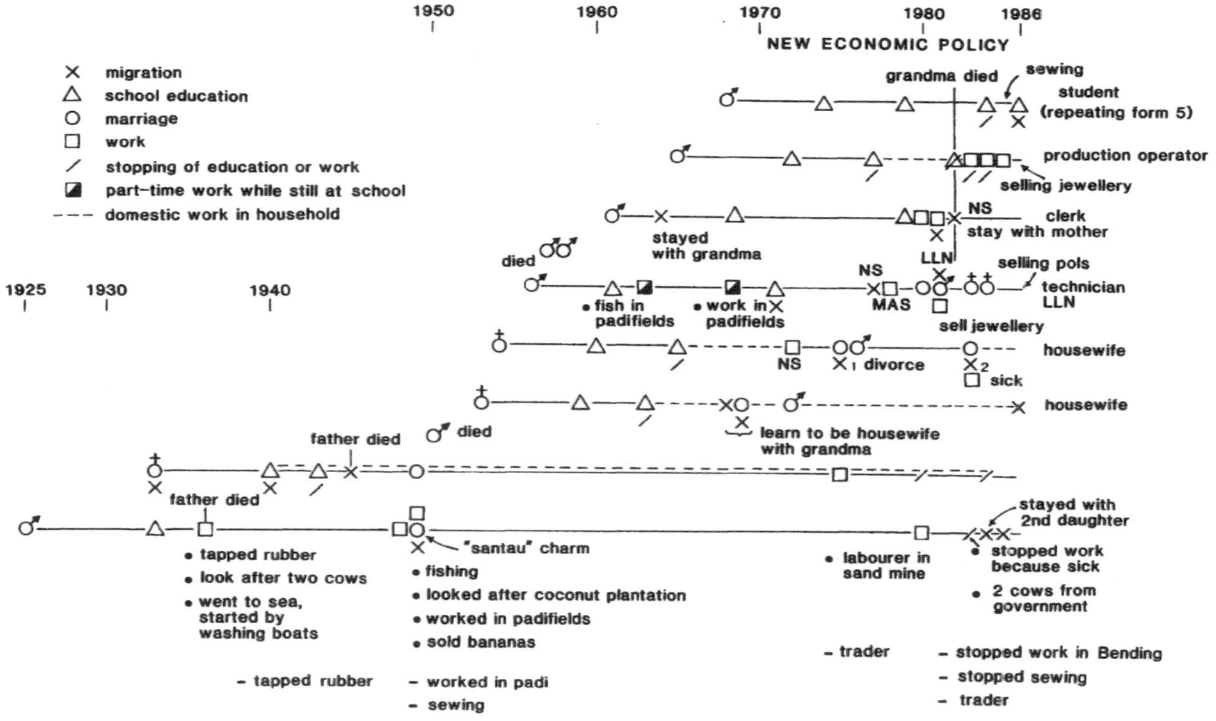

As in the previous case study, aspects of household adaptive strategies, income pooling, sibling and sex order all play a role in this household. The head of the household is not the husband but the wife. Pak Cik (the husband) is quiet, meek, and sickly - and quite peripheral to the family. His relationship with his wife started to degenerate with his failing health, culminating in his leaving the house between 1984-5 to live with his second daughter. Because of Pak Cik's inability to earn enough (due to his ineptitude), Mak Cik (the wife) started to work in padi fields and to sew clothes immediately after her marriage. She says that life was much easier prior to marriage. After marriage it became very difficult, especially when the children arrived. Her first child died during these initial years of economic pressure.

The first two girls were then taken out of school early to help with the housework and to look after the younger siblings. The economic situation eased somewhat with the second daughter working as a production worker in National Semiconductor from the age of 16. She was to be a major income contributor for eleven years, helping with her siblings' education (none had a scholarship) and the household expenses. All the other children helped in pooling income by working in the padi fields, sewing, and later, when the mother became a trader, assisting her in selling crockery sets, carpets, bedspreads, pots, and jewellery. Even after her marriage and the arrival of her son, the electronics worker continued working. This marriage was to end in divorce within the next year, so she and her son went back to her mother's household. In fact, she continued working until her second marriage six years later, at the time when her youngest sister entered the electronics factory and took over as the other major contributor to the household (other than the mother). The boys do not appear to contribute, except to help her sell her goods.

The younger children comment on how easy their lives have been compared to their two older sisters. For example, the eldest daughter is so poor today that her mother often gives her money to help her family. She was married off (arranged by her grandfather) after the second daughter had started taking over the housework. And it is the second daughter, the one who had contributed most to this household, who has quite a strong say in household decisions. Although the youngest sister is now the main contributor, certainly to most of the house renovations, she defers still to her eldest sister and mother.

CONCLUSION

The problem to which we address ourselves at the broader level is that of the processes of labour force formation in the periphery, as engendered by the rise of export-platform productive activity. In particular, we seek to analyse the transformation of the Malay household economy in response to world and national-level events affecting the processes of labour-force formation and reproduction. The micro-level case studies that we have analysed highlight the role of conjunctural events occurring during the life course of the different families, as well as through the eyes of individual workers. We hope to have shown the interplay between the structural and conjunctural processes. The structural processes are the internationalization of productive activity and the role of the state which determine the macro-level structures. The conjunctural events are the combined effects of the everyday (or short-term) economic, social, and political circumstances that affect the daily life situation of the individual and household economy. The individual and household case studies we have used in this paper are rich in depictions of micro-level life situations as they unfold through historical and family time. The difficult task is to relate these micro-events to the higher chronology of industrial time, some of which are captured in the impact of the recession.

The aim of this chapter is to seek a common theoretical and methodological framework to link micro-level processes (the conjunctural) with the macro-level processes (the structural), and to relate the culturally specific experiences in Penang with the depiction in other diverse parts of the world economy. The body of theory relating to the development of the capitalist world economy would seem to provide such an overarching framework but it needs to be amplified in areas which recognize the different articulations of structures at the local level in different parts of the world system.

NOTES

1. The study on household response to industrial change is a joint effort by the Universiti Sains Malaysia and the University of British Columbia (funded by IDRC). It covers about forty Malay rural and semi-rural families

where at least one daughter is on wage work. Most of the families had at least one daughter working in an electronics factory, the industry selected for this project to examine the impact it had on the newly emerging industrial Malay female workforce. Life histories were constructed for all household members and were then incorporated into family histories. The aim of this approach was to capture the changing configuration of the family through time, and to assess the impact made on it by individual members of the household.

2. Derived from intensive interviews between 1984 and 1986 by Chin Siew Sim, research assistant for the USM-UBC Household Response to Industrial Change Project.

3. All currency designated in Malaysian dollars, which is about $2.50 to one US dollar.

4. In fact, today (1987) automation in this section (wire-bonding) is quite advanced. That Mak Teh is vague about the machines and the impact on production is quite typical of the factory girls.

5. Village work which encompasses a variety of rural activities from work in the padi fields to labouring.

6. White man, a general term for Europeans, Americans, Australians, etc.

7. These production workers are particularly susceptible to direct selling. Amway products (like Arcoroc, Tupperware, and Avon cosmetics) are in the direct-selling line and are very popular among the factory girls.

8. It means uncle, and is used when addressing an elderly man, as a sign of respect.

9. Such elements of reciprocity still exist among the poorer kampung folk.

10. In fact, Yati was retrenched by Mostek.

11. Such peer-group influences are very strong among the factory girls.

12. It means half-and-half: half for the girls and half for the retailer.

13. Celebrations at the end of the fasting month in Ramadan.

14. Raised wooden floor used as an eating space.

Chapter Five

The acceptable face of self-help housing: subletting in Fiji squatter settlements - exploitation or survival strategy?

Jenny J. Bryant

INTRODUCTION

In the past decade it has become clear that the housing shortage in the Third World is reflected not only in the increasing number of squatter settlements and pirate subdivisions, but also in their intensification. As more people move into established settlements, households become increasingly overcrowded as families let and sublet rooms to others in an attempt to obtain additional income. As writers have become more aware of the increase in letting and subletting, there has been a revival in the commodity housing debates of the late 1970s and early 1980s. These debates revolved around the question of whether housing was only for consumption or whether it had income-generating properties (see Burgess 1978; Turner 1978; Conway 1982). These commodity debates were more concerned, however, with housing as a product for resale than for rent, and it was not until more recently that discussions turned to tenants and subtenants of squatter housing (see Amis 1982; Edwards 1982; Gilbert 1983; Gilbert and Ward 1985).

The revival of this debate has essentially grown out of a desire to extend the theoretical discussions on housing as an income generator, but it has also led to a more detailed examination of the notion of housing as an instrument of social control (McCallum and Benjamin 1985). The expanding knowledge of the role of women in household economies (Drakakis-Smith 1985), and of the importance of housing in creating and maintaining a stable middle class in Third World countries (Teedon and Drakakis-Smith 1986), has undoubtedly stimulated research on the importance of letting and the position of very poor groups which sublet houses or rooms within squatter houses.

The aim of this chapter is to examine the phenomenon of squatter renting in Fiji, and in particular the subletting of rooms within squatter households. Until recently, much of the research on squatter renting dealt with the relationship between income earned from letting and subletting and the upgrading process (Jimenez 1982; Struyk and Lynn 1983). In the so-called self-help housing process, squatters are thought to be able to afford the addition of rooms to their houses, or other structural improvements, as a result of 'windfalls', either in the form of rent or from family gifts. Although this research does examine the relationship between letting/subletting, upgrading, and maintenance, it will also examine mechanisms for survival undertaken in squatter settlements and will seek to discover whether those renting are in different circumstances from their landlords. Who are the renters and who are the landlords, is a crucial question. Is it true that tenants are poorer than owners, or are the former choosing to rent until they can afford something better (Gilbert and Ward 1985: 246)? Are the landlords exploiting tenants in order to build new homes, to upgrade them to make a house of better status, or simply to carry out basic maintenance work? Or are they merely taking rent in order to survive the depressed economy and high unemployment which is affecting most societies today (Edwards 1982: 55-6)?

SUBLETTING OR SQUATTER RENTING

Letting of houses and the subletting of rooms in squatter and pirate settlements is common practice in the Third World. Several of the writers examining this phenomenon highlight the situation of poverty and marginalization which forces the poor to rent as they have no other choice. They also show how it may possibly form the basis of a low-cost housing policy which provides legal rental space for the lowest income-earner (see Laquian 1983: 82, on Nairobi; Drakakis-Smith 1981a: 87, on Hong Kong).

In their study of Tondo in Manila, Struyk and Lynn (1983: 447) found a strong correlation between squatter 'lodging' (or subletting) and house upgrading. This led them to suggest that since housing improvements are carried out by squatters, governments can institutionalize this phenomenon by fostering letting and subletting in owner-occupied units (ibid. 453). Struyk and Lynn did not consider the implications of such policy on housing availability for the poorest group, nor did they take into account the fact that many squatters do not even have access to basic services and may need to take in subtenants to pay for these before even considering the upgrading process. The work of economists such as Struyk and Lynn, and also Keare and Parris (1982), does not consider the survival strategies undertaken by squatters and has tended to concentrate on legitimizing the informal housing sector. Fortunately, a number of researchers have been concerned with housing for the poorest group (Amis 1982; Gilbert 1983; McCallum and Benjamin 1985), and in recent years there have been suggestions of not simply institutionalizing the process but allowing subletting in squatter settlements to continue as an appropriate response to housing shortages (Laquian 1983: 82). It is important to note, however, that legitimization may mean deteriorating standards as landlords rent out more rooms for profit (Drakakis-Smith 1981a: 86). Examinations of the ultimate outcome of such a laissez-faire policy have not yet been carried out, but obviously the implications of a landlord-based housing market as a result of formalizing the informal sector are immense (Teedon and Drakakis-Smith 1986: 322).

This chapter examines the increasing incidence of letting and subletting in squatter settlements in Fiji. It seeks to gauge the extent of the practice, and the value which it has to those involved. McCallum and Benjamin claim that 'the poor obviously have a high tolerance for what outsiders perceive as overcrowding as long as it provides space which can be profitably rented out' (McCallum and Benjamin 1985: 283). They do not tell us how the rent received is used, but it is implied that the squatters have at least a choice about how they utilize their space and what they do with the money received; for example, they may upgrade their houses over a period of time, or build a new home when sufficient funds have been saved. McCallum and Benjamin also fail to examine the social and physical disadvantages of high densities.

Whether or not the tenants and subtenants are also making choices about whether to rent or buy should also be examined, because renting to them may be just a transition period before choosing to buy (Turner 1978). On the other hand, tenants may have no alternative, as land becomes more difficult to obtain and/or incomes fall (see Gilbert

1983: 454; Gilbert and Ward 1985: 129; Drakakis-Smith 1981a: 85; Edwards 1983).

A further dimension to the choices made by squatters about their housing futures is the role of the landlord. Publicity given to squatting in Fiji sometimes focuses upon the fact that certain groups of squatters are exploiting even poorer groups in order to pay off their assets of a house and land elsewhere. In this survey an attempt was made to discover how rental income is spent and what proportion of the weekly income of landlords and tenants is constituted by received rent. Whether or not the money involved is significant, and can be regarded as exploitative, is an important outcome of the survey.

This chapter is based on fieldwork carried out in a squatter settlement in Suva, the capital of Fiji. The subletting of rooms in Fiji squatter settlements is not universal, although the incidence is certainly increasing (Suva City Council 1983). Some background will be given on Fiji, its urban growth, and the history of squatting and subletting in squatter settlements. The results of the survey will be presented and conclusions drawn on the characteristics of squatter tenants and landlords and their ambitions. The link between subletting and upgrading will be examined in the light of the value of income received from subletting and the importance attached to it by both primary tenants and subtenants. The research has important implications for low-income housing policy in Fiji. If subletting becomes part of the acceptable face of self-help, it may provide the basis for government abdication of responsibility for the poor, as private landlords are left to provide cheap shelter. The long-term result of such non-policy would undoubtedly be to alienate further the poor from adequate shelter.

URBANIZATION IN FIJI

Located in the South Pacific, Fiji has the second largest population in the region (after Papua New Guinea), and also the second highest proportion of the population living in urban areas (after New Caledonia). Urbanization in the South Pacific is recent and a product of colonial settlement, unlike most other Third World countries, where there is often a long history of urban centres. In the South Pacific, cities and towns generally developed as administrative centres to serve plantation or mining regions and quickly became foci for employment and thus for rural-urban migration. Fiji, with the longest history of urbanization in the Melanesian islands of the South Pacific, is a good example of the colonial form of urbanization (Connell 1985: 71).

Ninety per cent of Fiji's population of around 715,000 lives on the two largest islands of Viti Levu and Vanua Levu. Of the 300 islands in the group, only about 100 are inhabited and the majority have no urban centres. More than 39 per cent of Fiji's population is now urban (Parliament of Fiji 1988), with over 54 per cent of urban dwellers now living in the capital, Suva (Connell 1985: 72), which has a growth rate more than twice that of rural areas (Walsh 1984: 185) (Fig. 5.1). Moreover, Suva's primacy is increasing, with it now being more than four times the size of the second largest city, Lautoka. More than half (51 per cent) of Fiji's population is of Indian descent; their forebears came to Fiji in the late nineteenth century as plantation labourers. Ethnic Fijians number 44 per cent of the total, and the remainder is made up of other Pacific Islanders, Europeans, and Chinese.

The growth of Suva was estimated to be 17.5 per cent in the decade 1966-76 (Chandra 1981: 30). Although urban growth in Fiji declined in the decade 1966-76 (Parliament of Fiji 1977), Suva - and in particular the 'Suva-Nausori corridor' - continues to expand with largely Fijian rural migrants (Connell 1985: 72-4) (Fig. 5.2). For the ethnic Fijians, however, rural-urban migration is a recent phenomenon, because absentee taxes, imposed by the colonial administration on those absent from their villages, were not lifted until the 1950s (Connell and Curtain 1982: 125). Many Indo-Fijians, on the other hand, have been urbanized almost since their arrival.

While the growth of Suva is not particularly rapid, it is significant that the population growth rate for poor groups in Suva is double that for the city as a whole (Walsh 1984: 185). Walsh does not define who are the urban poor, except in terms of their unsatisfactory housing in inner-city slums, housing authority flats, or squatter settlements (ibid.). It may be assumed from this that the poor are the 'socio-economic residue of an acquisitive, affluent, capitalist bourgeoisie' (Drakakis-Smith 1981a: 86) and as such may have little choice in housing. They must take what is available. It is this group, often destitute or earning minimal wages, which reflects the inequalities prevalent in Fijian society. It

Figure 5.1 Fiji: urbanization 1976

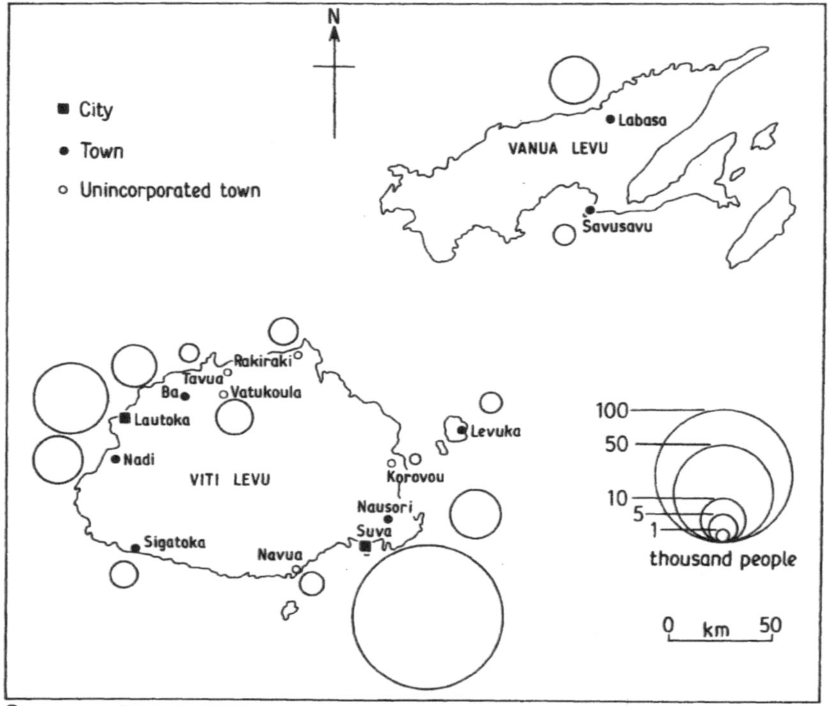

Source: Walsh 1984

Figure 5.2 Suva: distribution of squatter settlements

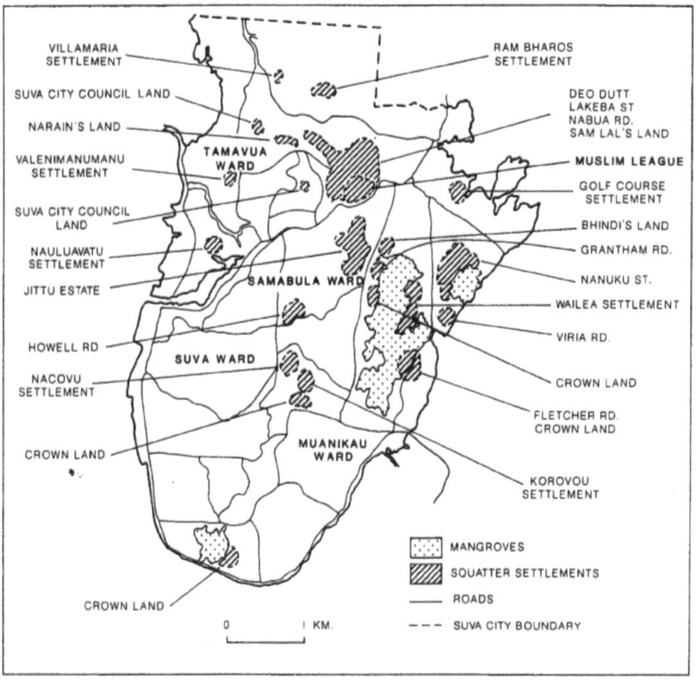

would be wrong to say, then, that the poor in Suva are increasing in number because of growing urbanization. Whether or not they have a choice about their housing future, the urban poor in Fiji are tending to concentrate in squatter settlements and inner-city slums, not solely because of a housing shortage, but because they can no longer have immediate access to the means of obtaining affordable housing. The poor are not the only group to inhabit squatter settlements and otherwise poor-quality housing. The presence of a landlord and wealthier class within these areas must be examined.

The Fiji Household Income and Expenditure Survey of 1977 found in urban areas that:

> 60 per cent of urban households live in single-unit dwellings, 30 per cent in flats, and 10 per cent in shacks, bures,[1] or temporary structures. Half of the

> homes are constructed primarily of concrete, a third of wood, 13 per cent of corrugated iron and the remaining 7 per cent are less substantial bures or shacks.
>
> (Fiji Central Planning Office 1980: 230)

In addition, the Fiji Central Planning Office survey found that more than 2,000 additional homes a year were needed between 1981-5, largely in the Suva-Nausori area. This was thought to be due to the increase (although slowing) of more than 3 per cent a year in the urban population in the decade 1966-76, and a trend towards smaller family sizes (Bryant and Khan forthcoming: 2).

The inability of the authorities to meet the housing demands of the poorest groups is partly responsible for the expansion of the squatter settlements which have been characteristic of Fijian towns since the 1950s. The question that must be asked, however, is whether there really is such a simple relationship between population growth, housing shortages, and squatting. Fiji is facing worsening poverty as the effects of declining world sugar prices and a downturn in tourism are felt, but the links between poverty and squatting are by no means clear (Drakakis-Smith 1981a: 86; Connell 1985: 85).

There has been no systematic survey of poverty in Fiji, although 15 per cent of the population is thought to be suffering acute deprivation (Cameron 1985: 488). One of the aims of this chapter is to demonstrate whether or not it is poverty which is forcing people to squat in increasing numbers and to sublet rooms in squatter settlements. The suggestion is made that both squatter housing and the subletting of rooms in squatter households are the inevitable attempts at survival by landlords and subtenants alike in the face of increasing social inequalities. Alternatively, renting in squatter settlements may be by choice, until a better life-style is possible (Gilbert 1983), while for landlords it may be an easy way of supplementing their inevitable movement out of the poorest class.

SQUATTING IN FIJI

The term 'squatter' is used loosely in Fiji to include all spontaneous or informal settlements which have substandard and unauthorized structures and a lack of basic services. It does not apply solely to those who have occupied land illegally. Using the above definition, surveys indicate that between 12 and 20 per cent of the population of Suva were squatters in the late 1970s (Walsh 1984: 185; Fiji Central Planning Office 1980: 232; Suva City Council 1983: 2). Other cities such as Lautoka, Ba, and Labasa also have sizeable squatter populations (Table 5.1). The exact figures are difficult to assess precisely, but most importantly, the squatter population is thought to be increasing at more than 500 households a year (Fiji Central Planning Office 1980: 232) (see Table 5.2).

Since 1972 the Suva City Council (SCC) and the Directorate of Town and Country Planning have carried out a number of surveys of squatting within the country. They show that the number of squatter households is increasing and reveal the changing tenure status of the squatters (Table 5.2). There is also an increasing emphasis given in the surveys to the employment status and access by squatters to other assets such as a house or land elsewhere. This reflects governmental concern that many squatters may not in fact be genuine, in the sense of being landless and homeless, but may instead be utilizing available land for squatting as a means of saving for a better standard of living in the future. This concern is reflected in the media which frequently reports that, for example, 'Squatters held good jobs such as bus driving and were civil servants' (Fiji Sun 3 March 1985: 3), or that, 'several squatters were earning as much as $100[2] a week' (ibid.).

Table 5.1 Fiji: estimated numbers of urban squatters 1976

Locality	No. squatter households	% households squatting
Suva	1,728	18
Lautoka	375	9
Ba	43	5
Labasa	66	9
Levuka	44	19
TOTAL	2,256	19

Sources: Fiji Central Planning Office (1980: 232); Bryant and Khan (forthcoming)

Table 5.2 Suva[1]: estimated number of squatter structures 1978 and 1983

Occupancy	1978	1983	% change
Owner occupied	818	1,090	33.3
Owner/tenant[2] occupied	139	163	17.3
Tenant[2] occupied	106	98	-7.5
TOTAL STRUCTURES	1,063	1,351	27.1

Source: Suva City Council (1983: 2)

1. The Suva area as administered by the Suva City Council includes only four wards and not the outer suburbs as included in the Housing Authority figures. Thus the figures here are lower than those found in Development Plan 8 (Fiji Central Planning Office 1980), but give the scale of increase in squatting.
2. No differentiation has been made by the Suva City Council between tenants and subtenants.

To support such statements, the newspaper concerned claimed to be quoting a 1983 Suva City Council survey report which few members of the public would be likely to read. The report in fact showed that 3 per cent of the 1,436 individuals employed and interviewed earned more than $100 per week (of whom only one-fifth were Fijian), that 47 per cent earned between $30 and $50, and that 31 per cent of individual incomes were between $50 and $100 per week (Suva City Council 1983: 7).

Attempts to challenge the genuine needs of squatters are undoubtedly a reflection of first, underlying ethnic tensions in Fiji (as many squatters are Indian); second, a worsening economy; and third, severe land shortages in urban areas; all of which, the authorities claim, make it difficult for them to meet urgent housing needs. Publicity given to squatters who 'make deliberate attempts to flout the law for personal benefits', and to the 'speculator' squatter who is usually a 'professional' to whom squatting may be a business venture (Fiji Ministry for Lands, Local Government, and Housing 1982: 1-4) does not take account of the fact that there was an increase of 16.4 per cent in the number of people living in squatter settlements in Suva between 1978 and 1983 - an increase of 3.3 per cent a year.[3] There was also a 17.3 per cent growth in the number of households which were 'owner and tenant' occupied, indicating an expansion in the phenomenon of subletting (Table 5.2). This may represent a response of the poor to worsening poverty and not merely to an acute shortage in public housing (Bryant and Khan forthcoming: 12). Since 1972 there has been an increase of 50 per cent in the number of squatter households which comprise either owners and tenants together or tenants alone. This rate of growth is twice as fast as the growth of squatter ownership during that period[4] (Table 5.3).

Table 5.3 Suva: squatter tenancy* 1972-83

Tenancy status	1972	1978	1983	% change (1972-83)
Owner occupied	865	818	1,090	26
Owner and tenant	95	139	163	72
Tenant	79	106	98	23

Source: Adapted from Suva City Council (1983: 8-11)

* The Suva City Council does not differentiate between tenants and subtenants.

SQUATTER RENTING IN FIJI

As indicated earlier, in Suva between 1972-83, there was a 50 per cent increase in the number of squatter households with tenants and subtenants (Table 5.3). Although detailed surveys of the entire population of landlords and tenants in squatter settlements have not been carried out, the practice of renting in settlements is recognized by government documents and the media. For example, the Ministry for Lands, Local Government, and Housing referred to tenants in a recent report as 'the poorest class of squatter' (Fiji Ministry for Lands, Local Government, and Housing 1982: 3).

The report was referring, however, to primary tenants (i.e. those who rent an entire house from a squatter owner) and not to subtenants, (those who rent a room from other squatter tenants). The media on occasion view landlords as 'predators...who put up shacks in a shanty town and then rent them out to desperate [sic] poor people' (Fiji Times, 21 February 1983).

While it is likely that certain tenants are benefiting from subletting rooms to even poorer tenants, the link betweeen a developing wealthy class of tenants and the practice of subletting is unproven. Indeed, to take the example of house upgrading as an indication of increasing wealth and social mobility, the evidence is minimal, according to Walsh (1979). Walsh has also tested Turner's views that squatter settlements are self-improving suburbs (Walsh 1979; 1984). He found that while 64 per cent of squatters had made improvements since they built the house or moved in, only 7 per cent of the housing could be considered adequate in terms of structure, size, water supply, and sanitation (Walsh 1979: 7; 1984: 193). He demonstrates that while squatters are willing and able to help themselves to a certain extent, without reasonable income and a better relationship with the larger urban society, they are unlikely to carry out substantial upgrading (Walsh 1979: 8). Although Walsh was not specifically studying landlords and the value to them of rental income, he did not find that squatters had substantial amounts of surplus income, although they did have the desire to improve their situation. The landlordism in Suva can thus be assumed to be petty and not on a large, commercial scale.

The value of rental income and the importance attached to it by both landlords and tenants is the theme of this chapter. If rent is contributing to the development of a wealthy landlord and tenant class able to divert rent into upgraded and even new housing, then the long-term implications for the poorest group in need of housing must be examined.

The Muslim League survey

Based on figures released by the Suva City Council, there are four settlements which have around 20 per cent or more of the structures occupied by owners and tenants, or tenants alone. These are Jittu Estate, Narain's Land, Deo Dutt Estate, and the Muslim League. Narain's Land and Deo Dutt are small settlements with fewer than seventy-five structures, and the extent of letting and subletting has not increased significantly since 1978 (Suva City Council 1983: 10).

Jittu Estate, the largest squatter, or informal, settlement in Suva had 224 structures in 1983, 58.5 per cent of which were occupied either by owners and tenants together, or tenants alone. This settlement of 17 hectares is long established, the privately owned freehold land having been let and sublet to settlers as well as settled by squatters over the past thirty to forty years. Because of its size, lengthy history, many well-established homes, and the presence of such a large number of subtenants, Jittu Estate would have been the ideal site for a survey of subletting. Unfortunately, during 1985 and 1986 the settlement became the target of several redevelopment plans and was caught up in an argument between the Suva City Council and the Fiji Housing Authority as to how best to redevelop the site. The uncertainty of these development plans, as well as the administration of a (poorly designed) survey by the Housing Authority, which had already taken up a considerable amount of the tenants' time, made it necessary to select an alternative site, that of the Muslim League, for this study.

The 22-hectare Muslim League settlement off Ratu Mara Road in Samabula, Suva, is owned by the Fiji Muslim League, an affiliate of Rabettat, Al-Alam Al-Islami in Saudi Arabia (The World Muslim League). The freehold property was acquired in 1964 for the development of an educational centre. To date, the Suva Muslim League High School has been built in the north-east corner of the site. Plans for the development of the remainder of the site were forwarded in 1985 to the Islamic Development Bank in Saudi Arabia, although the Muslim League is also considering an alternative development plan. This could include the possibility of co-operating with the Fiji Housing Authority to upgrade and subdivide the house sites and to give squatters security of tenure. At present, the Muslim League pays $6,088 a year in rates to the Suva City Council, none of which is recovered from the squatters (Dean, Pers. Comm. 23 April 1986).

There were 178 squatter structures on the Muslim League site in 1983 (Suva City Council 1983: 8), with a population of 1,162, and a household density of 6.5 persons per structure (Table 5.4). Although the people are not strictly squatters in the sense that they have the permission of the Muslim League to be there, the term is used to indicate that the settlement is informal, and that a number of the structures do not have adequate facilities.

Table 5.4 Muslim League settlement: structures and facilities 1972-83

	1972	1978	1983
No. structures	131	162	178
No. people per structure	7.2	6.8	6.5
Facilities:			
Pit latrine	129	153	157
Pour flush	1	4	17
Septic tank	1	2	6
Tap water	n.a.	89	147
Storage tanks	n.a.	1	12
Wells	n.a.	13	19

Source: Suva City Council (1983: 11)

The population of the settlement increased by only 5 per cent between 1978 and 1983, with its most rapid growth of 17 per cent occuring in the six years prior to this (Table 5.5). While household density has apparently fallen over the past fifteen years (Table 5.4), there has been a change in the ethnic composition of household units away from what was largely an Indian phenomenon of squatting to an increasing number of Fijian families moving into the settlement. This situation is reflected throughout Suva squatter settlements where there was an increase of 9 per cent in the number of Fijians squatting between 1978-83 (Suva City Council 1983: 2). While Indians continue to own the majority of the structures (99 per cent), they are increasingly subletting rooms to other people, which means that there are more household groups in one dwelling, in some cases with landlords, tenants, and subtenants sharing (Table 5.6).

The changes in tenancy status, while not of major proportions, were nevertheless occurring by 1983, and it was

Table 5.5 Muslim League settlement: population 1972-83

	1972	1978	1983
Adults	n.a.	n.a.	698
Children	n.a.	n.a.	464
TOTAL	946	1,108	1,162

Source: Suva City Council 1983: Appendices

Table 5.6 Muslim League settlement: tenancy status 1972-83

	1972		1978		1983	
	no.	%	no.	%	no.	%
Owner occupied	109	83	137	85	144	81
Owner/tenant occupied	14	11	18	11	20	11
Tenant occupied	8	6	7	4	14	8
TOTAL	131	100	162	100	178	100

Source: Suva City Council (1983: 8-11)

assumed for the purposes of this research that the phenomenon would continue into the mid-1980s and beyond, reflecting the worsening economic situation for landlords and poorer groups alike.[5]

As letting and subletting are also linked to the process of upgrading existing homes or improving one's economic circumstances (Walsh 1979; 1984; Jimenez 1982; Keare and Parris 1982; Struyk and Lynn 1983), it was considered essential to discover what other property was owned by the squatters in the Muslim League settlement. In its 1983 survey, the Suva City Council found that of the 178 owners of structures, seventeen (9.6 per cent) owned other houses or an area of freehold land elsewhere in Suva. Two of these owned two houses each which were rented out in the Muslim League Estate itself (Suva City Council 1983: 22).

Subletting in Fiji squatter settlements

In April and May 1986, after discussions with the Suva Muslim League and Suva City Council, it was decided to carry out a 20 per cent sample survey of Muslim League squatters to determine the extent and contribution of letting and subletting. The Muslim League had also carried out a partial (43 per cent) population survey in 1985 which provided a basis of comparison for the 1986 survey.

The sample of 178 dwellings was undertaken in 1986. The questionnaire was comprehensive, eliciting information on the structure (number of households, quality of dwelling, building materials, etc.), on the households (finances, tenure status, previous dwellings), and on the individuals (age, sex, employment). Questions concerning letting, subletting, and upgrading were repeated within the questionnaire in an attempt to cross-check responses.

Survey results

A most significant finding of the survey is that the number of people per dwelling has increased dramatically since the Suva City Council survey of 1983. In the thirty-five dwellings there were 303 people, or an average of 8.7 persons per dwelling. This compares with 6.5 people per dwelling according to the Suva City Council (1983: 8). In 1986 the 303 people were living in 86 household units: 38 of these units were interviewed. It can be assumed from these figures either that the squatter population increased dramatically after 1983, with no corresponding increase in the number of dwellings, or that the Suva City Council surveys interviewed only the main household (usually the dwelling owner), and that subletting is not admitted to officialdom.

The average length of residence of household heads in the survey was between eighteen months and two years. In other words, they had arrived after the Suva City Council survey had been conducted. This comparatively recent increase in tenants is also reflected in overcrowded houses, as there has been a total of only three rooms added during the period in which the people have been resident. Indeed, the majority of the houses in the estate are old, with an average size of four rooms, with rooms to let and sublet (Table 5.7 and 5.8).

Table 5.7 Muslim League settlement: age of house

Age of house	No.	%
< 1 year	0	-
1 - 2 years	1	3
2 - 5 years	1	3
5 - 9	3	9
> 9 years	30	85
TOTAL	35	

Source: Fieldwork (1986)

Table 5.8 Muslim League settlement: type of house

Type of house	No.	%
Family house	2	6
House with rented rooms	29	83
Group of structures	4	11
TOTAL	35	

Source: Fieldwork (1986)

Sixty-nine per cent of the houses were built by the current household using self-help methods, and 23 per cent were built before the current owners moved in, making the building history of the latter category an unknown element. Most are constructed of metal with timber foundations, the majority have kitchens under cover and an outside, mains water-supply system. The majority use wood or kerosene for cooking and have outside, shared, pit latrines, a situation which has changed little over the past few years (Tables 5.4, 5.9, 5.10). There are no households without access to cooking, water, or sanitary facilities.

Table 5.9 Muslim League settlement: building materials

Building materials	Walls	Roof	Foundations
Timber	2	0	27
Plyboard	7	0	0
Corrugated iron	18	35	0
Timber and iron	8	0	0
Earth	0	0	0
Concrete	0	0	2

Source: Fieldwork (1986)

Table 5.10 Muslim League settlement: household facilities

Facilities	Water	Sanitary	Kitchen
Inside	2	1	29
Outside	30	37	11
Both inside and outside	6	0	0
Shared	0	27	11
Not shared	0	11	27
Mains water	31	0	0
Rainwater tank	5	0	0
Well	4	0	0
Container	5	0	0
Flush toilet	0	1	0
Pour flush	0	7	0
Pit	0	30	0

Source: Fieldwork (1986)

Note: Ten of the households had more than one type of water supply.

Twenty-five of the thirty-five dwellings surveyed had tenants, seven of them with more than one tenant family. In fact, sixty of the households were subletting (70 per cent of the total number) (Table 5.11). The indications are, then, that according to this survey, letting and subletting in squatter settlements are of far greater importance than ownership, thus begging the question of the value and importance of letting/subletting.

Table 5.11 Muslim League settlement: tenancy status

Status	No. households	%
Owner occupiers	21	24
Tenants	28	33
Subtenants	32	37
Non-paying guests	4	5
Other	1	1
Total no. households	86	100
Total no. dwellings	35	
Total no. people	303	
No. people per dwelling	8.7	

Source: Fieldwork (1986)

Unfortunately, it was not possible to interview more than nine of the twenty-five primary tenants who were subletting rooms. Those nine primary tenants claimed to spend the rent money on food, school fees, and other everyday needs such as clothing and household fuel. Only one primary tenant said that he used the money for house upgrading. Certainly the sums of money involved may seem small, averaging between $6 and $10 per week (usually $7.50), with no increase in the past two years (Table 5.12). Compared with average wages in Fiji, however, they are substantial. The average wage in 1986 grosses at around $40 per week, and in 1976 the mean household income was

$58.66 for Fijians and $38.88 for Indian squatters (Connell 1985: 85, quoting Walsh 1979).

Table 5.12 Muslim League settlement: rent paid per week by tenants and subtenants

Amount	No.	%
$5.00	10	32
$7.50	15	48
$7.50-$10.00	1	3
$12.50	3	10
$15.00	2	7
Unknown	1	-
TOTAL	32	

Source: Fieldwork (1986)

The amount received from subletting would be inadequate for the purchase of building materials and the hire of labour unless the primary tenant had more than one household as subtenants. In fact fourteen of the thirty-five dwellings do have subtenants, and in three of those houses there are five or more separate groups of subtenants, each paying $7.50 per week, ensuring an additional weekly income to the primary tenant of almost double the average wage of squatters in Suva (adapted from Suva City Council 1983: 4).

Walsh's findings in 1979 showed that very little upgrading was then taking place in Suva squatter settlements. While this might imply that any additional income was used for survival, rather than for improvement in house standard, it would appear that the situation had changed substantially by 1986. Despite the claims made by people about how they spent their additional income, it appears that substantial upgrading is taking place in the Muslim League estate. A number of homes in the settlement have had additions and improvements made since the people have been resident. The improvements have not all been minor, and all have been carried out by self-help means, indicating a desire to upgrade, even where large amounts of surplus cash are not available (Table 5.13).

Table 5.13 Muslim League settlement: dwelling improvements

Nature of improvement	No.	%
Additional rooms	3	6
Better kitchen	7	13
Better toilet	2	4
Improved wall materials	1	2
Improved roofing materials	1	2
Improved flooring	6	12
Improved doors or windows	2	4
Paint	2	4
Total reconstruction	4	8
Electricity connection	2	4
Water connection	1	2
Other (e.g. verandah, shelving)	5	10
No improvements	16	31
TOTAL NO. IMPROVEMENTS	52	

Source: Fieldwork (1986)

A major improvement which has taken place is the connection of an outside mains water supply. This is not the result of upgrading by squatters, but is instead supplied for a fee by the Suva City Council. In structure, the majority of the homes (74 per cent) are constructed of corrugated iron, or a combination of timber and corrugated iron, and all have metal roofs.

The extent of upgrading which has been carried out on houses in the settlement does indicate the desire of occupants to improve their homes, and acquire a feeling of long-term security, despite an apparently insecure future. It may also demonstrate an expansion in landlordism and a movement away from the limited nature of subletting where the number of subtenants was small and the income received minor.

However, the upgrading does not necessarily mean improved living conditions for tenants and subtenants. As Walsh found in his work, very few squatter houses in Suva coud be considered adequate in terms of structure, size, water supply, and sanitation (Walsh 1979:7). This 1986 survey confirmed Walsh's findings, in that 79 per cent of the homes still have outside water supplies, 97 per cent have outside toilets, and 71 per cent of those are shared with other households (see Table 5.10). Of course, the sharing of facilities may be an indication of upgrading, for as more toilets, for example, are added to dwellings, the accommodation is able to be shared by more landlords and tenants. In Fiji, however, it would appear that people do not 'squat' in increasing numbers because there are more facilities. They move into a settlement despite the number of facilities available, and the landlords themselves continue to share the same toilets and water taps, until they can move away altogether.

Such behaviour supports Drakakis-Smith's argument that there is frequently general deterioration in squatter settlements, despite upgrading (Drakakis-Smith 1981a: 86). It is more likely that the rent received is going into the upgrading or building of a home elsewhere, and not into routine maintenance of the houses in the settlement. Unfortunately, only three landlords were prepared to admit that this was the case, although the Suva City Council found that seventeen of the squatters in Muslim League, of the total 178 dwellings, did own land or a house elsewhere (Suva City Council 1983: 4).

Employment and squatter upgrading

A number of the households are supplementing their incomes by letting and subletting rooms, but those interviewed claimed that the money received was for day-to-day survival. It would appear from the amount of upgrading, the number of homes owned elsewhere, and the substantial proportion of weekly income which rent contributes, that this is not the case. It is likely, however, that people do not so neatly compartmentalize their earnings and really cannot say how much money goes into upgrading or building.

There is little evidence in this survey of squatters upgrading houses after receiving 'windfalls'. Every household head interviewed was employed, whether he was a primary tenant or a subtenant, and the majority worked in similar occupations, largely in production, transport, or sales. Thus household income in the settlement does not vary to a great extent. Such a finding supports that of Edwards in Bucaramanga, where it was discovered that primary tenants were generally low-income earners, living on the same lot as their subtenants and sharing the same facilities (Edwards 1982: 56). In the case of the Muslim League, however, there is evidence that the movement towards capitalist landlordism is occurring and may well increase as land becomes difficult to obtain.

Since a long-term study has not yet been carried out, nor household income correlated with the degree of letting and subletting, it is difficult to predict a trend where household upgrading may be correlated with income. The high degree of letting and subletting does perhaps indicate a response to poverty, but it does not explain why people are renting rather than building their own homes. The most likely explanation is to be found in the availability of land and in incomes. The tenants and subtenants in the Muslim League settlement of Suva may be almost identical to their landlords in employment status, but they do differ in race (the number of Fijian tenants is increasing, while 99 per cent of landlords are Indian), and they are slightly younger and more recent arrivals to the settlement. There are also more single parents among the tenants.

The indications then are that first, land is simply no longer readily available for squatting or for housing of any type. This is certainly the case in the urban areas of Fiji, where both formal and informal housing developments are moving increasingly to the periphery, and where only 20 per cent of the urban employed can afford to buy Housing Authority homes (Bryant and Khan forthcoming: 3). Second, the more recent arrivals in the squatter settlements are renting, not only because there is no land available, but also because they have to. Their incomes will not allow them to live on the outskirts of towns where transport costs are high, and access to employment difficult. To continue Edwards' statement that 'people rent because they are poor; they are not poor because they rent', the landlords initially also had to rent rooms because they too were poor (Edwards 1982: 59). As the need for rental accommodation grows and with the limited availability of rooms, it would appear from the data presented in Table 5.6, that landlords are improving their situation by taking in more tenants. Whether or not they are using that income for housing developments elsewhere was difficult to ascertain in this survey, although the Suva City Council did find in 1983 that seventeen of the Muslim League landlords owned houses or land in other parts of Suva (Suva City Council 1983: 22).

As constraints on land become more severe, it will be more difficult for the authorities to provide housing for low-income earners. This in turn strains the availability of rental accommodation and could stimulate the growth of landlordism.

In the case of the Suva Muslim League, the possibility of better housing for the tenants of its houses depends very much on charity. If the League does decide to subdivide the land and allow the squatters to remain, then it will be the landlords only who benefit by ownership of the house and land, and by being able to continue letting rooms to supplement additional building projects and upgrading. While this may mean that the tenants are no longer in deteriorating houses, it will probably also mean that they must pay higher rents. The future of such an uncontrollable aspect of private housing development will contain useful lessons for the government Housing Authority, which is currently examining the possibility of legitimizing letting and subletting within its own schemes. A prediction would be that, like most housing schemes designed for the poor, legitimate letting and subletting will once again fail to benefit the poorest group.

NOTES

1. Bure is the Fijian word for a house, although not one used by a private family. It is more commonly used to denote accommodation for single men or strangers, and often as living quarters for domestic workers.

2. At the time of writing, F$1.02 = US$1.00. Since the establishment of the republic, this has worsened considerably.

3. The number of people per structure in fact fell from 7.8 to 6.8 in that time, and there has been an increase in the number of structures of 5.5 per cent a year (Fig. 5.2).

4. These figures for Suva represent slower growth in

squatting than that estimated by the Housing Authority which suggested that squatter populations were growing at 12 per cent a year (Parliament of Fiji 1982: 7). The difference could be accounted for by the fact that the Suva City Council deals with Suva City only, and not the outlying suburbs where most rapid growth occurs.

5. The assumption made was based on trends in other Third World countries where renting is increasing because of land shortages and other difficulties (see Gilbert 1983: 453; Edwards 1982: 51) and was also based on the present state of the Fiji economy (Fiji Employment and Development Mission 1984; Connell 1985: 3-4).

Chapter 6

The built environment and social movements in the semi-periphery: urban housing provision in the Northern Territory of Australia

David Drakakis-Smith

This chapter examines the links between urban housing provision and economic development, together with the role of the state, and social relationships in shaping these links. It is set in the context of Northern Australia. Before undertaking the empirical analysis, it is necessary to establish the conceptual framework within which this will occur in order to relate the nature of the investigation to the themes that comprise the principal foci of this book. There are two distinct sections to such a framework; the first relates to the spatial dimensions of development, whilst the second encompasses some recent ideas on the relationship between urbanization and economic growth, specifically the role of social movements in shaping the urban built environment.

In this context, it is particularly important to clarify the concept of the semi-periphery because, as the discussion will show, its characterization gives an opportunity to place more emphasis on the superstructural elements of the modes of production. If this is the case, then an empirical analysis based on the semi-periphery should provide an appropriate framework for this study. It is thus the task of this conceptual overview, first, to establish a character for the semi-periphery which will support a contextual analysis of social movements; and second, to discuss why these are important in helping to understand the development role of the built environment.

THE SPATIAL DIMENSIONS OF DEVELOPMENT: CORE, SEMI-PERIPHERY, AND PERIPHERY

All development, and underdevelopment, has spatial variation. Some is the result of inherent geographical differences - for example, in the location of resources or cultural groups - but much is the consequence of historical and social processes. Just as capitalism has varied in intensity and nature through time, so it has over space. Geographically uneven development can therefore be the consequence both of historical forces and of ongoing change in the production process. It must be noted that in this respect assessing uneven development involves not just the superficial measures of wealth but also comparisons of production, distribution, and consumption systems, including the varying role of the state, class relationships, technology, and so on (Soja 1984).

Within geography, classification and categorization are something of an obsession (Sibley 1983), stemming largely from the 'cartographic instinct' to bring order, at least in two-dimensional terms, to complex world or regional data and produce 'patterns' of relationships. So has it been in development studies, with geographers arriving relatively late on the conceptual scene and seeking to give respectability to the spatial perspective on underdevelopment (Rimmer and Forbes 1982).

Some, such as Soja (1984), have argued strongly and eloquently for the relevance of space within the radical perspective, but for many the task of categorization has seemed sufficient in itself to state the geographical case, leading to a substitution of classification for explanation (see Reitsma 1982a, b, 1983; N. Smith 1982; Doherty 1983). Such categorizations are often solely descriptive, but can vary widely in nature and value, from a concern with revealing geographical inequalities to simple spatial interpretations of development theory. Archetypal in this latter context has been the geographical interest in the core-periphery concept.

The notion of cores (or centres) and peripheries is not new and is an interpretative device which has been employed by all political shades of development strategists. Friedmann (1972), for example, used it to illustrate his normative, functional model of regional change in the 1960s. However, at about the same time, more radical theorists of the dependency and world-system schools were employing the idea as a framework to illustrate the permanent disequilibrium in the global economy (Frank 1969; Wallerstein 1976).

For many critics (see Browett 1981) this simple bipolarization contained too many contradictions to be analystically useful. The overemphasis on unequal exchange ignored not only the global movement of finance and production capital (Palloix 1977), but also the fact that internal social, political, and economic structures, together with their articulation with international capital (colonial or neo-colonial), have strongly shaped the nature of specific formations in even the remotest corners of the Third World. As Gorz has noted: 'the frontiers between development and underdevelopment, between dominant economic powers and dominated populations, between colonizers and colonized, do not pass simply between nations ... these frontiers exist also within each nation of the capitalist world' (cited in Armstrong and McGee 1985: 25).

Soja, who regards core-periphery theory as a fundamental, 'first-order' conceptualization of geographically uneven development, lays great importance on the hierarchical nature of its structure. Not only is there a global core and periphery, but also a series of nested regional and sub-regional expressions of the same relationship extending down, according to Soja, to the design of the built environment, household, and workplace. Such nesting occurs within all social formations at all levels, and suggests that spatial linkages between components and levels can be complex and varied. Galtung (1971) claimed that relations between the core and periphery were characterized by what he termed vertical interaction (or exploitation), but that the conditions which gave rise to and maintained such exploitation were the feudal relations within and between the peripheries which fail to interact and mobilize to bring about change through unified action.

However, the stylized feudal relationships postulated by Galtung are not a reflection of the operations of the global enonomy in the post-colonial period. The rigid control of commodity trade by metropolitan countries has been superseded by the multi-faceted operations of transnational corporations and financial institutions. Linkages thus occur between cores and peripheries across all levels of the nested hierarchy, a point to which the discussion will return below.

Another criticism of the simple dichotomy of dependency and world-system theory was that it failed to recognize and account for variations from the polarized core and periphery, such as the new industrializing countries (NICs) of Taiwan, Singapore, and the like, or the 'developed periphery' of countries such as Australia and New Zealand (see Browett 1981; N. Smith 1982). To counter this criticism, writers such as Galtung (1971) and Wallerstein (1976) introduced the 'go-between' or 'semi-periphery' as an intermediary stage of development. Neither, however, was able to define this category satisfactorily. Galtung, in particular, adopted a very crude definition in which the nature of the production process itself was assumed to be at some transitional stage within the semi-periphery. In his words, the 'go-between' stages would 'exchange semi-processed goods with highly processed goods upwards and semi-processed goods with raw materials downwards' (Galtung 1971: 104). Significantly, when put to an empirical test by Gidengil (1978), Galtung's criteria failed to indicate any clear-cut gradation in the non-metropolitan countries.

Wallerstein (1976) adopted a firmer approach, at least in terms of his historical validation of the semi-periphery, alleging that it comprised nations on the way up or down in the world economic system. He particularly highlighted the importance of periods of economic stagnation (as at present) in changing the economic ranking of nations. As with much of the world-system theory, however, the links with the real world were not pursued vigorously enough to establish firmly the existence of a semi-periphery.

One of the most intensive programmes which attempted to follow through Wallerstein's conceptualization of the semi-periphery occurred in the Fernand Braudel Centre; a programme recently summarized by Arrighi and Drangel (1986). Their conclusions offer a strange mixture of perceptive insights, detailed statistical analyses, and sweeping, often simplistic, generalizations. Their concern is to establish the validity and stability of the semi-periphery by an analysis of data over a fifty-year period. The results apparently verify the existence of a stable group of some twenty countries 'capable of selectively exploiting the peripheralizing tendencies of the world economy so as to prevent downgrading of their mix of core-peripheral activities but not sufficiently to attain core status' (Arrighi and Drangel 1986: 41). The outcome might be more convincing had the whole edifice not been constructed upon the foundations of data on per capita GNP which are patchy and, more important, very limiting. The results (Fig. 6.1) throw up obvious anomalies, such as Libya being part of the core, which are not really addressed. But in seeking to explain a changing world situation, any investigation should surely focus upon mobility and its causes.

Earlier observers, such as Cardoso and Faletto (1978),

Figure 6.1 The changing world-system categories

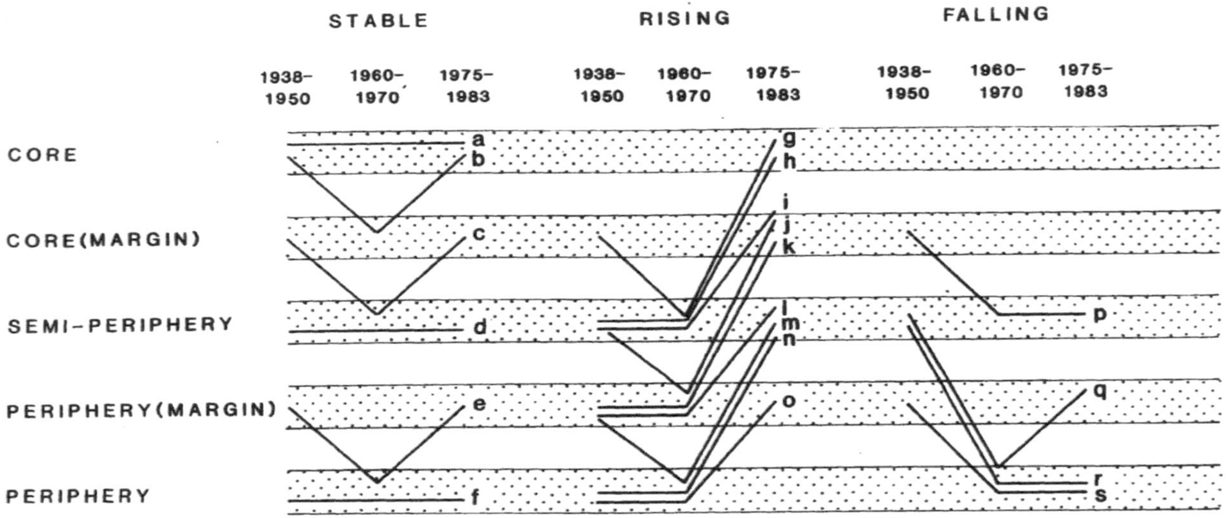

Source: . Arrighi and Drangel 1986

suggested that the semi-peripheral states can be distinguished by the extent of colonial settlements vis-à-vis direct exploitation; at least, that is the implication of their analysis of Latin American nations. Extended colonial settlement, it is claimed, led to the development of a domestic bourgeoisie and elite who were able to establish a degree of internal control over domestic resources and the nature of their exploitation, although overall there was still subordination to the interest of the core or metropolitan nations: a proposition further explored by Warren (1973). This characterization of the semi-periphery can be extended outside Latin America to other nations, such as Australia and South Africa, and perhaps even to the NICs where national control of domestic and regional elements within global processes, such as the New International Division of Labour, has begun to occur (Henderson 1986).

This emerges very clearly in what is one of the most vigorous attempts to identify the semi-periphery: Vayrynen (1983) has not only compiled a list of the major characteristics of expanding semi-peripheral nations, but also classifies them into four main types (Fig. 6.2). He attempts to test this empirically but falls into the common trap of selecting illustrative data on the basis of a priori decisions of major characteristics, thus selectively substantiating his own subjective judgement. Nevertheless, Vayrynen's work is important because it introduces non-economic variables into the classification system, although it does not go far enough in recognizing the importance of social movements in the generation of change. It is on these grounds above all that the system fails to incorporate some of the more developed semi-peripheral states, such as Australia and New Zealand. Vayrynen's classification is also limited in that, like many aspects of development theory, it does not seek to establish those characteristics of the periphery which persist in the semi-periphery (or begin to appear in declining core nations?). The concept of the semi-periphery has always contained a contradiction in that whilst the label itself emphasizes the 'periphery', it is always the features of the 'core' which are used to characterize it. Such is the Eurocentric logic of even radical development theory. It would be equally facile to identity a 'semi-core' by using only the characteristics of the periphery.

The pitfalls of employing such classifications for their own sake and with little understanding of the relationships

Figure 6.2 Vayrynen's semi-periphery

	Regional powers	NICs	Pariahs	Oil exporters
Economic				
Strong internal markets	★		★	
Middle income GNP	★	★	★	★
Rapid expansion of manufacturing exports		★		
Indebtedness	★	★		
Balance of payments deficit	★	★		
High level of state investment		★	★	
Social				
Continuing social crises	★	★	★	★
Growing gap between rich and poor	★			★
Political				
Unstable	★	★	★	★
Strong central government controls	★	★	★	★
Strongly nationalist	★	★	★	★
High military expenditure	★		★	★
Examples	Brazil India	Hong Kong Spain	South Africa Israel	Saudi Arabia Venezuela
			←—Iran	—→
		←—Taiwan	—→	

Source: after Vayrynen 1983

between the categories can be seen in Reitsma's (1983) attempt to produce an alternative model to dependency theory (Fig. 6.3). The 'explanatory key' that accompanies this model in the original text illustrates the problems inherent in separating classification from explanation. In particular, the varying strengths of the articulations of the components often bear little relationship to reality. Reitsma postulates, for example, that 'the majority of (peripheral) countries have only weak relations with (core) countries, so that they cannot be very dependent on them' (Reitsma 1983: 330). Such a statement totally ignores the web of financial and commercial relationships that ultimately ties ostensibly independent non-metropolitan firms to the economies of the core. The model, by its exclusion of core, peripheral, and intermediary linkages with the global levels, also fails to take into account Reitsma's own stated desire to further explore dependency theory (and therefore core-periphery relationships).

Figure 6.4 offers an alternative model which attempts to rectify this anomaly. It indicates the varying strengths of the complex relationships, including the expanding ties between the semi-peripheral cores and core peripheries (NIC plants in depressed regions of the UK, USA, and Europe), as well as the very weak links that characterize the relationships between the countries of the periphery. It should be noted at this point that such links are not only economic but also encompass social, cultural, and political ties, and perhaps offer a clue as to the apparent weakness of the Third World in the face of continued capitalist penetration. Certainly such points, however tenuous, also indicate what should always be the case, that model building is not an objective in itself but an aid to explantion.

With this in mind, it ought to be possible to identify features of a (modified) pre-capitalist mode of production within the semi-periphery. Thus, high levels of GNP and domestically generated and controlled exports might be accompanied by an extensive petty-commodity sector (as in, say, Taiwan or South Korea). But economic features must not be considered alone; the persistence of pre-capitalist cultural or ideological values are also important, a feature which is often related to ethnic factors. Such qualitative criteria would, perhaps, justify the inclusion of Australia and South Africa in the semi-periphery, and in many respects this takes the categorization much closer to the concept of internal colonialism that has recently been re-

Figure 6.3 Reitsma's alternative to dependency theory

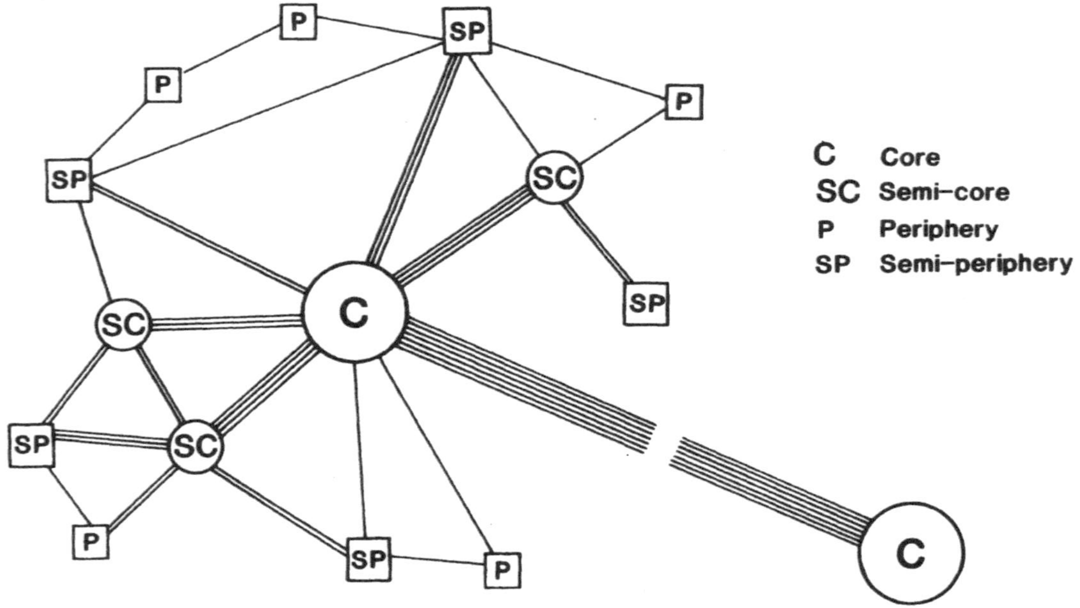

Source: Reitsma 1983

Figure 6.4 Linkages between the core, semi-periphery, and periphery

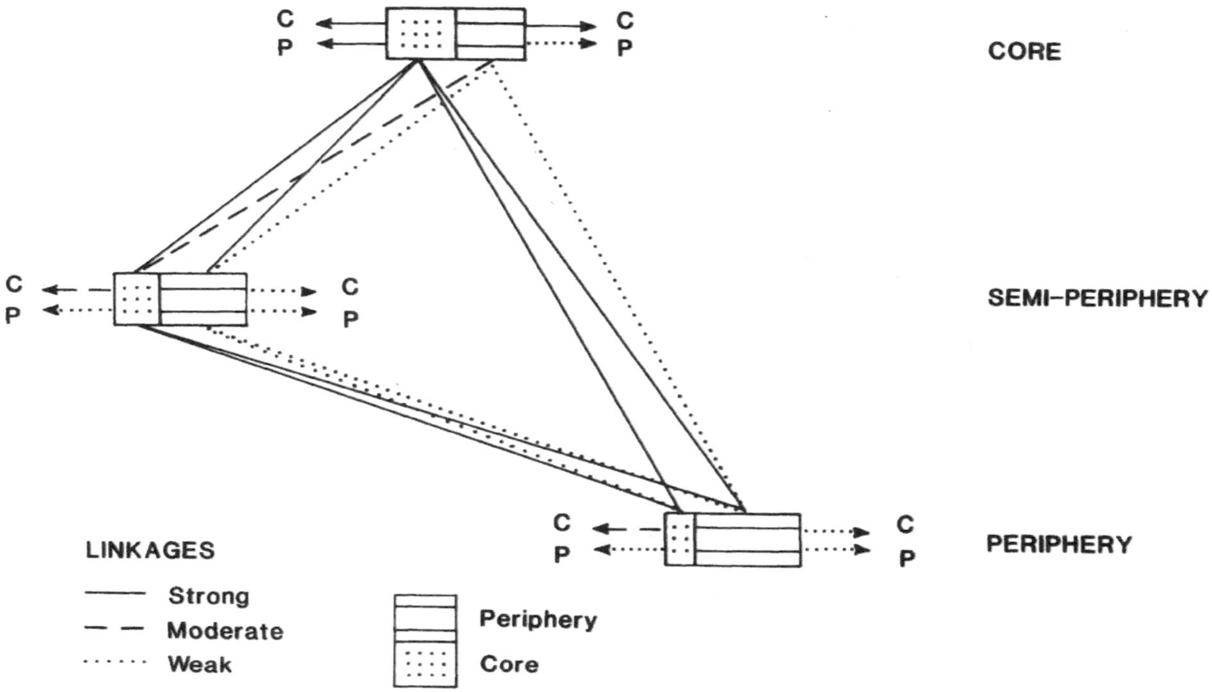

explored as a conceptual framework for explaining some aspects of uneven developments (Drakakis-Smith and Williams 1983).

It is not surprising in these circumstances that there has been no clear agreement on a list of nations or features that represent the 'semi-periphery'. In many ways, this must be expected, for the notion is more symbolic than real, representing as it does the full range of features between what Bromley (1979) has termed the 'crude buckets' of conceptual dualism. However, the discussion on the existence of a qualitatively, if not quantitatively, recognizable semi-periphery does serve a useful purpose in drawing attention towards the processes which gave rise to social formations that were neither core nor periphery. In this sense, it has often featured in the debates of geographers investigating the broader question of the spatial dimensions of development.

One of the most important aspects of such investigations, especially for geographers, is the spatial transfer of value - 'the mechanism or process through which the value produced at one location or in one area is, at least partially, realized and thus adds to the localized accumulation in another' (Forbes and Rimmer 1984: 23). There are two basic categories of geographical transfer of value (Hadjimichaelis 1980; Forbes 1981), namely the indirect or basic form, which is intrinsic to capitalism and is epitomized by unequal exchange at all levels; and the direct form, which encompasses various deliberate institutional efforts (usually by the state in concert with specific class interests) to redirect surplus value to designated groups or areas. Of course, both direct and indirect transfers of value vary through time and can thus produce not only inherited unevenness in development (the context of articulation) but also a created unevenness (the ongoing process of underdevelopment).

These variations and their complex interacting have been explored in a less deterministic way by Massey (1979), through an analysis of the spatial division of labour. Uneven development is here seen as a function of the changing priorities of capital accumulation and the constant restructuring of its organization within the various components of capital, whether vertical or horizontal. Different facets of capital can have different spatial consequences, giving rise to 'more localized complexes' (Cooke 1983: 164); interestingly, Massey also utlizes the concept of the semi-periphery to illustrate the non-polarized nature of this process.

The most important point which arises from Massey's work is that it brings theory, explanation, and the complexities of the real world closer together than in many previous attempts to grapple with spatially uneven development. However, some such as Cooke (1983), claim that this approach is still too economistic and underemphasizes the role of what neo-Marxists would regard as the more 'passive' elements of the superstructure: the state, class relations, ethnicity, ideology, etc. How does this criticism relate spatial perspectives to the urban focus of this paper?

THE URBAN BUILT ENVIRONMENT AND URBAN SOCIAL MOVEMENTS

All theories of uneven development assume some sort of economic accumulation within an urban focus or foci. Cooke has noted in this respect that 'it seems difficult to abstract regional processes from urban ... since all regional growth is theorized as occurring in urban concentrations of some kind' (Cooke 1983: 132). He claims that this represents 'a particular, spatially dominated, way of thinking about processes which are not themselves primarily spatial' (ibid.). It is, however, difficult to deny that such processes have a spatial dimension and even Galtung's model of early core-peripheral relationships reveals a geometrical patterning similar to that of Losch or Christaller. With specific reference to uneven development in the Third World even before the emergence, from the 1960s onwards, of the contemporary radical debates on uneven development, there was a concern with the contradiction between the city as a parasite or as a centre of change. Certainly both linear/modernization theory and its obverse, dependency and world-system theory, have accepted urban networks as diffusers of development or underdevelopment. Sector, marginalization, and petty-commodity theories too, have mostly stemmed from direct fieldwork in the city, whilst the later abstractions of neo-Marxist theorists, from NIDL to the role of the state or class conflicts, have been based largely on an analysis of urban change. Lipton's (1977) 'urban bias' was probably more true of development theory than development itself!

This is not to say that rural change, and analysis of it,

has not been important, but for the purpose of this discussion it is clearly evident that cities play an important role within the hierarchical patterns of spatially uneven development and within the geographical transfer of value. What undoubtedly has undermined the validity of much theory in the past has been the reification of geographical space, whether as nations or regions, as the perpetrators of underdevelopment.

It is, however, difficult to draw general theories from the processes operating in the cities of the periphery and semi-periphery, since they are by their very nature highly individual admixtures of varied and complex internal and external forces. Armstrong and McGee (1985) suggest that many of these forces are contradictory ones of divergence and convergence. The latter encompasses the way in which many of such cities are increasingly becoming part of a global production process, with large numbers of their residents adopting the consumption patterns of their western counterparts in food, dress, and the built environment. Divergence, in contrast, emphasizes the way in which much of this capitalist change is geographically specific to some countries, regions, or cities more than to others, and has given rise to what Galtung identified in the early 1970s as an increasing disharmony of interest within and between the peripheries at all levels.

Writers more specifically concerned with the role and nature of the city per se, such as Castells (1977) and Harvey (1973, 1982, 1985), see it as the heart of the process of capital accumulation. Harvey goes so far as to state that 'an understanding of how capital becomes urbanized and the consequences of that urbanization is a necessary condition for the articulation of any theory of the transition to socialism' (Harvey 1985: 226). Indeed, it is axiomatic that, in theoretical and real terms, cities are central to the structuring of space for controlling production, distribution, and consumption, as well as the reproduction of labour.

In this process, radical analysts allocate a crucial role to the state in manipulating urban change in order to mediate between conflicting sections of capital, between capital and labour, and between classes to bring about the optimum conditions for accumulation. Not surprisingly, the function of the built environment itself is heavily emphasized as central to this role of the state; so much so that many have interpreted the emergence of suburbia and the restructuring of the inner city, inter alia, as essential elements of a strategy to overcome a major crisis of capitalism, viz. underconsumption, by means of an investment shift from the primary (production) circuit, to both the secondary (built environment) or tertiary (service) circuits. Harvey (1985) claims that the shift towards the built environment creates its own contradiction of heavy investment into an inflexible form of capital which, almost as soon as it is constructed, constitutes a barrier to further changes in production, distribution, or consumption by tying up capital for long periods of time. To a certain extent this is true, but constant restructuring[1] of both internal and external aspects of buildings themselves enables this obstacle to be partially overcome and even, at times, turned to advantage, for example through subdivisions. Perhaps nowhere is this more evident than in the frantic capital accumulation process in Hong Kong, where the built environment continues to undergo astonishing cycles of overt and covert renewal and adaptation in order to accommodate the increasingly urgent demands for profit realization.

This type of analysis has been criticized by several writers (see Mingione 1981; Cooke 1983; and Williams 1984). The common denominators of this critique are first, that this approach artificially segregates the city from the remainder of the accumulation process, much of which has its own built environment; and second, that its economistic reductionism underplays the role of class relations and urban social movements in shaping the city through, for example, the medium of the planning process (see Pickvance 1976; Slater 1986).

The notion of urban social movements has proved to be a rather slippery concept to pin down and define, despite its increasing recurrence in the literature related both to advanced and peripheral capitalist development. It is not clear, for example, whether urban social movements equate with social movements or urban movements, or constitute an intermediate category (Slater 1985b: 20). In the latter context do they constitute a form of transitional action, or of composite action into which social movement is placed in its specific spatial context?

There is, of course, no reason for urban and social movements to coincide. Social movements may occur outside the urban milieu and not all urban movements share all features that are alleged to characterize new social movements. Nevertheless, there are sufficient similarities in the analyses of those examining new social movements and urban social movements for the two to be usefully fused together in searching for a set of ascribable characteristics suitable for empirical analysis.

Other factors to bear in mind must be that not all urban social movements necessarily lead to changes in the urban built environment. Indeed, as Ceccarelli (1982) has noted, not all urban social movements lead to social change and this fact itself has been central to the debate on the definition of urban social movements. Conversely, and equally important in this particular investigation, it is axiomatic that not all social and morphological changes are the result of urban social movements. Much change occurs through the medium of the dominant social group,' primarily the state, by means of the operations of the planning process.

This last point, however, again begs the question of what is an urban social movement, because it assumes that urban social movements function on behalf of subordinate and disadvantaged groups in society. As the discussion below indicates, this is still a moot point because the notion of the urban social movement has, over the years, been considerably broadened by some of its major proponents so as to incorporate non-proletarian elements. Indeed, the whole question of links with classes and class analysis, together with the role of political parties and the state, comprises the heart of the debate over the definition of urban social movements.

Some analysts have preferred to stay clear of attempts to grasp the nettle of definition. Slater (1985b), for example, in his introduction to a book of essays on new social movements, simply notes their immense diversity and that there is little consensus over definitions. Instead, he fruitfully concentrates his examination on the character of 'new' as opposed to 'old' social movements. Certainly the fact that such diversity exists makes for difficulties not only in attempting to define urban social movements, but also in comparative analysis. Not only do advanced and peripheral capitalist societies differ in the nature of their economic, social, and political contexts, but within these crude divisions of the world economy there is also massive diversity. One needs only to contrast the political economies of Cuba and Nicaragua to appreciate this point (see Slater 1985b: 16-17).

Within the attempts to characterize or define urban social movements, the trend over the last decade or so has clearly been away from a narrowly deterministic and dogmatic interpretation of urban social movements as the new cutting edge of proletarian socialism to one in which they are seen as examples of granulated socialism attempting to democratize contemporary global capitalism on a localized basis.

It is perhaps instructive, in this context, to review some of the major issues involved in this changing conceptualization of urban social movements, as it provides the basis for the conceptual assessment of the empirical section of this investigation. The empirical findings will then be used to inform the ongoing debate, in so far as it is valid to do so.

The springboard for much of the debate on urban social movements has been the work of Manuel Castells. Indeed, it is the fundamental changes in his own conceptualizations, together with the reactions of others to these positions, that constitute the heart of the literature. Lowe (1986) has divided Castell's work into three main phases.

In the early phase of this analysis, during the 1960s and early 1970s, Castells' theoretical position was conditioned by formal structuralism. The primary area of contradiction between capital and labour was seen as being at the point of production, so that any secondary contradictions in the consumption sphere would have to be linked to the dominant level (Jacobowicz 1984: 154; Lowe 1986: 19). This led Castells to argue that transformation of the urban structure could only come about by 'an articulation of the urban movements with ... in particular, the working class movement and the political class struggle' (Castells 1977: 453). The corollary of this position was that only if urban movements produced social change did they become urban social movements; mere protest was regarded as ineffectual and therefore encouraged by the dominant system.

The critiques of this position recognized the importance of Castell's identification of an alternative form of challenge to the dominant sytem. Several attempted categorizations of urban movements. Pickvance (1976), for example, also defined urban social movements not by their form or internal organization but by their effect: structural transformation of the power of the State earned the soubriquet of urban social movement; failure to achieve this resulted only in the designation 'participation' or 'protest'. Toucaine (cited in Slater 1985b) adopted a similar system in which less effective conflict became 'social struggles' or 'defensive collective behaviour'.

However, if the accolade of 'urban social movement' is bestowed on those conflicts or movements which effect a change in the social relations of production, attention is thereby diverted away from the means by which the dominant groups subvert potential urban social movements. Indeed, Lowe (1986) suggests that an important focus for investigation ought to be the failure of what he terms the mobilization process.

For many observers, the essence of this debate was whether urban social movements were indeed 'new forms of popular struggle' (Slater 1985a, 1985b, 1986) responding to new forms of subordination within more open and pluralistic forms of democracy which have increased the autonomy of the individual in this respect. Many others were not so sure and continued to insist on the paramountcy of the organized labour movement to bring about change in the social relations of production. Castells in his early work seemed to want to incorporate both of these views in his interpretation of urban social movements, but was also guilty of assuming that only urban social movements as he defined them could bring about change, thus ignoring a large cast of other urban actors (Pickvance 1976). In this context, the role of the state is of particular importance.

By the late 1970s, however, Castells (1978) was beginning to modify his position, admitting that links to the working class were not essential for action to occur and that consumption cleavages might produce autonomous political mobilization. Indeed, urban social movements began to be recognized more widely as 'pluriclass' when consumption cleavages began to lead to cross-class alliance. However, this is clearly not a precondition for action and can equally be divisive rather than unificatory.

The final phase in Castell's interpretation of urban social movements emerges in The City and the Grassroots, (1983) when the class dimension is further diluted by the recognition that urban social movements not only can coalesce around a variety of 'community' interests, akin to Cooke's (1983) 'local civil society', but also must exclude links with existing political parties. It is difficult to see in this context how urban social movements actually achieve social change, and both Castells and his critics thereafter begin to move towards outlining more general conditions which give rise to and nurture urban social movements, and presumably increase their chance of success.

Pickvance (1983), for example, identifies the importance of rapid urbanization, the intervention of the state in collective consumption, the ability of political parties effectively to express protest, and the particular combinations of social and economic conditions (in, for example, raising expectations) in any explanation of urban social movements. However, as Lowe (1986) notes, these general conditions cannot explain the specificity of any particular urban social movement or, indeed the absence or ineffectiveness of such movements when the urban chemistry appears to be otherwise promising.

Lowe (1986: 182) himself suggests that the crucial element in attempting to understand urban social movements is 'the nature and influence of corporate and state power acting to incorporate, marginalize, or repress urban conflict' and pinpoints, in particular, the presence or absence of key political actors in this process. He further specifies (Lowe 1986: 194) four major analytical themes in the general focus on the process of mobilization: the identification of social bases; the nature of local political power structures; the ideological structuring of the urban system; and the immediate and changing nature of the general economic climate.

Such a rationale seems to be an improvement on what has gone before, although Lowe's model is not too dissimilar from that of Pickvance in its overall scope. It is disappointing, therefore, when Lowe proceeds to define urban social movements as 'organizations standing outside the formal party system which bring people together to defend or challenge the provision of urban services and to protect the local environment' (Lowe 1986: 3).

The problem with this definition is that it is far too broad and, in particular, lacks any reference to the 'grassroots' element of urban social movements. In short, the contemporary use of the term seems to have strayed too far from the original concern with underprivileged urban groups seeking to increase their share of urban resources at the expense of the dominant section. Perhaps a more useful definition can be obtained by combining Lowe's with that of Castells (1983: 305), which is that an urban social movement is a collective conscious action, outside the formal party system, aimed at defending or challenging the provision of urban services, or protecting the environment, against the logic, interest, and values of the dominant class.

Providing that this caveat is incorporated, Lowe's

general themes of investigation do seem to offer a fruitful way to investigate urban social movements. But for such theorizing to be of practical use, it must be translated into an empirical context. Such is the intention of this chapter.

In recent years, increasing numbers of empirical studies of urban social movements have appeared, apparently adding a comparative dimension to the theoretical texts that have been discussed here. However, individual studies placed next to one another in a text do not really constitute true comparisons, so that although the theory of compatibility has featured prominently in the recent literature, empirical investigation of whether similar structural circumstances give rise to similar or similarly effective social movement has not been so frequent.

This chapter investigates recent changes in the residential built environment in two towns - Darwin and Alice Springs - in the Northern Territory of Australia. These are the only two substantial settlements in this thinly populated state which attained self-government in 1978. An investigation of the housing situation draws together not only forces from the public and private spheres, and the local and central state, but also their articulation with the household itself, with all the implications such linkages have for, inter alia, the reproduction of labour or consumption cleavages, and their expression in a miscellany of urban social movements coalescing around a variety of non-class interests, such as gender or ethnicity.

The following section places the analysis of the housing situation in the two towns into a historical, social, and political context which encompasses Lowe's fourfold mobilization factors. It illustrates clearly the need to understand the regional, national, and international dimensions of these elements in order to appreciate fully the reasons for the similarites and dissimilarities between the two settlements.

AUSTRALIA AND INTERNAL UNDERDEVELOPMENT

The opening section of this chapter has briefly reviewed several concepts that are central to the empirical analysis. It is the intention of this second section to present an equally short summary of the spatial and historical background to the study area - the Northern Territory of Australia. Of the concepts examined, the one which is most directly drawn into the discussion is that of the semi-periphery.

As noted earlier, this concept has often been less than rigorously addressed, despite its frequent incorporation into analytical discussion related to uneven spatial development. However, in the context of this chapter, a subjective justification for its existence was at least noted, namely to draw attention to the fact that the world cannot simply be divided into core and periphery. On the other hand, the few attempts to identify empirically the nature and extent of the semi-periphery have floundered because of its very diversity. Even Vayrynen's (1983) apparently detailed categorization cannot be accepted as complete because not only is it restricted to economic and political criteria, but also it omits any reference to what was earlier characterized as the developed periphery.

In many ways, it is the dominance of mode of production analysis, with its emphasis on the importance of economic substructure, which has clouded this debate. If the social, cultural, ethnic, and ideological elements of the superstructure are allowed more prominence, then the contradictions within the articulation of the pre-capitalist and capitalist modes of production become more apparent and understandable. Clearly, it is less than satisfactory to categorize countries by simple, reductionist labels, such as those Vayrynen employs, but by most criteria Australia must be recognized as part of the global semi-periphery, and has been categorized as such in some of the literature; for example, Forbes and Rimmer (1984).

To establish this objectively and conclusively could itself form a major investigation (if the task were considered to be worthwhile!). Subjectively, and for brevity's sake, it is possible to refer to several varied criteria in support of the statement. Economically, for example, Australia has a set of MNC relationships, inter alia, in which her firms are both subordinate and dominant, exploited and exploiting. Ideologically and politically, Australia is a corner-stone of that collective bastion of the semi-periphery, the British Commonwealth. Culturally, ethnically, and economically, Australia has a demographically small, but spatially extensive, pre-capitalist (albeit modified) mode of production embodied in its Aboriginal population, particularly in the non-metropolitan areas outside the southeast. Despite almost two centuries of White, capitalist domination and exploitation, resistance has ensured the survival and gradual reassertion of Aboriginal rights and interests (Drakakis-Smith 1981c, 1984). In short, given that the semi-periphery is itself a subjective concept, the subjective criteria used above establish Australia as part of it, at least to the extent needed for this analysis. A more important point is perhaps to emphasize the fact that, as with all countries, Australia itself has internal core/ peripheral features.

The identification and nature of the Australian periphery is not a concept or task that has received much direct attention from Australians over the years. Despite its place in the folklore of the Australian psyche, most Australians know little and care less about underdevelopment in outback or remote Australia: only relatively recently have substantive texts begun to appear that address the political economy of the northern section of the country (e.g. Courtenay 1982; Loveday and Wade-Marshall 1985a; Parkes 1984). Theories of uneven spatial development are rarely incorporated into such discussions, unless the related but distinct concept of internal colonialism is included amongst these (Drakakis-Smith 1983). However, recently Gerritsen (1985) has introduced an implicit spatial dimension to his examination of the place of the Northern Territory in the federal dualism that other observers have identified in the Australian political economy. In simple terms this comprises, on the one hand, a set of resource-rich 'outer' states and, on the other, a set of manufacturing, 'inner' states. The former are heavily reliant on the export of primary (usually mining) products, are often in conflict with federal economic policies, and are characterized by extensive and increasing reliance on multinational investment.

Specifically, Gerritsen assesses the position of the Northern Territory within this model, in comparison with the two 'outer' states of Western Australia and Queensland. He concludes that, in many ways, the Northern Territory is a more extreme example of the outer states, particularly in terms of employment patterns (Table 6.1). However, the recency of the state's political independence has left it still heavily reliant on federal funding, and also with a relatively large state bureaucracy. The local-national state relationship is, therefore, highly peculiar to the state itself, and sustaining and expanding the power and influence of the local state has been an important influence on development policy in the Northern Territory.

Table 6.1 Northern Territory: comparative employment percentages 1981

	Northern Territory	W. Australia/ Queensland	Victoria/ NSW
Agriculture	4.0	8.0	5.1
Mining	4.6	3.0	0.8
Manufacturing	4.6	12.8	20.4
Utilities	1.9	1.7	2.2
Construction	9.2	7.5	6.0
Wholesale/retail	11.9	18.1	17.1
Transport/storage	4.8	5.7	0.5
Communications	1.8	2.0	2.0
Finance, property business services	5.5	8.2	7.7
Recreation, personal services	5.9	5.2	5.1
Public administration	10.2	3.8	3.8
Community services	19.5	15.1	14.4
Unclassified	15.2	7.5	8.0

Note: excludes defence forces

Source: Gerritsen (1985)

In addition, the Northern Territory has experienced uneven spatial development not only as part of the Australian periphery, but also in terms of its own rural peripheries and urban cores. Although the latter are relatively few in number, they are surprisingly diverse in nature. Indeed, it is precisely because of the limitations of scale (in demographic or economic terms) that the process of uneven development, the transfer of value, the role of urban changes, and the nature of the built environment can all be examined a little more closely than in more complex social formations.

From the arrival of the British in 1788 to the 1930s, development and underdevelopment in Australia primarily related to the articulation of the pre-capitalist and capitalist modes of production; subsequently there has been an elaboration of these relationships within the modes of distribution and consumption. Inherent to the discussion is a growing influence from the social relations of the superstructure (ideology, ethnicity, etc.), not only in creating uneven development but in shaping the nature of the contradictions within, and the reactions to, such development. In this overall process, the role of urban centres did not become prominent until the 1940s, whilst that of the local state emerged strongly in the 1970s. The detailed nature of the pre-capitalist, Aboriginal mode of production and its penetration by white capitalism has been detailed elsewhere (Rowley 1973, 1976, 1978; Drakakis-Smith 1981c). What follows is a short chronological summary of the components this process entailed in Northern Australia (Figure 6.5).

On their arrival in 1788, the British declared all land to be the property of the Crown; no treaty was signed with any Aboriginal group and no compensation was paid. As the fragile ecosystem was broken, most Aboriginal communities in the southeast were rapidly destroyed by disease, anomie, and force. By the 1830s, increased immigration, together with the growing profitability of wheat and wool for export to Great Britain, had encouraged a rapid expansion of the agricultural frontier. Often this was in the form of illegal occupation of land by pastoral squatters, but it must be doubted whether more extensive government control would have changed the situation a great deal. What was occurring, therefore, was an indirect transfer of value from the Aboriginal mode of production to the British capitalist mode of production, through the medium of a colonial 'squattocracy'. Assumptions as to Aboriginal ethnic inferiority were used to justify both the appropriation of the land and the neglect of the communities.

As the nineteenth century wore on, the squatter pastoralists expanded into more remote areas in the northern half of Australia where conditions were harsh, facilities limited, and communications relatively poor. In these areas, the role of Aboriginal labour was crucial but seldom admitted to be so. Aborigines were forced off their own lands, directly or indirectly, and into the 'protection' of various white settlements, such as pastoral stations, missions, or government reserves. Essentially these were labour reserves where families were helped to survive whilst the men worked away, often for months at a stretch, on various pastoral stations. This was internal colonialism in full flow, where the dominant mode of production used institutionalized power (both temporal and spiritual) to establish a spatial division which was ideologically

Figure 6.5 Australia: a chronology of Aboriginal exploitation

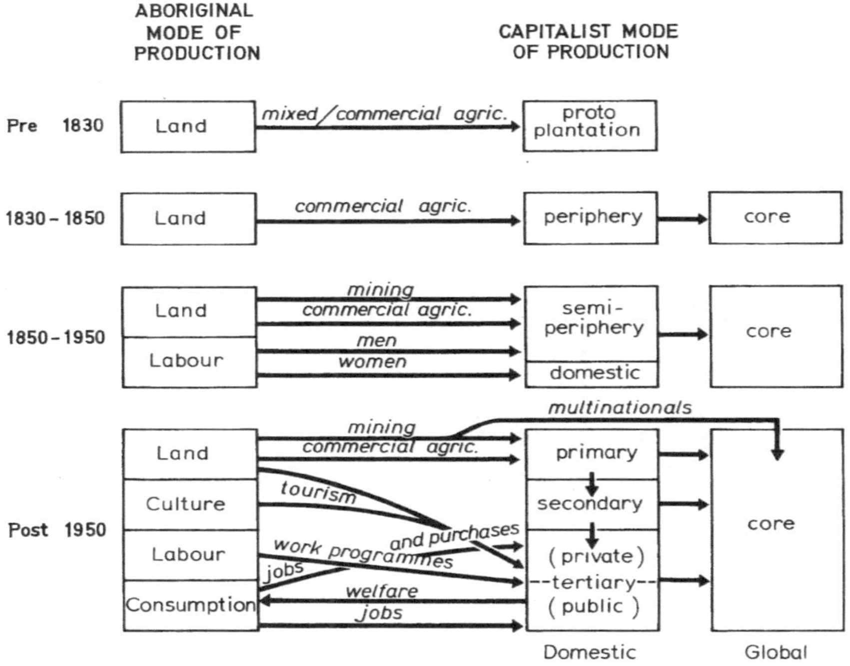

Source: Arrighi and Drangel 1986

buttressed by increasing ethnocentrism and racism on the part of the white population in northern and central Australia. During this period, therefore, the transfer of value began to become more complex, occurring through labour usage as well as from the occupation of land.

In 1901, Australia became a federation, and over the next few decades its population rose massively as a result both of encouraged immigration and a high birth rate. Most of this population increase was concentrated in the larger cities and, together with the expanded industrial demands of various war commitments, led to the emergence of a substantial domestic market, not only for the agricultural produce of the interior, but also for its commercial materials. It was during these decades, therefore, that mining production began to expand from the exploitation of precious to industrial minerals.

Despite the expanding capitalist penetration of the Northern Territory, this was still oriented around the enclave or dispersed settlement patterns of primary production. Communications were poor and related to the movement of commodities rather than of people. As a result, the two major urban settlements of the Northern Territory, namely Darwin and Alice Springs, although established in the nineteenth century, amounted to no more than around 1,500 and 400 inhabitants respectively on the eve of the Second World War.

Hostilities with Japan transformed the geo-political position of the Northern Territory from periphery to frontline and initiated an enormous federal investment in infrastructural and military projects (which has continued to the present day). The improvement in communications has had an important effect on the economy of the Northern Territory, particularly from the 1960s onwards, with the massive expansion of mining activities across the whole of northern Australia; usually through the medium of multinational corporations.

The same period has also witnessed the rapid growth of tourism in the Northern Territory, with total annual receipts reaching, by 1981, A$110m. The third, and in many ways the most important, factor in the expansion of the Northern Territory's economy has been the rapid growth in the welfare system associated with the Aborigines. The extension of cash benefits to the Aboriginal population has vastly increased collective consumption within the state and has resulted in the expansion not only of the service sector but also of the welfare and administrative systems themselves. Many white bureaucrats, many businesses, and almost all of the small nucleated settlements owe their existence to the presence of poor blacks in the Northern Territory (Drakakis-Smith 1980; Heppell 1979).

Although substantive urban growth in the Northern Territory really began only in the 1940s, many small nucleations have existed since the Aborigines first began to be deprived of their land and shunted into manageable pockets of labour on commercial properties, missions, or reserves. These have persisted as small service centres, dispensing the welfare facilities now sustaining their existence, and remain in census classifications as 'bounded rural localities'. Comprising less than 1,000 people each, they account for 10 per cent of the Northern Territory's population. Perhaps surprisingly, in view of the popular concept of a dispersed outback population, three-quarters of the Territory's population lives in 'urban centres' of more than 1,000 inhabitants; the great majority residing in Darwin and Alice Springs. In contrast to the 'bounded rural localities', the majority of the population of the urban centres is white, although within the various towns the proportion differs substantially.

Whilst all the urban centres in the Northern Territory are small by global and even Australian standards, there has been substantial variation not only in their economic functions, but also in their social and political development. Indeed, it is tempting to say that much of the urban built environment in the Northern Territory has been shaped primarily by these non-economic factors, but it is not easy to separate the various forces in empirical terms. However, it is certainly true that economic and urban growth in the Territory, over the last two decades at least, has occurred within an ideological and political framework that has proved both volatile and influential, and which has introduced considerable complexity and contradiction into the nature of the built environment of the state.

Of the many changes of this type which have occurred, two are of particular importance and will be briefly described below, namely, the civil rights of Aborigines and the Northern Territory's attainment of self-government. Capitalist penetration into the Territory must be seen in the context of these (and other) processes, because the consequence has been spatial unevenness, not only in economic development, but also in the nature of the built environment itself. Accordingly, the overview of these two prominent aspects of social and political change will be linked to a closer focus on the ways they relate to the built environments of Alice Springs and Darwin respectively.

ABORIGINAL CIVIL RIGHTS AND HOUSING PROVISION IN ALICE SPRINGS

Aboriginal rights

In Australia, the pace of change in the relationship between the black and white sections of the community has accelerated considerably over the last decade. In a world-wide climate, albeit uneven, of advancing civil rights for minority groups, the restoration of land rights, the conflict over mining procedures and royalties, and the resurgence of pride in their identity and culture, have all moved Aborigines to the forefront of important national issues. As a consequence, the Aboriginal people have found it necessary to reassess their position in this struggle for control over their heritage and their life-style, and have come to realize that central to their problems is their underprivileged and subordinate economic position within Australian society as a whole.

It is clear that, despite the efforts of the federal government, prejudice is still an important factor in determining Aboriginal employment and incomes, and that continued white control over the means of production will ensure continuity. Extensive provision of welfare payments does little to change this situation and, in fact, simply creates even more jobs for whites within the delivery system. Economic independence of Aborigines can therefore only come about through social movements which challenge the white domination of production in remote Australia. Two of the most important issues in this respect have been land rights and the control of mining operations.

However, it is not my intention to dwell on economic alternatives; instead I want to re-examine the underdevelopment of the Aboriginal people to establish the ways in which it, and the social movements reacting to it, have related specifically to the built environment. The discussion relates to empirical data from Alice Springs which, like many small towns, has increasingly become a focus for both the despair and hope of Aboriginal people.

Whilst much of the urban population growth in central and northern Australia has been the result of white migration, the Aboriginal component has also increased rapidly, even in towns which are not experiencing much overall expansion. Aborigines are traditionally a mobile people, but in the past their travelling occurred within a circumscribed area. Recently, however, there have been indications that the drift into towns is becoming more permanent.

Most of the limited evidence available on Aboriginal urban migration reveals that it is governed primarily by kin or tribal relationships, but that within this framework various factors have provided the initial impetus to move. Perhaps the most important of these is the increased physical mobility which has come from the provision of trucks and other forms of transport under various welfare schemes. The growing frequency of visits, particularly to those towns which are distributional centres for the welfare system, together with the slow spread of radio and television, has brought an increased knowledge of the attractions available. In addition, the extension of educational services to young Aborigines has induced many to seek what they regard as more appropriate employment opportunities in the towns; unfortunately this is often in vain.

One important point to remember with regard to the above generalizations is that the numbers involved in Aboriginal urban migration are still quite small. For example, although the Aboriginal population of Alice Springs virtually doubled between 1975 and 1980, this involved only 1,400 people, many of whom were born in the town. As a result, migrational trends are at times heavily influenced by local circumstances.

Aboriginal housing in Alice Springs

The town of Alice Springs was established about a hundred years ago as a supply and communications centre for the growing pastoral and mining population of central Australia. At the turn of the century, the white population was no more than a few score, but with the extension of the railway line to Alice Springs in 1929, the town became a railhead for the transhipment of cattle to southern markets. By the late 1930s, the white population of the town had increased to around 200, with a similar number of Aborigines.

After 1945 the character of Alice Springs began to

change. The precursors of this transformation were to be found in the Second World War itself, when the Japanese threat to Darwin caused many of its functions to be shifted south. With the establishment and expansion of welfare services for Aborigines in the 1950s, Alice Springs was further developed as an administrative centre for the whole of central Australia. In addition, the economy of the town received a considerable boost from tourism, which expanded rapidly in the wake of transport improvements.

Population growth from this economic development peaked in the late 1960s, when the average annual increase was around 11 per cent. Most of the increase comprised well-paid White employees of both the private and public sectors, usually on short-term contracts of two or three years - a typical colonial, expatriate situation. The Aboriginal population also expanded over the same period through both natural growth and migration. Although census deficiencies have made this figure difficult to quantify, Aborigines comprise about 20 per cent of the total population. Much of this Aboriginal increase has been accommodated in the camps that fringe Alice Springs and house about half of its black population.

Although the benefits of the recent economic growth of towns like Alice Springs have accrued primarily to white residents, it would be wrong to infer that Aborigines play no part in the economic system other than to receive welfare handouts. Indeed, as outlined earlier, the presence of Aborigines in the towns of central and northern Australia is the very foundation upon which much of the economy is constructed. Many service industries, such as taxi companies, are almost totally reliant upon direct Aboriginal patronage, whilst considerable proportions of the white-collar sector are linked in some way or other to the administration of services for Aborigines.

This reverse dependency of white on black gives rise to a mutual interlinkage within the current economic system. However, as in the underdevelopment of the Third World, the benefits of this interdependence accrue mainly to the dominant element: in Aboriginal Australia this is the white middle class, which is therefore struggling to maintain its position for a mixture of ethnic and class interests, within which the latter is becoming increasingly important. In order to examine how these interests, and the social movements they have engendered, have affected the built environment, I shall complete this case study by examining in a little more detail their relative importance to the field of Aboriginal housing (Drakakis-Smith 1981b).

Social conflict and social movements have shaped the residential environment of Alice Springs in many ways over the years. Initially, Aborigines were excluded from the town by legislation, and although by the 1950s this had been rescinded, the sheer poverty of the migrants forced them to live in peripheral camps rather than conventional housing. Most camps in Alice Springs are essentially close community groups based on kinship or linguistic/geographic origins, and have been established in accordance with Aboriginal law through agreements with the traditional Arunta owners of the land. However, most of the settlements were originally on Crown Land, and their inhabitants would, in the Third World, be designated as 'squatters'. In all there are about thirty campsites around Alice Springs, but only about half may have had a sizeable population at any point in time. Most dwellings are usually built of materials salvaged from the town rubbish dump, despite attempts by the municipality to prevent such recycling. Caravans and abandoned motor vehicles of all types are also used. Overall, camp dwellings provided insufficient protection against the extremes of cold and heat and the occasional heavy rain which Alice Springs experiences. Until as recently as the late 1970s, it was deliberate policy not to provide the camps with a service infrastructure, because the authorities were afraid that if improvements were made this would only encourage further illicit settlement.

The response to this situation in the 1960s was to restrict the residential space of Alice Springs to whites, through the indirect method of creating a special reserve at Amoonguna some 15km south of the town. The prevailing credo amongst administrators in the 1960s was that of assimilation, the gradual absorption of Aborigines into the white Australian way of life. However, this was only theory: most actions seemed designed to prevent any such assimilation (except a few 'deserving' instances) and to maintain Aborigines in a peripheral position in relation to the core of white society. The houses at Amoonguna were part of this process, being planned as a series of transitional dwellings which were intended to give the Aborigines some experience of living in a conventional white house. Unfortunately, these dwellings were so appalling that the Aborigines rejected them; which further confirmed the white view that Aborigines were not ready for conventional town residence.

Although no formal barriers existed within the conventional housing market of Alice Springs, almost 70 per cent of such properties were virtually inaccessible to Aboriginal households because of their cost or because a considerable proportion were supplied or subsidized by employers. The low Aboriginal incorporation into the formal job market was thus of fundamental importance in restricting access to residential space in the town itself.

Almost the only way in which the Aborigines have access to conventional housing is through the Northern Territory Housing Commission (NTHC) which accounts for just under 30 per cent of permanent units in the town. Applicants must fill in forms in the main office in Alice Springs and are then subjected to two separate screening processes by assessors from the Housing Commission. The first relates to the financial status of the applicants - not in terms of their having too high an income, but whether their financial resources are sufficient to meet the regular rental payments. There are additional, hidden costs involved in the move to a NTHC dwelling, such as rental and utility deposits and a rent advance, which can amount to several hundred dollars and must discourage many of the poorer families (who are most in need) from ever applying.

Not only is financial status assessed, but applicants are also screened as to whether they have had 'suitable experience' for coping with a conventional house. The condition of most camp and reserve housing is of course not conducive to a favourable verdict in this respect. Furthermore, almost all of the access procedure is undertaken with and by white officers of the NTHC - itself a discouragement to many Aboriginal families. As a result of these barriers, 80 per cent of NTHC houses are leased to white families.

Until the mid-1970s, therefore, the residential structure of Alice Springs was firmly the creation of the ideological forces, although the ethnic prejudice against Aborgines clearly had economic dimensions and was by then developing an element of class conflict. The inroads into this uneven access to housing resources and consequent spatial patterning have resulted not from changed economic circumstances but rather from the translation of the broader successes of social movements related to civil rights at the national level into purely local terms.

These changes began in the early 1970s, when it became

possible for Aboriginal campers to apply for the leases to the sites they occupied. Although several such leases have since been granted, applications for the larger and more prominent sites often met with strong opposition from entrenched White interest groups. Much of the success in the development of the leased sites has been due to the Tangatjira Association, a collective of Aboriginal leaders and organizers from the existing camps which was initially established to look after small-scale improvements. However, with government funding, Tangatjira has recently taken on the role of adviser to the camps in negotiations to obtain leases, and also to the various housing associations in their search for suitable housing designs.

It is evident, therefore, that the changing residential environment of Alice Springs has been substantially shaped by ethnic and class interests and by the social conflicts and movements associated with them. The very existence of reserves and camps and the ways in which these are changing are in themselves a major illustration of this process. Economic interests are involved, but in a subordinate capacity, and only tenuously related to the social relations of production and reproduction, since Aborigines are not a major element within the labour force. Although the state, principally through various welfare agencies and the Department of Aboriginal Affairs, has played an important role in the overall process, local circumstances and social relationships have clearly been paramount in the eventual outcome.

The localized nature of such relationships became very evident by comparison to the situation in Darwin, where Aboriginal residential space has been developed quite differently. But not only are social relations different in Darwin; so is their formalization into what might be termed a 'local state' per se. It is this which forms an important theme in the following examination of the built environment of the state capital.

THE LOCAL STATE AND HOUSING POLICY IN DARWIN

The Northern Territory became a self-governing state in 1978, and now has responsibilities comparable to all other states. The federal government, however, retains control of Aboriginal affairs and the two huge national parks, and has a considerable influence over decisions related to mining (particularly of uranium). Moreover, the Northern Territory is still reliant on federal funding for some 80 per cent of its revenue.

The stated development goals of the Northern Territory government are to provide the physical infrastructure and social capital to support the expansion of private enterprise, and to assist and direct the economy itself to ensure that the private sector does take over as 'the engine of growth'. As noted earlier, Gerritsen (1985) considers these objectives to be consistent with those of the other peripheral or outer states of Australia, but correctly identifies within the Northern Territory a third (implicit) goal, which is to assert and protect its own local political autonomy.

Given the history of federal control over the affairs of the Northern Territory, this objective is not surprising; but it does pose a contradiction within its development goals, because the desire to safeguard the political strength of the local state has resulted in a more extenisve influence over the structure of the economy and its likely future than the state government ostensibly wants. Perhaps nowhere is this more evident than in the employment patterns of the Territory (Table 6.1, see p. 217), which reveal the extent of the direct role of the state. The contradictions which this development dilemma poses for the Northern Territory are clearly evident in the nature of the built environment in the state capital and will be discussed with particular reference again to housing patterns.

Darwin is the most northerly city in Australia, located on a hot monsoonal mangrove coast some 4,000km from Sydney. Until recently its transport links with southern, settled Australia were poor, with neither sealed road nor railway extending over much more than half the 2,200km distance to Adelaide, its nearest neighbour amongst the state capitals. Not unexpectedly, this has pushed the costs of living and development to very high levels, so that most projects have to be very profitable or highly subsidized to justify investment. Increasingly, that level of justification is being dictated by world demand for minerals, such as manganese, bauxite, and uranium, or by strategic considerations related to defence on the part of the federal government.

With the development of mining activities and tourism in the 'Top End' (the northern part of the Northern Territory) since the 1960s, Darwin's population has grown to some 70,000 and the city has become the focus for considerable migration from other states, particularly of young people (see Taylor and Lea 1988). Given also the relatively high birth rate, this makes Darwin the most youthful state capital in Australia. However, the growth and form of the city have not been entirely the result of the economic fortunes of the north. The city is located in an environmentally hazardous zone, with the oppressive wet season prone to erupt into destructive cyclones. These environmental hardships, together with the high cost of living, have ensured a high level of state and federal investment in the development of the city through subsidies of various kinds, above all in the built environment, as expressed particularly through the housing construction programme.

The nature of the Darwin housing market

Darwin has a complex housing supply system. Not only is there considerable variation in the nature of the conventional housing market, but there is also a non-conventional supply system of Aboriginal camps, reserves, hostels, caravan parks, and boats which officially house some 5 per cent of the city's population. This proportion is much higher than in other state capitals but is much lower than in Alice Springs. Within the conventional sector, the principal contrast between Darwin (and the Northern Territory as a whole) and the rest of urban Australia is its relatively low proportion of home ownership (Fig. 6.6), a feature which seems to have obsessed the Territory's planners since independence. The preoccupation with raising the level of home ownership appears to have been a growing feature of housing policy in Australia as a whole for the last ten years. Paris et al. (1982), for example, alleged that there had been a steady decline in the finance available for rental programmes - a fall directly related to the reduction in the availability of federal funding. The rental sector in Darwin, therefore, constitutes an unusually large slice of the housing market, particularly in view of the strong capitalist leanings of the state government. It is an 'anomaly' of the built environment that requires explanation.

There are five main categories of rented property in Darwin (Drakakis-Smith 1984, Corbett 1985):

(i) accommodation rented to employees of the commonwealth (federal) government;

(ii) accommodation rented to staff of the Northern Territory Public Service;

(iii) accommodation rented to the general public by the Northern Territory Housing Commission;

(iv) accommodation rented or made available to employees by private sector firms;

(v) accommodation rented on the open market.

Figure 6.6 Australia: housing tenure by state 1981

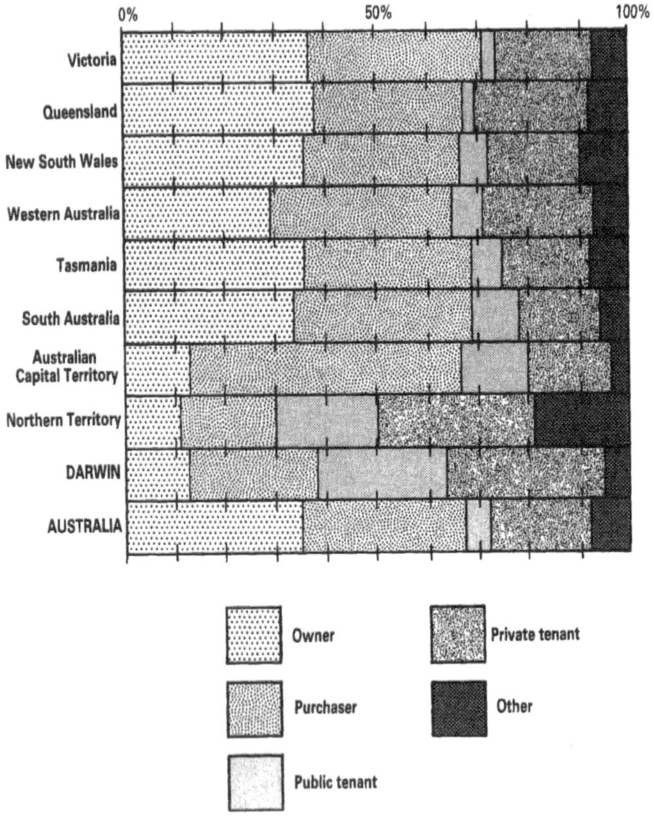

Perhaps the most important feature of the renter households in Darwin is the large number of single-person households. They comprise about one-third of the total, and of these well over 80 per cent are restricted to the open-market tenancies by virtue of their ineligibility for government housing and/or the nature of their employment (if any) which excludes them from job-related accommodation. Many single persons are, therefore, considerably disadvantaged in the housing market, even before their level of income is taken into account. Data on employment and occupation accentuate these differences, particularly as job-linked accommodation in the private sector tends to favour the higher-status occupations. Conversely, those who work in less remunerative occupations or who are out of the workforce tend to concentrate in NTHC general and open-market rented properties.

These employment-tenure relationships are, perhaps most convincingly relayed through the income data in figure 6.7, where it can be seen that those with accommodation linked to jobs tend to be the better off (through higher status jobs and working spouses). Such families are doubly favoured since they have both job and accommodation security and so have no need to enter the relatively expensive open-rental market or to wait (for lengthy, unsubsidized periods) for allocation of government housing. Moreover, many of those families in the private sector living in employer-provided housing tend to receive a rental subsidy. Indeed, over three-quarters of such tenant families receive a rental subsidy in this way, compared to only about 13 per cent in the open rental sector.

In contrast, those with lower incomes are obliged to seek accommodation either within the open market or, if eligible, with the NTHC (although the increasingly lengthy waiting-list forces many multiple-member families to wait for a considerable amount of time in the expensive private sector). There are, therefore, two populations who are 'in need' in terms of income available for housing: one, the multiple family, can be eventually catered for, although the waiting period often causes much hardship; the other, the single persons, are left largely to fend for themselves in the most expensive housing sector, despite having the lowest incomes. Rents in the private sector in Darwin are double those found in most other state capitals, and it is estimated that two-thirds of the families in the city are unable to afford the median rent. Although rent relief is available,

Figure 6.7 Darwin: tenure and household income 1984

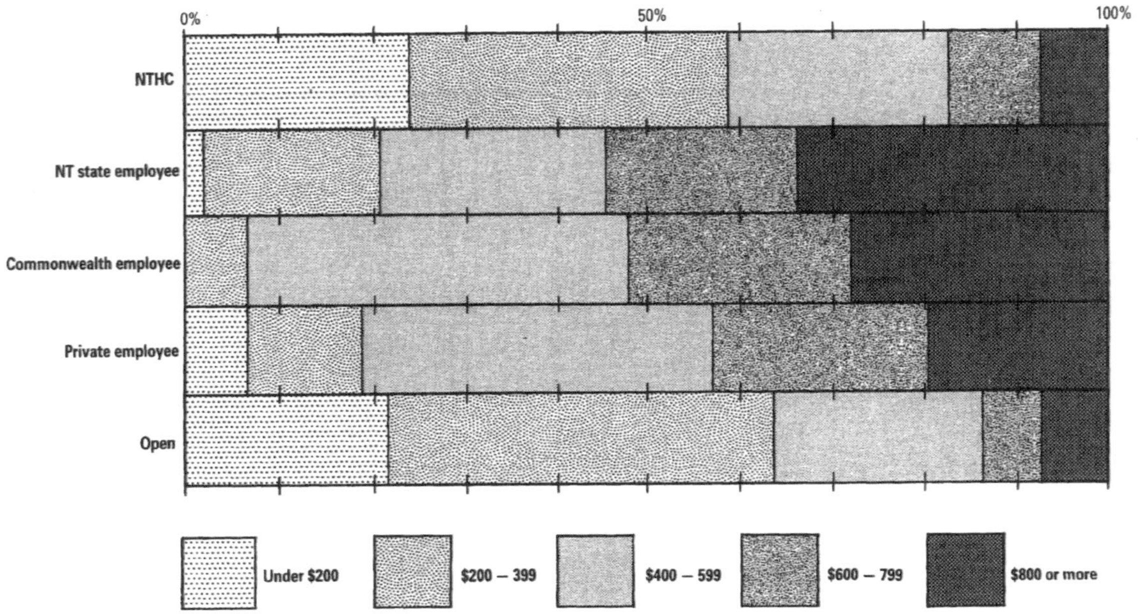

almost half of those in the open-market sector were unaware of the scheme and only 3 per cent were receiving assistance. The most common response within the housing market for those who cannot afford such high rents, other than to abandon it altogether by leaving Darwin or moving into non-conventional housing, is to reduce rents by sharing. Almost all such accommodation is open-market accommodation, where approximately 40 per cent of those interviewed were sharing.

Figure 6.8 indicates the weekly rents paid by each family unit and shows the expected differences in range between the controlled public and uncontrolled private sectors as a whole. The data represent rents actually paid, and therefore do not reveal lowered rents through subsidies, rebates, and sharing. Even so, the private sector indicates much high rental levels, with the real extent of these demands being clearly revealed by the size of the household rents (i.e. the rent per unit of accommodation) within the open-market sector, despite the generally smaller size of the units. Within the private sector as a whole, subsidies tend to exaggerate the differences between the two tenure types. On the one hand, the rent relief system, as noted above, is as yet ineffective in providing assistance; on the other hand, the rental subsidy provided by some employers is extremely effective in lowering rents. The overriding impression is once again of pressures being placed on those families living in unsubsidized accommodation in the private sector. For many this is increased by the need to provide their own basic facilities and furnishings.

In short, a sharply polarized situation exists in Darwin's rental market (which comprises the majority of its housing) between a large group of highly subsidized families on the one hand, and a substantially disadvantaged group on the other. Given the developmental context of the Northern Territory and its identification by Gerritsen (1985) with the Australian periphery, where laissez-faire, export-oriented capitalism is predominant, the degree of state involvement in housing provision could be regarded as unexpected. In this context, Paris et al. (1982) have drawn a clear distinction between public and welfare housing, with the latter comprising accommodation specifically for those 'in need' and the former being characteristic of efforts to provide a genuinely equitable choice between ownership and rent without prejudice to standards. Allegedly, welfare housing is more typical of right-of-centre governments, whilst public

Figure 6.8 Darwin: tenure and weekly rent 1984

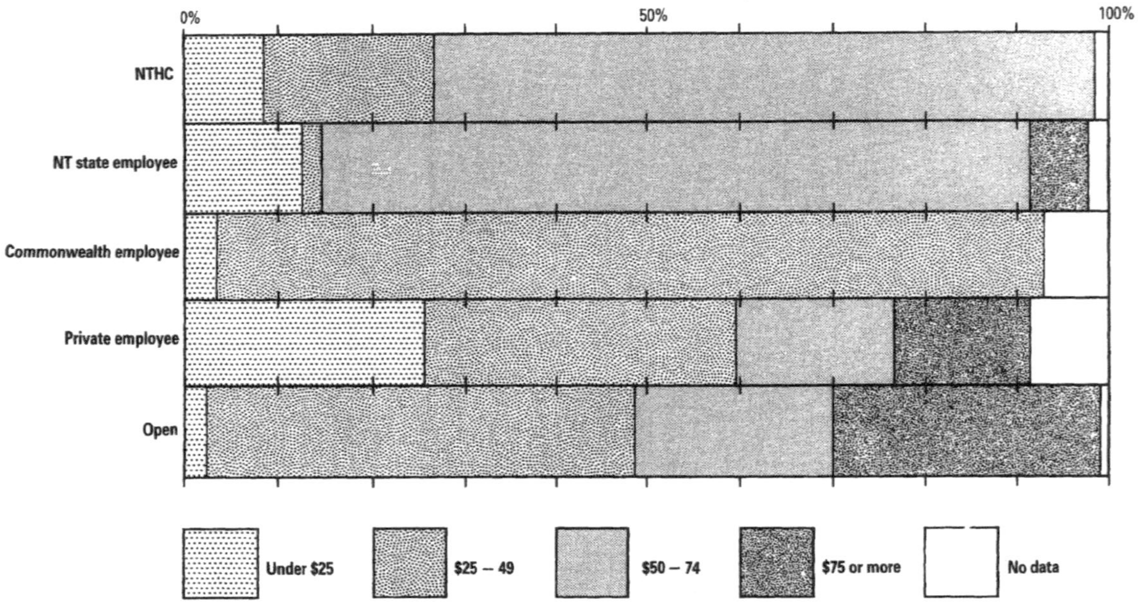

housing is the product of socialist administrations.

The current Australian situation shows several contradictions in this scenario. First, the national trend towards ownership is occurring under a socialist federal government (although some observers would question its socialist standards); second, the extensive public housing programme in the Northern Territory can hardly be said to be the product of socialist policies. These features suggest that the substantial shaping of the urban residential environment in Darwin by the local state needs alternative explanation.

It could be argued that the local state is carrying out its stated intent of promoting capitalist development by providing relatively cheap housing in order to lower the costs of the reproduction of labour. But there are several problems with such an argument. The first is that a large proportion of the Darwin workforce has secure jobs, with salaries fixed by national agreements and swollen by tax rebates for working in the far north, as well as housing subsidies (direct or indirect). In contrast, it appears to be the migrant, young, and/or blue-collar section of the labour force which is most adversely affected by high rents.

It might be argued in the light of Lowe's theorization that the extended role of the local state is a response to a consumption cleavage; that the state is acting to forestall the mobilization of an urban social movement. Clearly this is not the case in Darwin, where the underprivileged groups have failed to coalesce around their common social bases and present a challenge to the dominant faction. Indeed, the ethnic balance in the underprivileged groups in Darwin seems to have prevented the unification of protest, in contrast to Alice Springs where the ethnic factor was a strong motivating force. The consequence, in Darwin, has been that the intervention of the state in the collective consumption of the housing market has been to the advantage of those with high salaries already swollen by tax rebates for working in the far north.

A closer examination of the situation reveals that a powerful motivation for this policy has been the need for the new state government to secure its own power base. The generous subsidies are used as an added incentive to attract an experienced bureaucracy and bourgeoisie and induce them to put down permanent roots in the Territory, thereby creating a more loyal and solid political base from which to confront the federal government.

CONCLUSION

How does this empirical investigation inform the earlier theoretical debate on urban social movements and their impact on the built environment? The first point must be to emphasize the degree of contrast that exists in apparently similar situations. The comparative analysis was set in the only two substantial settlements in a thinly populated state. The national and international forces impinging on these urban centres were constant, the ideological structuring of the urban system was the same, and yet the residential changes in the built environment resulted from quite different movements.

On the one hand, in Alice Springs, there was a genuine urban social movement, coalescing substantially around an ethnic core, but whose strength and direction was strongly influenced by a wide range of factors from local collective leadership to the global resurgence of identity for indigenous peoples. Whilst the success of this movement owed little to local political parties, it was always strongly supported by the local Australian Labour Party (ALP) although this had little practical influence. However, it would not be correct to state that the Aboriginal urban social movement in Alice Springs was devoid of a class dimension. As noted above, the white economic position in Australia is sustained by a subordinate, welfare-supported Aboriginal population and the (private and public sector) mobilization of whites against the Aboriginal social movement was as much motivated by class interests as ethnic considerations.

In contrast, the evolution of the residential environment in Darwin has been quite different and, as yet, the dominant section has been unchallenged by any form of granulated socialism. Why have the underprivileged in Darwin failed to mobilize? Lowe (1986) is quite correct in emphasizing this as an equally important line of analysis, particularly when the structural framework of society seems so similar. Certainly the smaller number of Aborigines has reduced the potential for ethnicity to provide the necessary social basis for coalescence. Indeed, given the much larger number of young, low-income, white migrants to the city, ethnicity has been a divisive rather than a unifying factor amongst the underprivileged.

This lack of unity and direction has also been affected by the fact that Darwin is the state capital. This has resulted not only in a stronger political response but also in what Lowe has termed 'the osmosis of activists from urban-based involvements into the party political arena' (Lowe 1986: 197). The result has been, as Castells (1983) has noted, the uninterrupted use of urban planning by the State (in this case the Northern Territory) for its own ends. Urban planning in Darwin is the consequence of political rather than social goals.

Clearly, the comparative analysis of urban social movements reveals their complex nature and the difficulties involved in attempting to produce stereotypical models. In this context, insistence on rigid criteria, such as non-class or non-party bases, for urban social movements is unwise and a theoretical framework of general considerations for investigating the nature of the mobilization (or non-mobilization) process seems more useful. However, it must be remembered that urban social movements are not the only way that underprivileged groups can bring about urban change, and that class/production conflicts may still be more powerful sources of transformation in the longer term.

NOTE

1. See Knox (1988: 365) for an interesting but inconclusive discussion on the role of architects in 'adding exchange value, stimulating consumption, and fostering the process of capital accumulation'.

ACKNOWLEDGEMENTS

I am grateful to the following organizations and institutions for fieldwork funds: Department of Human Geography (ANU); Northern Australia Research Unit (ANU); the Northern Territory Housing Commission and Department of Lands; Keele University; the Australian Studies Centre (University of London); and the Royal Society. I am particularly indebted to Peter Loveday, Field Director at the Northern Australia Research Unit, for his organizational assistance and his invitation to become a Visiting Fellow at NARU. I am also grateful to Steve Williams and Joe Doherty for comments on some of the conceptual issues raised in this chapter.

Chapter Seven

Upgrading the 'matchboxes': urban renewal in Soweto, 1976–86

Charles Mather and Susan Parnell

INTRODUCTION

The call to South African urban geographers for research on African townships has generated a variety of informative work (see Beavon 1982). The plethora of research on aspects of the urban economy, most notably the informal sector (Rogerson and Beavon 1982, 1985; Tomaselli 1983; Beavon and Rogerson 1984; Rogerson 1985; Rogerson and Hart 1986), stands in stark contrast to the lack of information on the residential fabric of the 'locations' (Black township areas). Work on the built environment has either tended to be historical (Morris 1981; Western 1981; Mather 1985; Mabin 1986) or has emphasized impressions of the townships depicted through the literature of earlier decades (Hart 1984). Little research has been directed at the physical changes currently occurring in the townships. In this chapter an attempt is made to redress this situation by examining the emergence of middle-class housing as one dimension of the changes in the urban aspect of Soweto, the principal Black dormitory town of Johannesburg.

During the 1980s, major transformations have been wrought on the urban landscape of Soweto. The nature of this change has not, however, been uniform. On the one hand the deepening economic depression, and the drought being experienced throughout the country, have witnessed a massive increase in unemployment and poverty. In the city, people have been forced to share accommodation in an attempt to reduce monthly expenses. The success of recent rent boycotts in the townships suggests that even a shared rent is beyond the means of thousands affected by inflation, retrenchment, and unemployment. Physically, increasing hardship has been manifested in the increase in squatting in the large cities and the dramatic rise of overcrowding in the already small state housing units allocated to Blacks. While the increase in subletting and squatting is undeniably the dominant trend of the 1980s, affecting thousands of people, it has not been the most marked visual transformation in the urban face of Soweto. The most obvious change in the geography of Soweto has been the emergence of new elite residential areas (Tuswa 1983). In addition, scattered throughout Soweto, there are upgraded and renovated homes which stand in contrast to the rest of the monotonous, grey environment.

Through the description of the emerging middle-class housing market, the aim of this chapter is to highlight political and economic forces which have prompted the process. The first section of this chapter details some of the new housing initiatives in Soweto. The provision of middle-class housing assumes the existence of people who can afford to pay for the quality of shelter on offer. A brief historical survey of the Black middle class reveals that conditions under which housing is now made available differ sharply from previous epochs. The background to the current housing situation in Soweto is presented through an examination of state and capital responses to the 1976 uprisings. Finally, the implications of private sector investment in the creation of a Black middle-class residential environment are evaluated.

SOWETO TODAY

The urban fabric of Soweto, South Africa's largest Black township is dominated by row upon row of small, four-roomed houses. Home for almost two million Blacks employed in Johannesburg, the urban landscape is distinctive for the so-called 'matchbox' houses constructed during the 1950s and 1960s. Severe housing shortages and lack of finance during the period were largely responsible for the low costs of the housing units and their subsequent poor quality (Mandy 1984). The pace of construction, which peaked at over 11,000 units in 1958, also contributed to the poor quality and monotonous layout of the houses (Lewis 1966). Recent changes in housing construction has, however, made such generalizations of Soweto's urban fabric no longer current. As one newspaper reported:

> the monotonous cornfield outlook that four-roomed houses gave Soweto is fast disappearing as more and more residents are making alterations to their houses...and with new houses are giving the township a much needed facelift.
>
> (Star, 29 May 1984)

Soweto's 'facelift' is occuring in two different forms. On the one hand, there are the new elite suburbs on newly developed land such as Protea North (Fig. 7.1). The West Rand Development Board, overseer of land in Soewto, provides the sites and the basic infrastructure such as roads, street lights, electricity, and sewerage. Once the land is fully serviced, the sites are sold to private developers or construction companies who build houses on the sites. In some cases the houses are sold to private individuals by the developers. In other cases, however, large companies contract building concerns to construct houses for their 'junior management' employees. Houses in these suburbs are usually sold for upwards of R35,000[1], a middle- to upper-class market in Soweto (Star, 3 May 1986). The second aspect of the 'facelift' is the renovation of existing 'matchbox' houses (Rand Daily Mail, 10 July 1984). A large majority of the upgradings are of a substantial nature, where the original 'matchbox' core is no longer recognizable (Home Life, February 1985; Star, 3 June 1986). In recent years, therefore, the emergence of a middle- to upper-class housing market in Soweto has effected important changes in the urban face of the township. Attention now turns to the forces behind the development of a middle- and upper-class housing market in Soweto.

THE BLACK MIDDLE CLASS

The growth of an indigenous manufacturing sector after the Second World War dramatically increased the number of Africans in the labour force. Occupational positions and wage opportunities were, however, severely limited by job reservation, and as a result Blacks were concentrated in unskilled or semi-skilled jobs. By the late 1960s the situation had begun to change. The economic boom and subsequent growth of the tertiary sector allowed Blacks to penetrate 'middle-class' jobs where racial embargoes were

Figure 7.1 Soweto: 'elite' suburbs

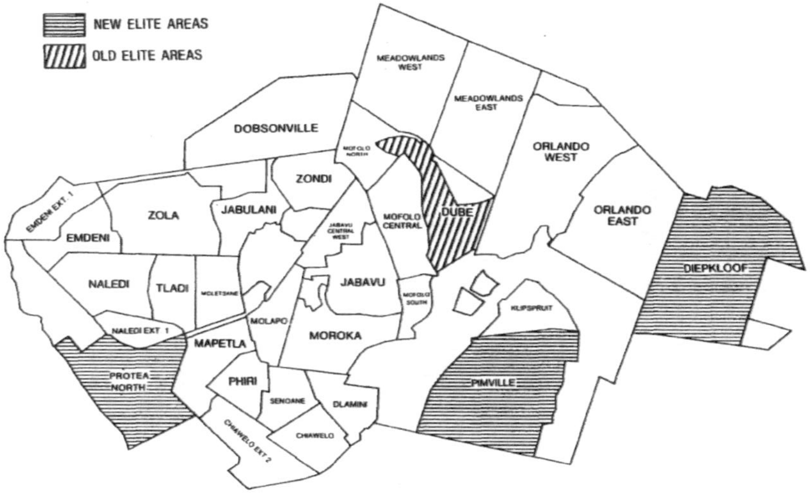

less rigid (Simkins and Hindson 1979; Scheiner 1983).

The size of this Black middle class has increased steadily from the late 1960s. In the period considered in this chapter, from the late 1970s, the rate of growth of the Black middle class was particularly high, higher even than during the boom years of the mid-1970s. More importantly, the increase in the size of the Black middle class in the post 1979 period, took place against a decline in the total number of Africans employed (Crankshaw 1986). While the Black middle class remains a small minority of the population, its presence and, particularly, the growth in its number, must be acknowledged as an important social transformation. The impact of this change is most important in Johannesburg which, as the nation's industrial centre, hosts 52 per cent of the headquarters of the top 100 companies (Rogerson 1984). As the focus of the nation's tertiary activity, it is home to many of the emergent Black middle class.

Although the past ten years have seen an increase in the size and importance of the Black middle class, there have always been at least some Africans whose status was markedly superior to those around them. In 1946, the state attempted to accommodate these 'better-off' Blacks, as well as other landowners, in Dube, which is a suburb of Soweto. The success of this venture to create a Black middle class suburb was limited, as few had the means to purchase their homes and Blacks were sceptical of state promises of security of tenure. By the late 1960s, when many Blacks with higher wages may have been in a position to purchase their homes, the policy of separate development had become entrenched. Home-ownership rights were withdrawn for Black in 1968 and construction outside the Bantustans almost ceased. Blacks in townships were demoted to the status of 'temporary sojourners' and were encouraged to build their homes in the Bantustan areas. As a consequence, waiting lists for existing housing stock grew. These houses, available only for letting, were allocated without reference to economic criteria. In fact, the only logic to the allocation process was that of ethnic origin (Pirie 1984b). As a result, the residential fabric of the Black townships of the 1970s did not reflect any of the class cleavages usually found in an urban settlement the size of Soweto.

THE SOWETO UPRISINGS AND THEIR AFTERMATH

The Soweto uprisings of 1976 drew sharp attention to the plight of South Africa's Black township dwellers. A government Commission of Inquiry into the dissatisfaction which led to the riots found that housing featured prominently among the grievances of township residents (South Africa 1980). The 'temporary sojourner' status held by Blacks outside the so-called 'independent' Bantustans was unacceptable. Housing in Soweto was reportedly drab and of a very poor quality. Overcrowding, high rents, and the supply of only minimal infrastructure were other grievances found by the Commission. The seriousness of the revolt and the findings of the Commission marked a watershed in the South African State's Black housing policy. The reverberations of the riots were also felt in the private sector. The responses of capital and the state to the 1976 riots are considered in turn.

Local and foreign capital's response spawned the Urban Foundation, an organization the aim of which was to foster the 'free enterprise' ethic among Blacks and to encourage private sector involvement in improving the quality of life for urban Blacks. These goals soon took the form of assisting Blacks to own a home, as the Foundation became a direct actor in housing provision, initiating housing schemes in various areas (Wilkinson 1984). The Foundation also became an important agent in pressurizing the state into legislative reform in the housing arena. Foreign capital responded in other ways too, and its response was generally more direct than its indigenous counterparts, with companies other than those involved in construction engaging in housing provision under their own names. A possible explanation for their more active involvement may have been the pressures these subsidiaries were feeling from parent companies with corporate responsibility programmes geared to assisting their Black employees in the housing arena. Other foreign multinationals were pressurized into the housing issue through business ethics such as the Sullivan Code (Housing in Southern Africa, April 1986). Clause six of this dispensation endorses the principle of 'improving the quality of employees' lives outside the work environment in such areas as housing, transportation, schooling, recreation, and health facilities' (Schmidt 1980). As a result of such pressures, a few of these large corporations provided housing for their Black employees in the elite areas of Pimville, Diepkloof, and Dobsonville. In addition, one large computer multinational constructed houses for its employees in Orlando West. The area is now ironically known as the 'Beverly Hills' of Soweto. Generally, however, these new 'elite parks' are very small and have not had a marked effect on the landscape of Soweto.

One of the major breaks in the state's new housing policy was the recognition of the permanence of urban Blacks. The 99-year leasehold legislation, passed in 1978, allowed Blacks legally resident in urban areas (in terms of South Africa's influx control laws) the right to own their home, although not the land on which it stood. Homes purchased under the new agreement could be rented, renovated, or torn down. With the new legislation, the state anticipated the rise of a Black home-owning group with middle-class values and aspirations. Government officials believed that a home-owning class would have a stabilizing effect on Black society; home-owners who had acquired a stake in the system were less likely to participate in riots, boycotts, or strikes (Hendler 1985).

The number of houses sold under the 99-year leasehold was by all accounts, very disappointing. By 1982, four years after the legislation was passed in parliament, only 1,700 leaseholds had been registered (Soussan 1984): 1,400 of which were in Soweto (Wilkinson 1984). It was clear that the state's anticipation of a growing, Black, home-owning class had been unfounded. In the first instance, there were very few Blacks who had the cash available to purchase a house. Since building society loans were almost impossible to secure, owning one's home was an unlikely reality for most Sowetans. A further factor in the failure to create a private housing market was the suspicion many Blacks felt towards the new legislation. Many remembered how, after the 30-year leaseholds had been revoked in 1968, they were forced to sell their houses to the state at prices far below their market value. Up until 1982, therefore, the state's drive for providing the basis for a Black home-owning group had failed. Since then, however, renewed government efforts to encourage Black home-ownership have met with some success, at least in Soweto, and it is to these efforts that attention now turns.

PRIVATE SECTOR INVOLVEMENT IN BLACK HOUSING

The failure of the 99-year leasehold scheme was responsible for the appointment of the Viljoen Commission of Inquiry into private sector involvement in African housing (South Africa 1982). Published in 1982, the Commission's main finding was that private capital could play a vital role in alleviating the shortages of housing for Blacks and addressing the issue of home-ownership. The Commission observed that legal restrictions had prevented private building companies from becoming involved in Soweto where they could realize large profits in the housing market. According to the Commission, employers had an active role to play in assisting their employees to buy and then possibly upgrade their rented homes. Some of the larger companies could become involved in providing housing for their employees, as had some of the foreign multinationals. The Commission found that employers had been unaware or confused as to how they could assist their employees in the housing sector. In addition, many companies had schemes which catered only for their White employees. As a further incentive to home-ownership, the Commission recommended that the state sell all its existing housing at discounts to Blacks legally resident in South Africa (Wilkinson 1983).

The state responded in the affirmative to these recommendations, and in April 1982 private building companies were permitted to take out 99-year leases on land in Black townships. This legislative move was soon followed by the announcement that building societies would be permitted to award bonds to houses purchased under 99-year leasehold agreements. In the following year, the government announced that it planned to sell 500,000 of the houses it owned at discounts of up to 40 per cent (Rand Daily Mail, 4 March 1983; 5 March 1983). In Soweto, almost 90,000 houses would be up for sale (Sowetan, 24 January 1983). Announcement of these sales was followed by a tremendous advertising campaign by the state, asserting the role of the private sector in Black housing, lauding the merits of home-ownership, and explaining ways in which employers could assist their employees in owning their own homes. The government argued that it could no longer be held responsible for providing houses for Blacks as it had in the past. According to the state, the backlog in housing was due to the non-involvement of the private sector; it was now up to private capital to build houses for their employees or assist them in buying a house (Housing, January 1983; Soweto Today, May 1984). On the merits of home-ownership, government publications were full of information on the desirability of buying rather than renting a house (New Horizons, February 1985). 'Buy now, improve, and feel secure' was a catch-phrase often used to entice would-be home-owners (New Horizons, January 1985). Information on the role of the employer, and on ways in which employees could be assisted, also featured prominently in State publications. In most cases, the employer was urged to assist employees by offering collateral to building society bonds (Housing, January 1984).

Up until 1982, the registration of 99-year leaseholds had been very slow, and only 1,400 had been awarded in Soweto. Since then, however, the number of leaseholds registered has increased rapidly and in 1985 alone, 6,295 were registered in Soweto (Hansard, 27 February 1986, col. 237) both for new houses and for existing 'matchbox' houses. New houses being built by construction companies range in price from R35,000 upwards (Star, 3 May 1986). Construction companies have the added incentive of moving into Black housing as a very profitable alternative to the oversupply and sluggish market in White Johannesburg and the virtually insatiable demand for houses between R40,000 and R50,000 in Soweto (Business Day, 20 June 1986).[2] Companies that provide their employees with housing also appear to be on the increase: one local supermarket chain recently constructed fifty houses for its 'junior management' employees in Soweto (Housing, January 1985). Forty-seven houses were built by a large South African insurance firm for their 'long-standing, high-salary earning' Black employees in Protea North, a suburb of Soweto (Sowetan, 11 July 1986). These houses range in price from R40,000 to R80,000 - at least four times the cost of constructing and servicing a 'matchbox' (Hendler 1986). Upgrading of purchased 'matchboxes' has also taken a marked upward swing, and since 1983 over 10,000 units have been renovated at an average cost of well over R2,000 (West Rand Development Board, pers. comm.).

Not all the upgradings necessarily represent gentrification. Many extensions occur with a view to subletting as a means of surviving the harsh economic reality of township life. One building company has simplified upgrading for many Sowetans and provides pre-cast units which can be attached to a 'matchbox' in a matter of days (Sowetan, 30

April 1986; Du Plooy, pers. comm.). The unit comprises 'two rooms and a garage', now a catch-phrase for renovations in Soweto. Bonds are also available for the pre-cast units which cost upwards of R10,000. The appearance of newspapers such as Home Life and home-ownership fairs catering for the renovator, are further indications of the extent to which upgrading is on the increase in Soweto.

Prior to 1982, one of the major obstacles to home-ownership was the scarcity and cost of building society bonds for Blacks. Since 1982, however, the granting of bonds for houses under the 99-year leasehold legislation has been permitted and building societies have been encouraged by the state to grant bonds to Black buyers. The availability of loans to Blacks has therefore played a major role in the upward trend of a home-owning class in Soweto. Very few Blacks, however, approach building societies privately, and 90 per cent of loans are employer-assisted (Star, 5 July 1986). Not all employees qualify for this assistance and, on the whole, it is white-collar or management level employees who are being supported. In many cases, the worker's pension fund is used as insurance for the company, so that only long-standing employees with large pensions qualify for assistance.

Since 1982, through legislative changes, the state and the private sector appear to have made important progress in fostering a Black home-owning 'class' in Soweto. It has been demonstrated that the divisions which home-ownership forges within the community reflect in turn more fundamental social cleavages (Mabin and Parnell 1983). To assume that the impact of the new urban policies has been uniform is simplistic, though the current situation in the township makes it impossible to establish what changes have been wrought by the new urban policies. Ten years after June 1976, the resurgence of township violence and the imposition of a national state of emergency, despite the improvements in the townships, begs the effectiveness of the responses of the state and capital to the riots of the previous decade. In the next section of the chapter, the gains the state or capital have made from their respective post 1976 urban Black housing ventures are examined.

ASSESSING THE IMPACT OF MIDDLE CLASS HOUSING FOR BLACKS

For the state, the political returns on its investment in middle-class housing have been limited. It is true that Soweto, once the name which encapsulated the horrors of the apartheid evil, is now used to illustrate the material benefits for Blacks as a result of reform initiatives such as the lifting of job reservation. The state's most important political objective in encouraging home-ownership and improved housing conditions, that of undermining the potential for unrest, has not been successful. The extensive involvement of middle-class Blacks in leadership roles in oppositional politics, is evidence that living in superior housing has not lessened the abhorrence of, or resistance to, apartheid.

Although the returns for the state on housing activity directed at the elite Black market have been limited and uneven, this is less true for private capital. Benefits from increased private sector involvement in the Black townships have accrued to both local and foreign capital. Furthermore, capitalists in a range of sectors have profited from the middle-class housing boom in Soweto.

It has been argued that for foreign investors to adopt the principles of the Sullivan Code justifies their continued participation in the South African economy which generates profit by its dependence on cheap, largely Black labour (Schmidt 1980). In this regard, the impact of these firms in the housing arena is informative. In the companies under discussion, the general practice has been that only permanent, high-salaried staff qualify for housing assistance. Recently the number of Blacks eligible for such perks has increased, largely as a result of the affirmative action applied within the firms. The provision of housing for management employees, however (whether Black or not), in no way ensures the adequate housing of the rest of the workforce. The intervention of foreign companies in the housing process has been characterized by provision for white-collar workers only. It may, therefore, be argued that their assistance has in no way transformed the South African economy, nor has it susbtantially altered the limited housing access afforded to Blacks.

In so far as local capital's involvement in Black housing has not differed fundamentally from its foreign counterpart, the above points also apply to the South African-based firms. In addition, certain sectors of local capital have enjoyed direct advantage as a result of the construction for Black home-owners. Most notable among those who benefit from the new housing initiatives have been the construction industry and the building societies. At a point when the South African economy has been experiencing one of the most intense crises ever, the construction of houses and the establishment of services such as electricity for the township of Soweto have provided extensive investment opportunities (Hendler 1986).

CONCLUSION

Urban renewal in Soweto, although perhaps not the dominant trend in the changing urban fabric of the township, does represent a significant shift that demands the attention of urban geographers. It has been shown that an explanation of the changes that have occurred in Soweto over the past decade may be found in the urban policies spawned by the state, under pressure from capital in the form of the Urban Foundation, following the upheavals of the 1976 uprisings. The most important aspect of the policy shift is the granting of the 99-year leasehold, thereby formally acknowledging the permanence of Blacks in the city.

Initially, attempts to create a home-owning Black middle class met with only limited success. The degree of support for the push to home-ownership increased dramatically after 1982, once loan finance was made available to prospective purchasers, and once companies were allowed to be registered as the owners of property. A result of these changes has been the appearance of new luxury homes, as well as extensive alterations to existing township houses, thereby transforming the built environment of Soweto, which, until the 1970s, had been characterized by the monotony of one house design: the 'matchbox'.

The first phase of geographic research on the South African city focused mainly on historical developments in the built environment. More recently attention has been paid to particular aspects of the social and economic fabric of the city, such as the informal sector, while contemporary changes in the physical fabric have been under-researched. The changes in the landscapes of Soweto, outlined in this chapter, highlight the importance of investigations of this nature. It has been the intention to document and account for upgrading, one of the most visible changes in the urban landscape of South Africa's best-known township.

NOTES

1. In 1987, S.A. Rand 1.00 = US$ 0.48.
2. For R50,000 a new house may be purchased in suburbs intended for White, first-time buyers.

ACKNOWLEDGEMENT

The authors would like to thank Gordon Pirie for comments on an earlier draft, and Wendy Job for the cartography.

Chapter Eight

Economic considerations on the renovation of the historic centre of Salvador (Bahia), Brazil

Johannes Augel

The ageing and the 'decadence' of a town are not natural processes, they are bound to historical framing conditions which characterize the town as a whole; in one or several of its main functions or one of its important inner-urban functions.[1] The need for urban renovation only arises if the continual process of maintenance, conservation, re-structuring, modernization, destruction, renovation, and adaptation to the changes in external conditions is interrupted, and the society feels that it is necessary to overcome the stagnation to make up for neglected adaptation, or to save buildings or whole sections of town that are in danger of disappearing (see Albers 1980). Furthermore, it has to be considered that an historical building authentically restored, or a renovated quarter will not be re-integrated into the historical context of its epoch with its human beings, customs, ideals, and needs, but is designed for present-day people and present-day needs.

A town is only conceivable as a continually changing object that is dynamic and able to survive if its various functions meet the needs of its inhabitants and their structural setting in an historical context, and if it is able to continue developing its complex functions and adapt them to new circumstances and conditions. For a renovation process, this may mean that an overemphasis on any particular factor, such as tourism, may raise the danger of 'monocultural' dependence.[2] Every partial renovation can, and must, tend to lead to problematic distortions in the urban structure with repercussions on other functions of the town and on the different social classes and groups involved. It is true that such socially differentiated repercussions arise in all places and at all times, but in a slow and natural process they are not thought of as much as a social problem as in an accelerated process induced by external criteria: as brought about, for example, by tourism or the goal of urban modernization.

The considerations outlined above concerning the historical and structural framework of urban ageing and the restoration of old sections of towns, point to the need to analyse our example of the city of Salvador as part of a broad historical process, as well as a manifestation of the dynamics of a modern, rapidly growing, industrial, and administrative centre. The decadence of the old city, in its physical-architectural and social aspects, can only be understood as part of the modernization which directed all public and private attention and nearly all investment to the modern sector. The old city remained on the edge of the development, and the former city centre became peripheral to all the dynamic interests. It became forgotten, stigmatized, and put to one side by public opinion; cut off from investment in maintenance and restoration by private owners and the state. The former importance of Salvador as the capital of a colony and as the largest urban colonial centre south of the Equator rapidly deteriorated into a reminiscence found in literary and artistic works[3], while respectable sòciety avoided the former upper-class district as a den of iniquity (or, at least, gave this impression), and all investment and speculation in real estate concentrated on the new, modern sections of the city (see Simas 1978; Mattedi et al. 1979; Augel and Augel 1984).

SALVADOR: 'SHAKESPEARE'S MAGNIFICENT WIDOWED QUEEN'

Even today, Salvador is still considered to be the most Brazilian of all Brazilian towns. It was founded in 1549 as the capital of the Portuguese colony, retaining this position for more than 200 years (until 1763). As a port, and as an administrative and residential city, it became the prototype of an urban colonial society (Russell-Wood 1968; Schwartz 1973; Alden 1968; Lugar 1980); it was the heart of the colony, and its link with the transoceanic world, in continuous communication with Lisbon and Coimbra, where the sons of the rich studied, and also with Goa, Macao, Luanda, and all the other markets and commercial centres of the Portuguese overseas empire (Boxer 1965). The role of the city as a harbour for the import of slaves, its large population, and the important role of the civil, military, and clerical urban society (with a disproportionately high number of African slaves), even today make Salvador the 'darkest' Brazilian metropolis. It has retained more African traditions than any other town outside Africa, and these have developed independent cultural, and especially religious, forms.[4]

Innumerable travellers in colonial times, especially in the nineteenth century (after the opening of the port in 1808), have described the city and expressed their surprise and admiration (M.P. Augel 1980). The Englishman Thomas Lindley calculated in 1802 that all the fleets of the world could fit without difficulty into the bay of Salvador (which was comparable with Constantinople, the mouth of the Tejo river, and Naples). Charles Darwin visited the city twice, in 1831 and 1836. In 1855, Robert Christian Berthold Ave-Lallemant admired the 'City of Saint Saviour of the Bay of All Saints, which extends toward the east of the bay as broadly as the length of its printed name across the page'. For the nineteenth century alone, there are more than one hundred travel descriptions, such as those of Prince Maximilian zu Wied-Neuwied (1815/17), the painter Johannn Moritz Rugendas (1821/25), Prince Adalbert of Prussia (1843), and Archduke Maximilian of Austria, later the Emperor of Mexico (1860), just to mention a few prominent German travellers. Besides effusive descriptions of the tropical environment, these accounts share the astonished observation that the population was divided into a very thin social stratum of administrative officials and businessmen; an almost equally thin social stratum of artisans and other poor but free inhabitants, and a large mass of slaves and former slaves. Even at this time, there is an impression of a geographical distribution of the residential quarters according to social strata, and a corresponding process of change (UFBa 1980, Azevedo 1969, Santos 1959).

The transfer of the functions of the capital city to Rio de Janeiro, the decline of the sugar-cane industry even during colonial times, and the stagnation of Bahia and its capital Salvador during the nineteenth century and the first four decades of the present century,[5] caused the city and its centre to change very slowly, while the southeast, especially Rio de Janeiro and Sao Paulo, consolidated their positions as the regions of new development. Between the years 1920 and 1940, the annual population growth of Salvador fell to its lowest level, only 0.2 per cent.[6] The industrialization and modernization of Salvador and its hinterland only began during the Second World War, and continued with the subsequent implementation of governmental measures for economic promotion in the 1950s (Kruger 1978; Espinheira 1985a; Kraychete 1986; Faria 1980; Singer 1980). This led to increasing immigration and the rapid growth of peripheral urban settlements, which these 'invasions' have made densely populated and poor (Brandao 1978; Mattedi 1979; Sarmento 1984; Salvador, Prefeitura 1976). The city which, during the nineteenth century and up to the Second World War, had grown very slowly from its original centre to the south and from the Bay of All the Saints to the Atlantic coast (the Vitoria and Barra quarters), now rapidly expanded to the north, showing a very clear division according to social strata. Along the coast, some fishing villages, farms, and small suburbs became upper-middle-class residential districts, while the middle- and lower-class suburbs between the Atlantic coast and the bay developed into huge settlements almost totally without infrastructure.[7] With the construction of wide arterial streets, which provide access to the numerous valleys of the urban district, the latter expanded in a remarkably discontinuous way (Carvalho 1985), reaching the region between Itapoa and the airport in the early 1970s and has since spread into the neighbouring community of Lauro de Freitas, approximately 40-45 km away from the city centre (Salvador, Prefeitura 1985).

The broad expansion of the urban settlements, the old and new 'invasions' often involving several tens of thousands of persons, and the well-planned, park-like residential settlements of the wealthy classes, nearly always remain unknown for foreigners. For them and for the general consciousness of the population of Salvador, the city still mainly consists of the centre: the historical city with its baroque churches, its colonial architecture, its African rites, and its folklore, strongly influenced by the Black population, and encouraged by the tourist industry (Mattos 1978). Although the infrastructure of tourism has been developed especially along the Altantic coast, Salvador's beaches are no better than other Brazilian beaches or those of several other tourist centres. Even today, the cultural identity of the city remains, without doubt, in the city centre. 'Shakespeare's magnificent widowed queen' was the name Stefan Zweig (1984: 61) gave to Salvador, the oldest city in Brazil, the head of the colony, and once the most powerful city south of the Equator. The many-floored mansions, some from the seventeenth century and many from the eighteenth century, former residences of the urban upper classes, of high-ranking military, clerical, and civil functionaries, as well as the gentry, testify even today to the past wealth of the city.

PROBLEMS WITH THE RENEWAL OF THE OLD CITY CENTRE IN SALVADOR

It is true that nowadays only a small part of the old sector of the city corresponds to the usual idea of historical testimonies that are worth maintaining. Just a few rows of houses that are still being used for commercial or residential purposes or for public institutions are kept in something approaching good repair; all the other street rows show a dreadful picture of architectural as well as social decay. At the beginning of the 1960s, for the whole of the old centre of the city, investments in maintenance and renovation were made almost only for buildings used for public and commercial purposes, whereas the residential structures increasingly fell to ruin. Only the facades and the exterior walls of many former middle-class houses remain standing, while the number of former residential palaces which have become rubbish tips, heaps of rubble, and parking lots is increasing rapidly. Six houses collapsed during the heavy rains in March/April 1985.

The increasing decay has had the same effect on the resident population as have the restoration measures: eviction. The population of the Maciel and probably also the population of the other old sectors of the city (for which no statistics are available) have decreased continuously since the beginning of the 1970s, partly due to 'clean up' operations (Espinheira 1985b). The number of ruins is growing constantly. In some cases, houses were abandoned room by room as they became uninhabitable. During one of my visits to the house situated at Frei Vicente 16 in 1983, there were two (incomplete) families with eleven persons living on the second floor. Two years later, this floor was abandoned and only the first floor was occupied. Even here, one room was so badly damaged that it was only used to dry clothes or as a store room. Neither the public nor the private sector invests money in preserving or renovating these houses.

The restoration work carried out by the Office for the Preservation of Monuments (<u>Instituto do Partrimonio</u>

Artistico e Cultural da Bahia, IPAC) generally has the same effect; that is, it reduces the resident population. Only two houses have been restored for residential purposes; all the other residences and palaces restored by private or public institutions are no longer used as private dwellings but have been converted into public offices or tourist buildings.[8] The massive presence of public institutions in the quarter (which at the beginning of the 1970s was still considered to be the city's largest district for prostitution and criminality) also contributed to the considerable reduction in the earning possibilities of the informal sector of the population living in this area. Public opinion, as stated in the local newspapers and especially in Salvador's largest and most important conservative daily A Tarde, pleads unequivocally for the 'cleaning' of the ancient 'den of iniquity', although the reputation of this quarter has been ill-founded for many years. Today, the Maciel is, above all, a poor and decaying quarter, almost forgotten by the state, with a marginalized population which has lost its faith in any positive results of state intervention and sees no solution to the rapidly deteriorating living conditions. The measures undertaken up to now by public institutions have been of little real value. The several hundred houses in the historical centre of the city decay faster than the state is able to preserve or restore them, despite considerable increases in funds. Not only are funds lacking, but also the conception of renovation that has been followed seems to be inadequate. In general, the measures realized up to now for the restoration of the old city could, at most, be useful for the creation of a stage set, but do not lead to the restoration of a functional quarter. Until recently there has been a complete lack of models of co-operation between the state and private initiative in the housing sector.

While public authorities do not contribute to the preservation of housing conditions, and their activities lead to the direct and indirect eviction of the residents (despite protestations to the contrary, see Fundacao... 1979; Salvador, Prefeitura 1985), the private property owners either have no interest in, or little likelihood of making, such investments. The private property in the old part of the city generally belongs to old established families, who have little interest in these houses and are unwilling to invest in them; or they belong to private or public institutions, such as the Orphanage of the Sacred Heart of Jesus, the Association of the Bread for the Poor of Saint

Francisco, the Home of Forgiveness, the Venerable Third Order of St Francis, and many other similar institutions (see Santos 1959; Bahia, Governo 1978), which do not have the economic strength to preserve, let alone renovate, their property. Even in those cases where the inhabitants pay rent, it is generally so low that it does not justify investments. In these rented houses, the subrenters are often the ones who profit the most. They rent a complete house and sublet for relatively high rents individual apartments, single rooms, and even rooms that are divided up into smaller units with wooden partitions.

Government offices have neither a legal base nor an interest in subsidizing private property. The lack of resources complained about on all sides is only part of the truth, because both the city of Salvador and the state subsidize projects which are just as expensive and unproductive as a comprehensive programme of restoration for the old city. According to statements by local politicians, there is a lack of political decisiveness in favour of large-scale programmes, as well as a lack of the necessary 'emotional engagement', i.e. the creation of a strong public opinion in favour of the preservation of the old part of the city. But it is also evident that the enormous public investments of the last twenty years, which were placed in the expanding sector of the city along the Atlantic coast, complied with the interests of the local property business in a very favourable way; a situation which could never be attained for the restoration of the old city (CONDER 1977, 1979; Augel 1986: 369-81).

Besides property speculation, which could profit considerably from public investment, there is the important problem of the structure of the building trade. In Brazil, the sector of organized small-scale trades involved in repair and maintenace is almost non-existent. Although the great boom in the building trade has fallen back because of the general economic crisis in recent years, there still seems to be an apparently unlimited demand for new buildings. This has produced a largely rationalized, specialized, and industrialized building trade.[9] Despite high unemployment and short working, the building trade's craft section is underdeveloped. One is lucky if one can find qualified tradesmen to repair houses in Salvador or in the whole northeast of Brazil. At the present time, the building trade is not at all interested in, nor capable of, taking on the kind of work necessary for the preservation and restoration of old buildings like those in the Maciel. While in talks with representatives of the building trade, it is continually repeated that only the state is able to take on this kind of work, such statements are indicative not only of the supposed lack of rentability for the repair and renovation trade, but also of the lack of specialist qualifications and the organizational structure of firms specialized in a form of construction suited to big buildings, apartment blocks, and large estates, not to a holistic restoration of old buildings.

The frequently voiced criticism of the state's lack of interest in a political decision in favour of the renewal of the city centre, in favour of the preservation of the residential value for the population, or in favour of the rescue of the irreplaceable cultural values of this colonial residential district, has to be seen in this context. The interests of local capital, that have already been mentioned in connection with public investment, have up to now been strong enough to avoid large-scale investment in the relatively unproductive and, in many senses, problematic sector of city centre renewal.

The tenants and, to an even greater extent, the subtenants, have no interest and no possibility of investing in other people's property. In addition to the experience and the fear that improvements lead to higher rents, or to the omnipresent call that the state should be made responsible for the preservation of cultural values in the old city, there exists another factor which is less evident. In the ruins, one can find very interesting and unusual examples of personal initiative. Walking through the streets in the darkness, one can find lights and voices behind several housefronts which seem to hide only rubble and rubbish. These ruins are sometimes used in a rather rural way, or in a way that is like the suburban invasions. In these ruins one can find wooden huts and other forms of adapted housing. Sheep and chickens are kept, even in apartments on the upper floors. Adapted forms of housing construction, or the restoration of parts of ruins to make a new home, can also often be seen.

It is my conviction that in order to ascertain the reason for the contradiction between the evident lack of interest of the tenants in the preservation of the substance of a dilapidated house and the existing forms of self-initiative, it is not enough to refer to the above-mentioned relation between proprietors, subletters, and subtenants. Even in those cases in which it would not be possible to carry out small repairs gratuitously or with low costs, which surely would have no consequences on rents but would only be advantageous for the tenants and subtenants, a positive response is unlikely to be elicited. A comparison with the efforts of other inhabitants to build a wooden hut on a ruined site makes one want to ask what kind of general relationship the inhabitants have with the houses in which they live. In the discussion that followed a lecture I gave on the problems of the renovation of old towns, a politically active listener told me that the desire to preserve such quarters as the Maciel and to declare its former palaces to be cultural goods is part of the ideology of the ruling class. In my opinion, the population of this quarter have no kind of positive relation to the buildings, which were neither built by them nor, evidently, built for them. A comparison with the neighbouring quarter of Santo Antonio Alem do Carmo (Frosch-Asshauer 1986), which has a very traditional character, seems to confirm this statement. Living is here more a wearing out, an act of consumption and exploitation, and not a form of continual production and consumption of living. Most residents feel that a house that does not belong to them, with a structure and dimensions evidently designed for a way of life that is completely alien to them and with which they cannot identify, is merely destined for consumption: to be used until it is useless.

PERSPECTIVES

There are several conceivable solutions to the described problem. The most likely one is the continuation of the prevailing policies, with only slight variance, resulting in the perpetuation of the prevailing process of decay with only piecemeal intervention by the Office for the Preservation of Historical Monuments. This will have the consequence that the direct and indirect eviction of the population will continue, and the old city, despite being a very active and densely populated quarter, will become a mixture of ruins and restored administrative buildings. Because of tourism, with its general justification of saving cultural values, it is certain that the restoration policies will be continued and the quarter will become even more sterile than it is already, especially compared with the times when the Maciel was a decaying, but still popular and highly frequented amusement district (Espinheira 1985b).

There is no alternative in sight that can oppose such a development. But in no sense can these policies succeed, if they do not take into account the residential function of a city district or the role of the people who live in it. Therefore, the question arises, whether it is possible to combine the interests of all concerned - the property owners, the inhabitants, and the state. This would make a variety of strategies possible, as well as requiring flexible solutions to allow for each individual case. Legal measures for investment subsidies, rent allowances, tax privileges, and also different types of collective property, including the proprietors, tenants, and public authorities, together with planning aids for preservation and modernization measures would all be necessary. In all this, however, the critical point is always the incorporation of the resident population.

The potential of self-help as part of a strategy for finding a solution to these problems can only be interpreted as a goal of research in the described context. Neither the economic situation of the population, nor the patterns of ownership, nor the state of repair of the big patrician residences, allow autonomic self-help to be a realistic solution. Nevertheless, without self-help and without surmounting the present situation (in which the residents, who are the object of official and private measures, have virtually no capacity for opposition or for organizing themelves), no solution seeking to preserve even a little of the authenticity of a living urban quarter is conceivable.

The course followed up to now in the restoration of some fine examples of colonial architecture has proved far too costly: frequently it has reflected the professional ambition of the architect or restorer rather than a concern for the preservation of practical values for the population of a functioning urban district.

NOTES

1. Besides structural changes which were analysed to a large extent within the urban ecology approach, mainly by the Chicago School and its followers, much has to be done to understand the effects of historical, regional, national, and global changes on inner city transformation. Monographs such as Petshek (1973) and Rosenthal (1980) consider such factors. From a world-system perspective, Timberlake (1985) represents great progress towards the understanding

of global factors on urbanization and changing urban hierarchies, whilst Daus (1983) gives some interesting insights on the dependency of coastal cities within the rise and fall of the Portuguese Empire.

2. There are innumerable examples of monocultural or monofunctional cities such as Manaus, Potosi, Brasilia, Wolfsburg, or Fatima; each one with special kinds of one-sided dependencies and disequilibria.

3. For example, consider Jorge Amado in some of his earlier novels, as well as artists like Hansen-Bahia, Carybe, and many other modern artists of Salvador.

4. There is a long tradition of Bahian racial studies, from the physical anthropologists such as Nina Rodrigues and Arthur Ramos, the social scientists like Edison Carneiro and Thales de Azevedo, to the Departments of Anthropology and of History and the Centre for African and Asian Studies of the Federal University of Bahia in Salvador/Brazil. For a broader history of slave importation and African influences in the colonial period see Russell-Wood (1982).

5. In spite of varied and sometimes considerable progress in Salvador's nineteenth century urbanization, as well as in its civic and scholarly development, the former capital turned out to be clearly provincial, especially in comparison with Sao Paulo and Rio de Janeiro. A valuable economic history of nineteenth-century Bahia is Calmon (1925); see also Bahia, SEPLANTEC - CPE (1978-80).

6. De Azevedo (1969: 236) indicates an annual increase in population of 1.8 per cent between 1900-20 and only 0.1 per cent from 1920-40. Similar to Rome's physical beauty not having been violated thanks to the municipal poverty (Lottman 1976: 13), Salvador's conservation of its historical centre seems to be related directly to the urban and regional stagnation.

7. While the development of the coastline is one of the priorities of public urbanization policies, the planning institutions have only recently begun to pay attention to the miolo, the inner areas of the peninsula of Salvador (Salvador, Prefeitura 1985).

8. Until 1979 the Monuments Foundation completed nine reconstruction projects in the Pelourinho area, twelve others were still under construction (Fundacao do Patrimonio Artistico e Cultural da Bahia 1979); by 1983, thirty-seven of the 161 buildings of the eight inner slum streets had been restored; thirty-five of them are for administrative use and only two are used for housing.

9. Fifty per cent of the construction firms registered in 1982 were founded between 1975 and 1980. The eleven largest (6 per cent) of the 179 contractors employed 47 per cent of the Bahian construction workers (in 1976), obtaining 56 per cent of the turnover and owning 65 per cent of the fixed capital (Borges cited in Franco 1983: 128-30). See also Sindicato da Industria da Construcao Civil da Cidade de Salvador (1980).

Chapter Nine

Temporary trading for temporary people: the making of hawking in Soweto

Keith Beavon and Chris Rogerson

INTRODUCTION

The cry bo-Mkozi has been heard by thousands upon thousands of Sowetan workers in the smoggy early morning atmosphere and again in the dusty dusk that envelopes their city like a grey blanket. The cry, meaning 'friends', comes from the hundreds of elderly, and usually destitute, women hawkers calling out to passers-by to purchase from their stalls and pitches. Collectively and somewhat affectionately known as the township 'aunties', the sandwich 'aunties', or simply the bo-Mkozi, the women are part of the many thousands who seek a livelihood in the ranks of the petty producers or the 'informal sector', as it is more popularly known. This army of casual workers represents one of the outworkings of an ill-conceived concept that not only gave birth to their 'city' but forced them, and generations like them, into temporary and tenuous income-earning niches.

The concept that has indelibly drawn the bounds to the Sowetan life-style was the product of two commissions set up by the Smuts government as a prelude to the drafting of the 1923 Natives (Urban Areas) Act. The commissions had reported that the towns of South Africa should be regarded as 'European' areas, in which there was no place for 'natives', and stated that it should be a recognized principle that 'natives' should only be permitted within municipal areas for as long as their presence, and their labour, was demanded by the wants of the White population (Davenport 1960). So was born the concept that the Black African people would be mere temporary sojourners in segregated locations (townships) adjacent to urban areas that required their labour. The concept was subsequently entrenched in the 1923 Natives (Urban Areas) Act, which was the foundation on which regulations for the administration of the residential areas for Blacks was to be based. An administration that, as the years passed, was to become harsher before it became any easier and which was to restrict the opportunities available for entrepreneurial opportunities unless they were conducted in the Bantustans. So emerged a situation that can be aptly described as one of temporary traders for temporary people. It is against such a background that the making of hawking in Soweto was fashioned.

Several years ago, research for a study of contemporary street sellers in the 'White' city of Johannesburg (Beavon 1981) soon prompted additional research into the antecedents of such activities (Rogerson 1983). The results of those subsequent enquiries (Beavon and Rogerson 1980, 1984, 1986a, b; Beavon 1982; Rogerson 1986b, c; Rogerson and Beavon 1985; Tomaselli 1983a, b) greatly enhanced an understanding and concern for those who necessarily had sought a living from the pavements for almost a hundred years. Despite a growing literature that is related to aspects of Soweto and its people, including some contributions from geographers (Mashile and Pirie 1977; Morris 1980; Pirie 1977, 1982, 1984a, 1984b), the social and economic history of the townships remains the great unwritten history of the working class of South Africa (Bozzoli 1979). So, although it has been possible to present a picture of contemporary hawking and petty production in Soweto (Rogerson and Beavon 1982), the antecedents of the contemporary scene remained a terra incognita for urban geographers (Beavon 1982). Fortuitously, in the pursuance of research material for other aspects of the 'informal sector' project on which a group of human geographers at the University of the Witwatersrand has been engaged for some years, data relating to hawking and trading in the formative years of Soweto are now to hand. The data inform us not only of the aspirations and frustrations that were experienced by 'formal' traders, aspirant traders, hawkers, and a variety of petty producers, but also of the attitudes, designs, and actions of those who responded to the ever-changing agenda determined by the ruling hegemony.

The content of this chapter covers some thirty years of trading activities in the townships on the southwestern side of Johannesburg (Fig. 9.1). A period that spans the formative years of the Vukenzenzele townships and extends into the 1960s before the townships, soon to be known as

Figure 9.1 Soweto: location

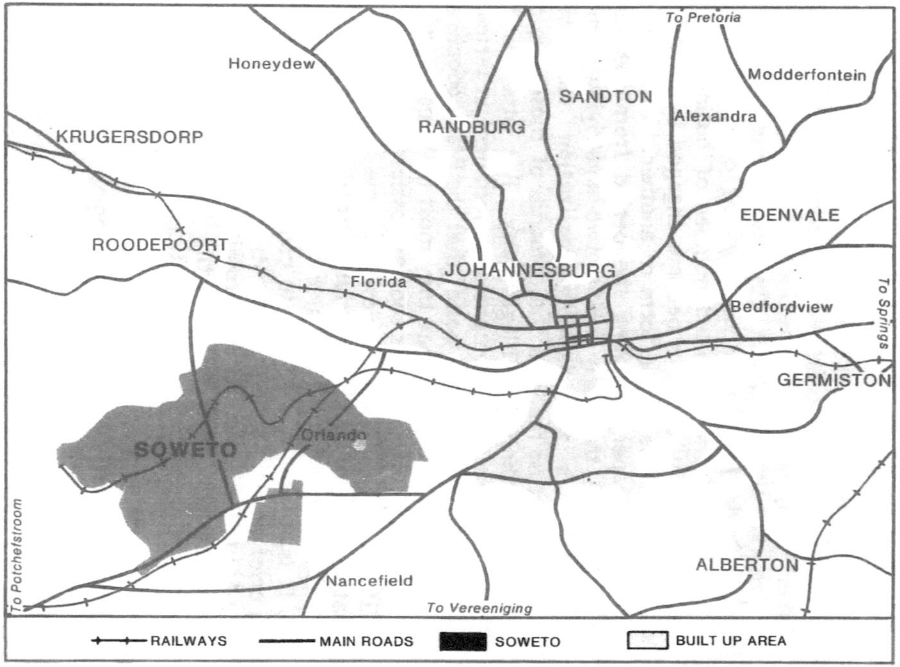

Soweto (an acronym for **So**uth**we**stern **to**wnships), passed out of the control of the Johannesburg City Council (Frankel 1979; Pirie 1984b). After a preliminary section which is related to the framework of broad restrictions imposed by acts of parliament, there follows a discussion divided into two sections. In the first, attention is paid to the genesis of hawking in the townships during the late 1930s through to 1950, a period that is marked by two struggles. On the one hand, there are those who are charged with the administration of the Urban Areas Act and its associated regulations, struggling to enforce the letter of the law. On the other hand, there are the local inhabitants, many of whom are not employed, who find themselves denied the right to provide for themselves anything other than bare daily essentials from a limited number of business sites. Consequently, many of them become engaged in a struggle for the right to trade in one form or another. In the second section of the paper, spanning the period from the early 1950s through to the early 1960s, two underlying themes emerge. First is the dawning realization that some concessions must be made to the demands of those who are trading in a technically 'illegal' way. The other is the increasing pressure from the central government in terms of its own agenda based on the belief that the Black people are temporary sojourners, and that they must be 'encouraged' to move out to the Bantustans as soon as possible.

Within the above framework, the discussion is illustrated by empirical material that demonstrates the consequences of official attitudes and actions in human terms. It is, however, essential to commence the discussion with a brief sketch of the legal constraints that provided the muscle for the suppression of 'formal' business development in areas for Blacks, thereby promoting the pressures for the emergence of petty forms of production.

THE MAKING OF THE 'TEMPORARY PEOPLE'

Street selling in the 'Black' townships has always been intimately related to the strict legal shackles clamped on the Black residents both in respect of their rights to live in the vicinity of Johannesburg (or indeed any urban place), and their right to trade with their own people (Kuper 1965; G.P. Hart 1972; Kegan 1978; Kitchin 1978; Southall 1980). The Native (Urban Areas) Act No. 21 of 1923 (subsequently replaced by the Native (Urban Areas) Consolidation Act No. 25 of 1945) formally laid down the rules that were to govern the segregated urban areas in which the Black people would be housed on the basis that they were merely temporary sojourners until their labour was no longer required by the White population. In terms of Section 22 of the Urban Areas Act, the (White) urban local authority controlled the allocation of all 'formal' trading sites within a Black township, and also had the right to prohibit hawking (and later peddling) if it so wished. Furthermore, the 'formal' trading sites could only be used for those types of businesses that would provide nothing more than the daily essentials for living (Kuper 1965; G.P. Hart 1972; Southall 1980). For example, dry-cleaning shops, bookshops, garages, and even chemist shops[1] (pharmacies) requested by the people could not be legally established in the townships. Such activities and all other kinds of businesses catering for more than the barest daily necessities would remain the preserve of White entrepreneurs and would be located in the 'White' areas. As a consequence, for about fifty years there have been continuous pleas from Black residents for more trading licences to be made available to potential shopkeepers. Denial of those requests has in turn encouraged applications for hawking and peddling licences, which have also largely been unsuccessful.

Provisions were made in the Natives Urban Areas (Section 17) for dealing with 'idle, dissolute or disorderly' natives in urban areas. 'Idle natives' were those regarded as habitually unemployed and not possessing the means of an 'honest livelihood'. A licensed trader or even a licensed hawker would, by definition, possess the means of earning an honest livelihood. The penalties for being classed as 'idle' were severe: removal from an urban area and expulsion to the Reserves, or being dispatched to a farm colony or work colony. Because possession of a trading or a hawker's licence offered considerable protection from removal in terms of the law, and was, moreover, an escape from proletarianization and a means of earning a legal, albeit meagre living, such licences were keenly sought. The fact that only a modicum of surplus value might be realized in providing for the needs of the poorest consumers living in Soweto and working in Egoli (Johannesburg) did not act as a significant deterrent (Rogerson and Beavon 1982).

THE GENESIS OF HAWKING IN SOWETO

The outworkings of the Natives Urban Areas Act, coupled with the rapid growth of impoverished Black townships on the distant margins of Johannesburg, fall within what might be termed the early phase of hawking, a period which runs from the establishment of the Vukenzenzele townships through to 1950. Thereafter, the designs of grand apartheid introduced by the Nationalist government after their electoral victory in 1948 begin to impact more forcibly both on the conduct of 'formal' trading and the sphere of 'informal' trading.

As the city of Johannesburg grew from mining camp to mining city and as it became the metropolitan centre of the industrializing Witwatersrand, it attracted and drew to itself an increasing population of both Black and White people. Between 1936 and 1946, as the region's industries became increasingly engaged in war effort production, the population in the Johannesburg municipal area increased from 229,122 to 384,628, with 57 per cent of the increase coming from the 'Black' rural areas (Morris 1980). In part, the great surge of Black families to the metropolis of Johannesburg in 1946 was due to a rumour that the city was providing land for people to build their own homes (Stadler 1979). This was not, however, the case; rather the council was simply allowing squatting to take place in the 'temporary' emergency camp of Moroka (Fig. 9.2) as the authorities fell behind in their capacity to provide sufficient housing. The imperative for increased shelter facilities was heightened by the implementation of urban racial segregation in the 1930s, which occasioned clearance of the multi-racial inner city slumyards and the resettlement of a growing proportion of their former Black populations in municipal townships (Koch 1983a). By 1939, the first township of Orlando had an estimated population of 38,000.[2] With the cessation of hostilities in Europe, the demands on the South African war industry declined and urban unemployment began to rise inexorably. Consequently, as the population increased and was funneled to the southwestern portion of the municipality, these areas became increasingly characterized by high and growing levels of unemployment (Stadler 1979). It was in these classic pre-conditions for the existence and growth of an 'informal sector' that significant numbers of Black people turned to the possibilities offered by a range of petty forms of production.

Figure 9.2 Soweto: growth and main townships

It is necessary to distinguish between the different types of business premises and street trades which now became features of the growing townships. At the top of the 'business hierarchy' were fixed sites licensed for a particular type of approved business and on which the council would erect the premises. Such businesses are referred to as 'legal' and 'formal' businesses, although they might be conducted in wood and corrugated iron shacks, or in units built out of breeze-blocks devoid of electric power or even piped water. Next down the hierarchy were approved 'stalls', which varied from single 'counters' in the open, to counters under a roof, and to selling points in shed-like buildings. Such 'stalls' were 'legal' in the sense that they were approved by the local authority and in several instances erected by the council. The lowest two rungs of the 'business hierarchy' comprised a variety of 'illegal' traders (Table 9.1) who operated either as itinerant hawkers simply storing their goods in private dwellings or shacks, or as petty producers and pedlars, functioning from the shacks and houses without licences or official approval. It is the stallholders, petty producers, pedlars, and other street traders who collectively compose the broad 'hawker' group referred to in the title of this chapter.

Table 9.1 Types of commodities and services offered by petty producers, hawkers, and pedlars in the southwestern townships c.1948

Food items		Services
Milk	Vegetables	Laundering
Cool drinks	Bread	Hairdressing
Poultry	Meat	Basketry
Sacrificial goats	Fruit	Bootmaking
Vetkoek	Maas	Carpentry
Porridge	Groceries	Tailoring
		Dressmaking
		Herbalist

Source: Letter C. Wilson to Town Clerk, with memorandum re conditions in Orlando township, 30 November 1939, A8624, Box 3, Orlando, JCHD.
Letter Acting Medical Officer of Health to Manager Non-European Affairs Department, House to House Survey Report attached, 18 March 1949, File 12/17/9, JCHD.

The extremely rapid growth of the Black townships' populations soon began to outstrip the ability of the local authorities to organize and license a sufficient number of 'legal' traders, and to erect the business premises for those traders as required by the Urban Areas Act. During the years of the Second World War, the Council officials had tended to turn a blind eye towards illegal trading.[3] and 'when it became apparent during the war that the authorities were not taking action against illegal traders, many Natives gave up their normal employment to become shopkeepers. Some erected stalls; others had shops in their homes' (Sunday Times, 23 October 1949). Accordingly, by 1946 there was a sizeable community of persons selling daily essentials such as bread, milk, fruit, vegetables, fried fish, vetkoek (batter cakes), and maas (thick soured milk). Frustrated applicants for licences, and others who simply decided to hawk items in the townships and to sell directly from their houses and shacks, swelled the mass of 'illegal' (street) traders; by 1948 this was precipitating official concern. In a 1948 survey of authorized and unauthorized businesses (Table 9.2), inspectors reported that in both types of operations the conditions under which food in particular was being prepared, stored, and sold was very unsatisfactory. For example, there were eighty-seven 'butchers' shops' functioning in wood and corrugated iron shacks where:

> the handling of meat renders it liable to infection and contamination, by dust and flies, and as the premises are not rodent proof, they afford attraction and harbourage [sic] for rodents. In all instances hawkers operate from these premises.[4]

In a report to the Town Clerk, the Manager of the Non-European Affairs Department stated:

> The position in regard to illegal trading in the Native locations has progressively deteriorated over the past year or two. The problem has been accentuated recently by the arrival of large numbers of ex-squatters in the Municipal locations; these squatters depended to a large extent on illegal trading in the erstwhile uncontrolled squatters' camp, where control was non-existent and where trading was one of the few means of earning a livelihood.[5]

Table 9.2 Authorized (A) and unauthorised (U) or illegal traders in the southwestern townships 1948

Township	General dealer		Butcher		Milkshop		Fish frier		Fruit and veg.	
	A	U	A	U	A	U	A	U	A	U
Orlando East	18	104	17	19	0	9	0	2	0	9
Orlando West	50	33	4	10	0	6	0	0	0	8
Orlando Shanties	21	114	3	24	0	4	0	5	0	11
Pimville	41	5	22	2	0	4	0	4	19	35
Moroka	14	199	5	43	0	23	0	6	0	17
Jabavu	47	?	5	?	1	?	1	?	0	?
TOTALS	191	455	56	98	1	46	1	17	19	80

Source: Report to Town Clerk: Illegal Trading in Native Locations, Annexure II, January 1948, JCHD

As early as 1947, the council was aware that the number of illegal traders operating in the burgeoning townships had reached a level where any attempt at 'eliminating illegal trading [would] inevitably engender resentment and organized demonstrations'.[6] Indeed, there had been a significant disturbance in the Moroka Emergency Camp in August 1947, caused by unhappiness over the allocation of trading licences. During a confrontation with the authorities three constables had been killed. Nevertheless, it was felt that 'the evil [of street hawking] must be suppressed in the interests of health and law and order'.[7]

Notwithstanding the dangers to public health, it was in fact the cumulative results of the policies of segregation and cheap labour that were beginning to overwhelm attempts to make those policies work in practice. The major single force, however, galvanizing the growing number of 'illegal' traders remained the woefully inadequate provision of 'legal' trading sites. As reported by the Manager for the

Non-European Affairs Department and the Medical Officer of Health:

> The consequences of the Council's inability to provide trading facilities in the new townships and to augment existing facilities in the older locations has had an immediate and unfortunate result, namely, the emergence, on a very large scale, of illegal trading.[8]

For example, the township of Moroka in 1949 sheltered a population[9] of 53,927 and yet had a mere eighty approved or 'legal' businesses, of which fifty-nine were general dealers and eleven were butchers (Table 9.3). Strikingly, there were only five milkshops for almost 54,000 people, and no fresh-produce dealers at all! The officially recommended numbers of different types of businesses for Moroka, as for other townships, was based on the ratio of one shop per thousand families (approximately 1 per 6,000 people). The exceptions were general dealer shops and butcheries, which were allocated on a ratio of 1 per 250 and 1 per 400 families respectively.[10] The situation for the townships existing in 1949 is shown in Table 9.3.

Although there were patent inadequacies in the shop allocation ratios, the local authorities continued to concern themselves with the letter of the law. More particularly, they worried over the absence of adequate health controls on the hawking of food items. As such, the focus of their attention centred upon the activities of the sellers of milk, meat, and offal, the fish-friers, and (to a somewhat lesser extent) the vendors of fresh fruit and vegetables. The meat and offal sold in the butcher 'shops' of the townships was usually obtained from the Johannesburg abattoir in the downtown area of the city. The hawking of offal, however, was done under less hygienic conditions and hawkers would be supplied with unwashed offal by a wholesaler. The dirty offal, consisting mainly of the heart, lungs, oesophagus, and intestines of slaughtered animals, would be washed and cut up in 'unauthorized wood-and-iron sheds' before distribution by the employees of the principal hawker, in wooden boxes or wicker baskets.[11] Not surprisingly perhaps, milk (with a high potential for contamination) was a special source of concern for the health authorities. The sale of fresh milk occurred either from recognized milkshops or through itinerant hawkers. Although the official approved milkshops were tolerated by the municipality they were nothing more

Table 9.3 Recommended (R) and existing (E) shops in the southwestern townships 1949

Township	Population	Milkshop		Fish-frier		Gen. dealer		Butcher		Fresh produce		Other	
		R	E	R	E	R	E	R	E	R	E	R	E
Orlando East	34,993	6	0	6	0	24	30	15	21	6	3	6	0
Moroka	63,135	12	5	12	5	48	59	30	11	12	0	12	0
Jabavu	37,400	5	0	5	0	20	0	13	0	5	0	5	0
Orlando West	40,998	7	0	7	0	28	61	17	7	7	2	7	0
Pimville	25,080	4	0	4	0	16	38	10	19	4	0	4	0
TOTALS	201,606	34	5	34	5	136	188	85	58	34	5	34	0

Source: Joint report by Manager, Non-European Affairs Department and Medical Officer of Health to the Public Health and Social Welfare Committee and Non-European Affairs Committee, 17 January 1950, Annexure 1, JCHD.

Report by Medical Officer of Health to Non-European Affairs Committee: Meeting: 29 September 1949, JCHD.

than wood-and-iron shacks and also apparently unhygienic.[12] While some milk was purchased from dairies in the municipal area of the 'White' city, a large percentage of the milk was obtained from uncontrolled producers on the outskirts of 'Black' townships where inadequate stabling and washing facilities were the norm. The 'unsafe' milk was then placed in cans or non-standard bottles and distributed by hawkers on pedal cycles.

In their endeavours to wrest some degree of control over the 'illegal' trading, the main line of action by the council took the form of arrests and prosecutions. The courts, however, as they were to continue to do, took the line that the accused were the victims of unreasonable circumstances. The local authorities were not satisfied, believing that the courts were not giving 'adequate support', for

> at the instigation of certain Attorneys [they take] the view that insufficient trading facilities are provided in the locations and in consequence, either dismiss the charge or impose a nominal penalty, which does not act as any deterrent.[13]

The Chief Magistrate had also indicated that on the general question of trading in the locations, he was very loath to impose severe penalties on any unlawful trader by virtue of the fact that, in his opinion, the council had not provided special and adequate trading facilities for the inhabitants.[14] Certainly there is evidence to suggest that the hawkers simply returned to their 'business' activity to be rearrested at some later date (see Table 9.4).

APARTHEID AND TOWNSHIP TRADING

In the 1950s, the procedures for screening became stricter, reflecting the tightening of the apartheid legislation affecting 'Black' areas. Prospective 'formal' traders had to satisfy a screening committee in respect of several major criteria: notably their ability to read and write and keep books and records in either Afrikaans or English; their previous experience in the type of business to be conducted; their soundness of character; and, most important, their sufficiency of capital resources.[15] Given the general educational and social deprivation that suffused the lives of

Table 9.4 Moroka: illegal traders and prosecutions 1947-50

Year	Number	Prosecutions	
1947	426		
1948	442	1948-9	280
1949	330		
1950	270	1947-50	610

Source: Extract from minutes of Non-European Affairs Committee, Illegal Trading: Moroka, 6 April 1950, pp.1-2, JCHD.
Report Medical Officer of Health to Non-European Affairs Department (NEA3/1949), Moroka Emergency Housing Scheme, 29 September 1949, p.5, JCHD.

most potential traders, it is scarcely surprising that many failed to qualify. For example, of 176 applicants for some trading sites available in 1950, only 30 (or 17 per cent) qualified for recommendation.[16] The officials of the local authority increasingly began to realize that some concessions had to be made in the townships to allow a greater number of people to become traders. One way of achieving such an objective was to increase[17] the number of 'open trading stalls' and 'market stalls', particularly for destitute women. It was known that 'a large number of hawkers were elderly women who were compelled either to support themselves or to supplement the family income'.[18] There is evidence that stalls had been made available to such persons during the war years of the 1940s. The stalls were usually located near schools and at points of entry into the township. Most of them (104) were positioned in Orlando East, with additional sites in Pimville (74), Orlando West (7), and Jabavu and Moroka (28).[19] The regulations governing the use of open stalls restricted traders to selling only fruit, cooked potatoes and mealies (corn on the cob), hard-boiled eggs, peanuts, and sugar cane. Each trader was required to have an approved storage box on wheels: in theory the goods would be sold directly from these barrows. Market stalls could be used for the sale of the same items, including fresh vegetables. With the sole exception that the council would provide tables or counters on which goods could be displayed, these stalls and the requirements for storage of goods were identical to those of the open stalls.[20] Thus, although the authorities were prepared to make some concessions towards the demand for additional approved trading points, they were not prepared to yield on what they regarded as critical public health issues. Storage requirements therefore continued to present a problem for the hawkers, who in many cases could not afford the barrow.[21] In addition to the above concessions, attempts were also made to accommodate offal sellers. The council was particularly concerned from a public health viewpoint that adequate facilities should be provided for the washing of the offal and the disposal of the dirty water. Consequently, it was only prepared to approve offal stalls that it had designed.[22] Yet in their desire to control hawking it was stated: 'we do not particularly want refrigeration in these stalls because that would increase the holding capacity in the very small quarters and would encourage hawking'.[23] The offal stalls, even without refrigeration, proved to be relatively expensive and too small,[24] and only a limited number were built. The potential offal traders who could not afford the stall rent simply continued to operate illegally.

The reference to impoverished women made in the previous paragraphy deserves a little more attention. Women hawkers have been one of the most common and persistent features of the petty producer scene over a long period of time. In part, their presence in this particular income-earning niche was related to the paucity of jobs for women in the formative years of the townships. It was also associated with a resistance towards domestic employment (Beavon and Rogerson 1986b) and, in the case of widows, with the need to earn enough money to pay the housing rent and thus to avoid the possibility of eviction and endorsement (back) to the Reserves. As stated by a Mr Masupha of Orlando:

> They are widows and have nothing except selling vetkoeks, they sell to these school children ... they are not sitting idle but trying to do something and if a privilege they have had for over twenty years is removed, it will create a hardship.[25]

The official attitude of relative tolerance towards these women hawkers, township 'aunties' or bo-Mkozi, is in part underscored by the lack of a meaningful social security system. As stated by the Medical Officer of Health in 1960:

> This Department has relaxed the general public health aspect of street trading to meet the genuine case of hardship where, for example, an elderly woman has no other source of livelihood, by agreeing to trading stalls being set aside for these persons where they can trade legitimately under reasonable control and with trading facilities provided.[26]

In any event, they are still a common feature in Soweto today, despite a hardening of official attitudes that became particularly marked in the 1960s, when the central government policy was more closely followed in the townships.

One other aspect of the street trade in food items deserves mention, namely the coffee carts from which a variety of cheap refreshments were offered for sale. The coffee carts remain the single most important category of street trading yet witnessed in the Johannesburg area. Yet despite the fact that most of the coffee-cart operators and owners were Sowetans, the carts were never concentrated in the townships. Almost from the outset of the trade, the coffee carts were located in the industrial areas of Johannesburg, where they were supported by the army of Black workers who daily arrived at work without having had a proper breakfast and before the limited number of native eating houses had opened for business (Rogerson 1983, 1986b). In the early days of the coffee cart boom, some 11 per cent of the 622 carts then known to be operating did so from positions outside railway stations and bus termini in Soweto. As the trade began to grow, and as the number of carts in the Johannesburg area increased, so both the percentage and the absolute number of carts in Soweto declined. By 1960, just prior to the final 'assault' on the carts in the last years of the 'coffee cart war' (Rogerson 1983; Beavon and Rogerson 1986a), only 3 per cent of the 1,407 carts operated at sites in Soweto.

Formal traders were precluded from dealing in soft goods and there is unfortunately little in the way of documentary evidence to gauge the extent to which hawking of soft goods took place. Yet there are clear indications that there was a sizeable trade and that it had direct links with soft goods businesses and wholesalers in the city of Johannesburg. Newspapers at the present time have columns of adverts in their classified small ads that offer special prices to hawkers on a variety of items from soaps, cosmetics, and jewellery, through patent medicines to clothing in particular. This tradition of soft goods firms acting as wholesalers to hawkers can be traced back through Black newspapers and periodicals to the 1960s and 1950s. The popular magazine Drum in both the 1950s and 1960s carried illustrative display advertisements directed at hawkers, offering such varied stocks as framed pictures of 'Queen Elizabeth II, African Chiefs, Religious Pictures, Medicines, Pills, Hair Oils, Soaps, and Face Cream', all at wholesale prices and at the same time informing hawkers that these were 'fast-selling, money-making lines' (Drum, July 1962, February 1955). The council was opposed to the hawking of soft goods in general, largely on the grounds that it was opposed to 'illegal' street trading. Yet, once again, it was perceived health dangers that provided the basis for opposition to the sale of clothing in the townships.

> In the case of clothing, the hawker or pedlar usually carries his wares about in a suitcase. The clothes are examined and handled and are often tried on by prospective purchasers in their homes or flats. If unsatisfactory, the article is returned to the hawker who moves to the next prospective customer where the process is repeated. At the end of the day's work the soft goods are usually taken home by the hawker and stored at his home. In the case of Non-Europeans particularly, this usually means that the goods are stored in the bedroom, often under the most unclean and objectionable conditions.[27]

The stance of the council is interesting, particularly as it was not until 1976 that laundering or dry-cleaning establishments or their depots were officially allowed to operate in Soweto.

By the late 1950s, there were already indications that stricter controls were going to be applied to both formal and 'informal' traders and hawkers in the townships. The Minister in charge of Bantu Administration and Development, the body that was ultimately in charge of all townships in South Africa, made his intentions clear in October 1959. He stated:

> I am considering the practical application of a policy to make it clear to these traders (Africans in the urban areas) that their trading facilities are temporary; that

> they must carry on their businesses in their home areas; and that the condition on which they are allowed to stay is that they are there only to build up capital and gain experience, for which the opportunities in their home areas are extremely slight.
>
> (Star, 24 October 1959)

The Minister is reported to have intimated that the petty traders in the townships were a potential threat to White businessmen. He argued that as Black traders began to accumulate capital, they would start selling commodities such as radios, stoves, and clothing, and they would offer competition to White-owned businesses near the 'locations'. The essence of the statement made by the Minister eventually became official policy and was included in a document[28] circulated to all local authorities charged with township affairs. In terms of what became known as the One-Man-One-Business policy,[29] Black traders were henceforth limited to only one business, and were precluded from forming companies and partnerships with the object of combining their resources in order to embark upon larger business ventures. Opportunities for economies of scale with the possibilities of reduced prices in the townships were thereby denied. It was categorically stated in the policy document (original emphasis here retained) that:

> Moneyed Bantu and Bantu companies and partnerships ought to establish themselves in the Bantu homelands where they can invest their capital on a permanent basis to the advantage of their people and own homeland.

In conclusion it was said:

> It will therefore be appreciated if moneyed Bantu, Bantu traders, and other trained Bantu who are able to contribute to the development of the Bantu homelands, could be traced and where necessary encouraged and assisted by the local authorities to settle and establish businesses and other enterprises in such homelands.[30]

Furthermore, it became clear that trading licences would only be granted to persons operating as grocers, butchers, fish-friers, greengrocers, eating-house owners, dairy owners, or wood and coal dealers. Consequently, anyone wishing to provide any other commodities or services for the Black community resident in the townships would have to be an 'illegal' hawker. The controls were aimed at winkling out those persons whose services were not essential for the benefit of the White population, and who could therefore be required to return or be relocated to a Bantustan or 'Homeland'.[31] As success in a formal business venture was now likely to make a person a candidate for removal, it must be assumed that it acted as a deterrent to some persons who might otherwise have taken out a licence. It is possible that the threat of being 'watched', if one was a 'legal' business person, could have driven more people into the grey world of 'illegal' petty production. The policy was advocated despite the fact that with the passage of time the ratios of number of traders (both legal and identifiable as 'illegal') to population had altered little in the fourteen years since its inception. To concerned people, there was still an inadequate supply of businesses, let alone businesses that operated from suitable premises offering services and commodities that were really in demand. The ratios that pertained in the early 1960s have been computed (making due allowance for the absence of some data) and are set out in Table 9.5 alongside the ratios for 1948. At first glance, the 1962 ratios may appear to be very reasonable. The 1962 survey, however, was a gross survey undertaken during a period of 'explosive growth', when the council was 'unable to erect buildings fast enough'[32] and includes businesses of both the approved and 'illegal' categories. On that basis, there seems to have been little improvement on the situation which had existed fourteen years before. Consequently the ratios, with the possible exception of that for fruit and vegetable shops, appear to be undesirably high. Clearly the concept that the Black population was composed merely of temporary sojourners, who as such did not warrant a full spectrum of businesses to serve them in the townships, was persisting.

As the controls on Black business activities in the townships became more closely policed, they spawned increasing resentment between the licensed, formal traders and the hawkers. The resentment had been simmering for years.[33] Part of the basis for resentment on the side of the 'legal' traders stemmed from a combination of their small inadequate premises and the need for them to pay rent and adhere to shop-hour regulations and health controls, all of which pushed up their overhead expenses. The chairman of

Table 9.5 Shop to population ratios for selected food shops in the southwestern townships 1948 and 1962

Recommended 1948		Actual 1962
Meat	1 : 2,400	1 : 1,743
Milk	1 : 6,000	1 : 6,694
Fruit and veg.	1 : 6,000	1 : 2,198
Groceries	1 : 1,500	1 : 1,150

the African Chamber of Commerce complained:

> Pavement salesmen, dealing in every conceivable commodity from candles and patent medicines to clothing, operate freely. They have no licences, and because they have no overheads they can undersell us.
> (Rand Daily Mail, 17 September 1962)

The average size of a Soweto shop was less than 40 square metres (Morris 1980), and certainly in the 1960s few shops had access to electricity. Consequently, shopkeepers were unable to attract customers through the normal attributes of a bright, attractive, and clean shop, nor were many of the food stores able to make use of refrigerated counter units. The cameo below illustrates, albeit in an amusing style, the tragedy of the trading scene of Soweto in the 1960s.

> Cattle by the hundred are being bought by the illegal butchers, slaughtered in their homes, then cut up and sold in the township streets from bicycle carriers. A Meadowlands butcher told me ... 'Customers of mine shout into my shop: "Your meat will go bad - we buy fresh meat from the bicycle boys".'
> (Star, 15 February 1960)

Packed tightly into the above quotation is a wealth of unwritten detail that resonates the feelings of despair, desperation, and hopelessness. It is only with effort that the mind of the outsider can comprehend the circumstances of desperate poverty that compel people to use their dwelling place as a slaugher-house; can share the despair of neighbours who find themselves, of necessity, 'living' [sic] amid such environs; or can appreciate the hopelessness of the 'formal' butcher trying to sell meat from a dingy small establishment lacking refrigeration and unable to compete against the fresh meat on offer from the 'bicycle boys'. Such a scenario is a microcosm of the conditions of temporary trading for temporary people forged by the apartheid planners.

CONCLUSION

The task in this paper was to unravel some of the ramifications of South Africa's unique contribution to the literature on urbanization: the notion of 'temporary sojourners'. More especially, the nexus of formal and informal trading networks as they evolved in the formative years of the settlement of Soweto constituted the essential axis of discussion. The genesis of hawking in the urban townships of South Africa was shown as inexorably wedded to the underdevelopment of formal trading facilities in these areas. In turn, the woeful condition of formal trading was a reflection initially of the manifold problems experienced in creating circumstances for 'normal trading' during a phase of very rapid expansion. The continuing inadequacies of the formal township system must be laid squarely at the doors of apartheid planners, who during the 1950s sought to fashion an urban environment in which its residents were still regarded as temporary sojourners, present only as long as they ministered unto the needs of Whites. The prevailing ideology behind the forging of township life in the 1950s and 1960s precluded the expansion of formal 'Black' trading enterprises, creating, as a consequence, a wide open niche for the operations of casual hawkers. That Black South Africans responded to the hostile economic conditions which they confronted in the townships belies the racist myth of their lack of business acumen. Indeed, in the final analysis, it is perhaps highly significant that the vernacular term that stylizes the townships of Soweto, Vukenzenzele translates into English as 'Get up and do it yourself'.

NOTES

Documentary material in the form of notes, reports, minutes of meetings, and other records were housed in the record store of the Johannesburg City Health Department. The records store has now been closed to the public and is in the process of being transferred to a temporary archive in Johannesburg, prior to being sent to the State Archives in Pretoria. All references to these materials are identified by the abbreviation JCHD. The reference codes are those that were applicable before the material was prepared for dispatch from the records office of the Medical Officer of Health to the closed temporary archive.

1. Letter C. Wilson to Town Clerk, with memorandum re conditions in Orlando Township, 30 November 1939, p.6, A8624, Box 3, Orlando, JCHD.
2. Letter C. Wilson to Town Clerk, with memorandum re conditions in Orlando Township, 30 November 1939, p.5, A8624, Box 3, Orlando, JCHD.
3. Extract from Report of Works Committee, Native Hawkers of fruit, communication from Messrs Helman and Michel, 29 September 1947, p.1, Box 62, File 3261, Offal Hawkers in Native Townships, JCHD.
4. Trading Native Townships, Annexure 1, 30 June 1948, Schedule A, File 8/3/2, Tea and Coffee Vendors, Vol.1, JCHD.
5. Report from Manager Non-European Affairs Department to Town Clerk, Illegal Trading in Native Locations, Annexure II, January 1948, p.1, JCHD.
6. Letter Manager Non-European Affairs Department to Medical Officer of Health re Illegal Trading in Locations, with Annexure, 20 November 1947, p.1, D1958, JCHD.
7. Ibid.
8. Joint Report by Manager Non-European Affairs Department and Medical Officer of Health to the Public Health and Social Welfare Committee and Non-European Affairs Committee, with Annexures, 17 January 1950, p.1, JCHD.
9. Report by Medical Officer of Health (NEA3/1949) re Moroka Emergency Housing Scheme, 29 September 1949, p. 1, JCHD.
10. Joint Report by Manager Non-European Affairs Department and Medical Officer of Health to the Public Health and Social Welfare Committee and Non-European

Affairs Committee, Annexure 1, 17 January 1950, JCDH.

11. Trading Native Townships, Annexure 1, 30 June 1948, pp.2-3, File 8/3/2, Tea and Coffee Vendors, Vol.1, JCHD.

12. Trading Native Townships, Annexure 1, 30 June 1948, Schedule B, File 8/3/2, Tea and Coffee Vendors, Vol.1, JCHD.

13. Report from Manager Non-European Affairs Department to Town Clerk, Illegal Trading in Native Locations, Annexure II, January 1948, p.2, JCHD.

14. Minutes of Meeting re Trading on the Perimeter of Moroka and Jabavu Native Townships, 4 December 1951, File 12/17/9, JCHD.

15. Joint Report by Manager Non-European Affairs Department and Medical Officer of Health to the Public Health and Social Welfare Committee and Non-European Affairs Committee, with Annexures, 17 January 1950, p.4, JCHD.

16. Memorandum Assistant Manager Non-European Affairs Department to Manager, Lease of Sites in Locations and Privately Owned Shops, 6 September 1950, p.2, File 12/17/9, JCHD.

17. Minutes of Meeting between Manager Non-European Affairs Department and Medical Officer of Health re Proposed Amendments to the Native Location Regulations, 20 March 1951, pp.1-2, File 12/17/9, JCHD.

18. Ibid.

19. Trading Native Townships, Annexure 1, 30 June 1948, p.4, File 8/3/2, Tea and Coffee Vendors, Vol.1, JCHD.

20. Letter from W. Carr to City Medical Officer of Health re Storage Facilities: Native Location, 15 December 1953, File 12/17/9, JCHD.

21. Minutes of Meeting between Manager Non-European Affairs Department and Medical Officer of Health re Proposed Amendments to the Native Location Regulations, 20 March 1951, p.3, File 12/17/9, JCHD.

22. Minutes of Meeting Non-European Affairs Department, 13 March 1952, p.1, File 12/17/9, JCHD.

23. Letter Deputy Medical Officer of Health to Medical Officer of Health re Offal Stalls: Native Locations, 28 October 1963, File 12/17/9, JCHD.

24. Minutes of Part of Special Meeting of Moroka Advisory Board with Representatives of City Health and Non-European Affairs Department, 20 March 1952, p.1, File 12/17/9, JCHD.

25. Notes on Meeting re Street Traders in Orlando, City Health Department, 18 August 1953, p.1, File 12/17/9, JCHD.

26. Letter Medical Officer of Health to Manager Non-European Affairs Department re Female Hawkers (Native Areas), 13 June 1960, p.1, File 12/17/9, Licences, Trading in Native Townships, Vol.2, JCHD.

27. Draft Report to Licensing Committee re Hawking of Soft Goods, undated, p.1, Box 62, Offal Hawkers in Native Townships, File 3261, JCHD.

28. Department of Bantu Administration and Development, Circular Minute No. A.12/1-A8/1, 14 February 1963.

29. Extract from the Report of Non-European Affairs to Management Committee Meeting, Trading: One Man One Business Policy, 17 September 1963, p. 1, File 8/3/2, Vol. 15, JCHD.

30. Department of Bantu Administration and Development, Circular Minute No. A.12/1-A8/1, 14 February 1963, pp. 3, 5.

31. Ibid.

32. Extract from Report of Non-European Affairs Committee Meeting, Allocation of Trading Rights to Bantu and Provision of Cinema Facilities in Urban Residential Areas, 6 June 1963, p.1, File 8/3/2 Vol. 15, JCHD.

33. Letter from Godfrey Trevor and Trevor on behalf of Emphunzeni Cash Bazar Orlando to Location Superintendent Orlando East, 20 November 1958, complaining about women hawkers, File 12/17/9, JCHD.

ACKNOWLEDGEMENTS

The authors wish to thank Philip Stickler for preparing the illustrations for the paper. The Human Sciences Research Council in Pretoria and the University of the Witwatersrand, Johannesburg provided financial assistance towards the costs of attending the conference of the Working Group held in Madrid.

Chapter Ten

Consumerism, the state, and the informal sector: shebeens in South Africa's Black townships

Chris Rogerson

INTRODUCTION

Scholarly observers and policy analysts of development and urbanization processes in the global periphery and semi-periphery have acknowledged only recently the crucial role assumed by the sphere of consumption (Armstrong and McGee 1985). The appearance of a copious volume of literature and the continuing fascination with issues of production within, what Walton styles, 'the third "new" international division of labour' (Walton 1985b: 3), have failed to be matched by rigorous analysis or debate on a host of parallel issues surrounding consumption and economic development. Curiously, this neglect of matters pertaining to consumption seemingly is shared by disciples of both liberal and radical paradigms of analysis in development studies (McGee 1984). Indeed, critical theoretical explorations on the role of consumption in peripheral capitalism have been pursued by a group of Latin American scholars associated with Raul Prebisch (see Prebisch 1976, 1978, 1981; Filgueira 1981; Rodriguez 1981; Tokman 1982).

The academic silence on the role of consumption in peripheral economies, has been breached, however, by a small handful of empirical investigations. During the early 1970s, Barnet and Muller deftly dissected the influence of multinational corporations in transferring consumption ideology to the underdeveloped world, arguing that 'stimulating consumption in low-income countries and accommodating local tastes to globally distributed products is crucial to the development of an ever-expanding Global Shopping Centre' (Barnet and Muller 1974: 173). Undergirding the promotion of imitative consumerism in peripheral economies is the enormous power of advertising which attempts to change cultural habits towards the adoption of 'modern' ways of living, encouraging a preference for industrial manufactured goods over traditional or indigenous commodities (Jouet 1984). Beginning in the 1960s, mass advertising in Brazil was linked to the greater availability of hire purchase credit in an effort to incorporate the working classes into the consumption patterns of modern transnational capitalism (Wells 1983). In the Philippines, efforts mounted by multinational corporations to mould local consumption patterns have been so successful with respect to tastes in food, drink, clothing, and beauty that it could be observed 'foreigners practically dictate to the ordinary Filipino what to use every hour of the day' (Lagboa and Pe 1981: 92).

The developmental ramifications of product taste-transfer from core to peripheral economies are teased forth in an important series of works on Kenya contributed by Steven Langdon (1974, 1975). In the experience of the local soap industry, the impacts of an inappropriate taste-transfer were to establish consumption patterns very difficult for indigenous enterprises to meet, except by dependence on government aid or by manufacturing multinational-like branded standardized products (Godfrey and Langdon 1976). In research also undertaken on Kenya, Kaplinsky reveals how organized attempts by manufacturers of breakfast cereals to influence consumer patterns combined together with the demonstration effects of expatriates' consumption styles to generate 'demand structures which have facilitated the introduction of inappropriate products, and hence of inappropriate techniques' (Kaplinsky 1979: 90). The crucial interplay between, on the one hand, proletarianization processes and on the other hand, the creation of consumption needs for commodities only purchasable with increased monetary incomes is shown by Terry McGee (1983) in the contemporary case of women workers in Malaysia's export-processing zones. Nevertheless, the connection of creating new 'felt needs' and the stabilization of workforces was recognized in Southern Africa almost a century ago by mining capital which deployed the commoditization of beer, guns, and even sex in order to extend the periods of its Black migrant labour force on the mines (van Onselen 1976).

The common threads which bind together these seemingly disparate writings are concern for the spread of consumerism, the implications of changing consumer tastes, and the world-wide homogenization of consumer preferences. Behind the international diffusion of 'consumerism', defined as 'the mass participation in the values of the mass-industrial market' (Ewen 1976: 54), have been major technological advances in transportation, facilitating the easier movement both of commodities and of labour, and an explosive growth in information flows through television, the press, and rising levels of education and literacy (Armstrong and McGee 1985). A pivotal role in the global dissemination of new life-styles and consumption patterns is accorded to the expanding 'middle classes' in peripheral economies (McGee 1984). In short, 'as well as adopting patterns of behaviour similar to those of the middle classes in the developed countries, these life-styles are also propagated towards the lower strata' (Filgueira 1981: 72). The result of this international diffusion of consumption styles is a growing degree of convergence in the consumption patterns, especially of the urban elite and middle-class groups in peripheral economies with those enjoyed in core or advanced capitalist nations (McGee 1984; Armstrong and McGee 1985). Osvaldo Sunkel maintains that this convergence in life-styles and consumption patterns is forging a 'transnational community', integrated at a world-wide level in spite of the fact that its members live in geographically separate territories (Sunkel 1973; Sunkel and Fuenzalida 1977).

It is the intention in this chapter to draw together strands from the nascent literature on consumerism and that now well-established 'growth industry of studies' (Moser 1984: 135) concerning the informal sector (see Portes 1983; Moser 1984; Rogerson 1985 for reviews). More particularly, it will be suggested that consumer diffusion and the convergence of consumption patterns may facilitate conditions for restructuring and ultimately the survival of spheres of petty production. The setting for this investigation is the semi-peripheral economy of South Africa, where currently the state is reversing its former policy of repressing informal sector activities (Rogerson and Beavon 1980, 1985; Rogerson 1983, 1986a; Beavon and Rogerson 1986a) and is embarking instead on programmes for 'developing' this part of the urban economy as an integral component of a 'reformist' package of measures vainly seeking to furnish apartheid South Africa with a more human face (Rogerson and Beavon 1982; Wellings and Sutcliffe 1984; Rogerson and da Silva 1986; Rogerson 1986b, 1987). Specifically, the study concentrates upon the shifting pressures which have triggered moves towards the formalization of shebeens, one of the most common and long-established elements of the informal economy in urban Black townships (Rogerson 1986c; Rogerson and Hart 1986). Shebeens may be likened to the American speakeasy of prohibition days or to the British 'pub' (see Sampson 1957; Harrison 1973; Lotter and Schmidt 1974, 1975; Bailey 1978; Lotter 1981; M. A. Smith 1983; Bramwell 1984) not only as outlets for the sale and consumption of liquor, but also as vital social and cultural institutions: focal points for popular recreation and culture in the drab ghetto environs of townships (Laubscher 1977; Kies 1982; Hart and Rogerson 1985).

The chapter unfolds through two major sections of discussion. First, the relationships between the spread of consumerism, changing patterns of liquor consumption among South African urban Blacks, and the rise of pressures supporting the survival of the shebeen will be explored. It is argued that shebeens moved to a position at the cutting edge of a taste-transfer process in the alcohol preferences of Black consumers, which propelled the shebeen into the wider nexus of the distribution network of the leading South African liquor concerns. The focus turns in the second section to address the issues surrounding the state's shifting policy post-1976 towards the distribution of liquor in Black townships. More especially, attention narrows to the setting of Soweto, South Africa's largest urban Black centre, examining the directions of contemporary struggles over the 'spoils' of formalization. The choice of Soweto for detailed focus on the liquor trade is particularly appropriate because it is clear that the state is using Johannesburg's 'unwanted city' (Kane-Berman 1978: 56) as a 'guinea-pig' or proving ground for evolving procedures to deal nationally with the question of shebeens in South Africa's Black townships (Rogerson 1987).

THE CHANGING NATURE OF SHEEBEENS - PROHIBITION AND BEYOND

The shifting trajectory of the township shebeen is inseparable from the historical vacillations in legislation concerning the supply of liquor to South Africa's Black populace. Between 1890 and 1962 this legislation turned a full cycle, from full availability through total prohibition to full availability once more (Stein 1981; Scharf 1984, 1985). The imposition of prohibition was rationalized by the view that Whites were protecting 'uncivilized natives' from the dangers of 'moral destruction' associated with alcohol, and thereby assisting them in 'advancing' towards civilization (Laubscher 1977; Stein 1981). Further justification was found in the racist beliefs that Blacks were unable to exert self-control on their alcohol intake and that excessive consumption inevitably precipitated a 'craving for the flesh' of White women (South Africa 1913). More significant, however, was the argument of White employers that free availability of liquor was detrimental to labour productivity and hence to profits (Stein 1981). Indeed, it was at the behest of mining capital that a total prohibition on Black access to alcohol was enacted during 1897 in the Transvaal (van Onselen 1982).

With two notable exceptions, prohibition on Black access to alcohol remained in force until as late as 15 August 1962. The first exception was the decision taken to provide Blacks with access to municipally brewed traditional kaffir or sorghum beer in Natal from 1908 and in the urban areas of the Transvaal from the late 1930s (Rogerson 1986c). The supply of this municipally produced beer took place through the network of municipal beer halls, the profits of which were to be a major source for financing improvements in township services (La Hausse 1984). The second exception to prohibition was a system of granting special permits to select 'Natives' deemed as having reached an acceptable level of civilization adjudged by White eyes. Such permits, which were renewable annually, allowed the limited purchase of liquor intended for home consumption. Nevertheless, a savage indictment on the operation of this iniquitous racially tiered system was furnished by the statistic that, at the close of prohibition, only 50,000 of a total Black populace of 9 million had been granted such permits for liquor purchase (South Africa 1960: 7).

Shebeens and prohibition

Denied direct access to the formal outlets of liquor supplies, many Black drinkers turned to supplies offered illicitly in the township shebeen in preference both to the ignominy of the permit system and to the inhospitable environs of the hated beer halls, whose appearance one observer likened to 'an overgrown lavatory' (Blumberg 1959: 9). The years after the Second World War witnessed an enormous expansion in the township shebeen trade and a shift in emphasis away from the former production of home-brewed beers and liquor 'concoctions' (see Koch 1983a, b; Beavon and Rogerson 1986b; Rogerson 1986c; Rogerson and Hart 1986) towards the selling of the so-called forms of 'White liquor', which remained unavailable at beer halls. This metamorphosis in the trade was so complete that by 1960 it was estimated that shebeens were responsible for an extraordinary 60 per cent of all liquor supplies to Blacks (South Africa 1960: 3).

What is critical to appreciate is that, by the close of the prohibition years, the 'informal sector' shebeen trade was geared increasingly to markets provided by the expanding 'middle classes' of the townships. Indeed, during the immediate post-Second World War years, most of the preconditions were fulfilled for the creation in South Africa's Black townships of the beginnings of a 'mass market' and the corresponding spread of consumerism (see McGee 1984). Not only did there occur a massive concentration of wage-earners in urban areas (Smit 1985) but also a vast expansion in a new class of township African, based outside of industrial employment, engaged in commerce, education, health, or state employment (Wolpe 1977). The group of teachers, nurses, social workers, doctors, clergy, and so forth enjoyed growing disposable incomes and reached towards ever-higher levels of both education and literacy. The emergence of this 'new petite bourgeoisie' (Wolpe 1977) furnished an ideological environment in urban Black townships wherein advertising, for the first time, could utilize status as a vehicle for creating a demand for mass-produced goods. If, in 1946, it could be observed that:

> the South African businessman has not been sufficiently imaginative to educate the very large African market into feeling the need for his products. He has shown little readiness to create in the African the urge to know and the desire to have the things that make life more pleasant[,]
>
> (Ngubane 1946: 139)

a decade later it was evident that a major transformation had occurred. The 1950s were major growth years in the South African advertising industry, with increasing efforts being devoted to the specific capture of mass markets offered by township consumers (Anon 1956). Advertisers sought to persuade the new middle classes of the townships of the benefits and 'wonders' of the products of the modern economy (see Fig. 10.1).

Despite the fact that during prohibition years advertising did not extend to the promotion of liquor products, there surfaced a growing status differentiation between shebeens and beer halls (Themba 1961). In particular, an enormous gulf opened between the drinking of 'White man's liquor' within the club atmosphere of shebeens, endowed with exotic names such as the 'Back of the Moon', 'The Thirty-Nine Steps', 'Falling Leaves', or 'The White House', and the quaffing of kaffir beer in the sterile social environment of municipal beer halls (Themba 1961; Hart and Rogerson 1985). One of the best-known Black creative writers, Nat Nakasa, draws forth the contrast, appreciating shebeens as 'hospitable homes...not like the municipal bar lounges with their business atmosphere and inevitable high fences which gives them the look of cages' (Nakasa 1975: 57). It is striking that surveys undertaken of the characteristics of Johannesburg shebeens in the early 1960s reveal these status differentials (Ntshangase 1961; Themba 1961). One survey recorded:

> Shebeens are patronised largely by the middle-income group: professional people like teachers, nurses, doctors, or social workers avail themselves of the facilities offered by shebeens.... People of the lower income group are rarely found in shebeens; their place is in the beer halls... The majority of shebeen frequenters do not care for beer halls - they say they do not go to beer halls because they do not drink 'mud' - beer halls are for low people.
>
> (Ntshangase 1961: 1, 6)

Beyond prohibition

The decade following the revocation of prohibition was a dramatic phase in the evolution of the shebeen trade. It was a period which the state felt would spell the death-knell for shebeens, as township residents were to be lured away by the opening of new, municipal-run outlets for liquor sales in townships, paralleling the municipal-run beer halls (Stein

Figure 10.1 The image of black consumerism in the 1960s

1981). None the less, shebeens thrived in the face of the hopeless inadequacies of the alternative drinking attractions afforded by beer halls. In terms of supplying the mass of Black consumers, the geography of township shebeens confers considerable advantages relative to that of the municipal beer halls. Shebeens are widespread in most urban Black townships of South Africa, with the majority of their customers drawn from within a ten minutes' travelling radius from home (Motumi 1979; Lotter 1981; Rogerson and Beavon 1982; Rogerson and Hart 1986). The ready accessibility of shebeens contrasts with the siting of beer halls, typically either close to railway stations (from which residents still have some distance to travel before reaching the safety of home) or in the environs of the ugly, single-sex, migrant worker hostels that scar the residential landscape of townships (Motumi 1979; Pirie and da Silva 1986).

Consequently, the locational patterns of shebeens and beer halls strengthened the growing status differentiation between the two drinking institutions, beer halls becoming increasingly reliant upon the patronage furnished by the lower-income residents of worker hostels. Other considerations, however, further reinforced the attractions of shebeens for the nascent township middle class. Unlike shebeens, beer halls are governed by rigid hours of opening, the absence of credit facilities, and a regimented rather than a convivial atmosphere for drinking. Finally, because of their ownership by local authorities, municipal beer halls became identified, alongside the new township bottle stores, as symbols of apartheid oppression (Motumi 1979).

But the relative prosperity of shebeens was assured ultimately by two further, albeit interwoven, factors. First was the continued expansion of a mass consumer market in the townships, led by the growth of a middle class. South Africa's trajectory of capitalist development in the 1960s and 1970s resulted in a considerable blurring of the former 'racial division of labour' (Davies 1979), with the penetration of Black workers into a host of new skilled manual and supervisory categories of work. Equally important, however, was the opening of new employment avenues in the mushrooming tertiary sector (Wolpe 1977; Davies 1979; Crankshaw 1986). A consequence of the progressive march of Blacks into new economic spaces was an expansion not only in the membership of the Black middle class, but also in their relative significance (Crankshaw 1986: 23). Because of the concentration of new tertiary income opportunities within urban areas, increasingly the mass market was concentrated geographically in the major Black urban townships (cf. Smit 1985). Wedded to this growth in the mass market was the enormous surge of advertising to Black consumers, seeking to advance the cause of consumerism (Financial Mail, 17 April 1970). In particular, the overwhelming message of advertisers was the superiority of 'modern' over traditional ways and life-styles, and the superiority of status attached to switching to new forms of food, personal care, and drinks.

Beginning in 1962, campaigns were launched for the first time by the leading brewing companies, distillers, and wine producers, seeking to capture the tastes of Black consumers. The prime thrust was to engineer a taste transfer towards the acceptance of malt rather than traditional sorghum beers and in favour of greater consumption of wines; a sphere of perennial over-production in the South African liquor industry and a major force behind the lifting of prohibition legislation (Bunting 1958; Stein 1981; Scharf 1985). While malt beers had been available in shebeens to Black consumers, they had never been a major constituent of the trade prior to the close of prohibition because of their general bulkiness and the higher profits to be made from selling hard liquor (Rogerson 1986d). The new advertising campaigns for lager beers emphasized the status attached to the drinking of 'European' as opposed to traditional 'Bantu' beer, always portraying a western way of life with successful and wealthy businessmen or happy middle-class families living in comfortable surrounds (Fig. 10.2). Beyond advertising per se, the aggressive marketing campaigns of the brewing and wine companies were boosted by a network of informal linkages to township shebeens, in order to promote the sale of particular products. The reinforcement of such linkages tied the profits of the capitalist enterprises ever more closely to the fortunes of the informal sector shebeeners, who were effectively absorbed as 'disguised' sellers of these large business enterprises (Rogerson and Beavon 1982; Beavon and Rogerson 1986b).

By the end of the 1960s, the efforts of the large brewing companies to penetrate the Black consumer market were manifest in dramatic downturns in sorghum beer sales and in findings by market researchers that sorghum beer was only preferred amongst the elderly, the less educated, and

Figure 10.2 Advertising malt beers to Black consumers 1965

those in the lowest income groups (Rogerson 1986d). Essentially, what occurred was the successful remoulding of Black consumption patterns for alcohol from a phase of home brews and liquor concoctions, through the monopoly of municipal beer, to a preference by the early 1970s for the formerly forbidden fruit, 'White man's liquor'. In terms of engineering this 'taste-transfer' process, township shebeens played a formative role in becoming de facto integral parts of the distribution chain of South Africa's powerful liquor concerns. By 1970 the major liquor corporations could hide no longer the vital significance of the shebeen as a marketing channel (Scharf 1985: 56). Accordingly, the activities of 'promotional teams' were stepped up in order to 'advise' and 'inform' shebeeners of their products (see Star, 27 January 1971). Covert assistance granted to the shebeens by South Africa's leading liquor enterprises became translated into overt support in the wake of events set in motion by the 1976 township uprisings.

TOWARDS THE FORMALIZATION OF SHEBEENS

It is highly ironic that the beginnings of a process leading to the formalization of shebeens should be catalyzed by the mass rioting and destruction that swept South Africa's Black townships during 1976 and 1977. The ambiguous position of alcohol in the lives and culture of working class Blacks was graphically underlined by the destruction in 1976 of hated apartheid symbols of oppression, such as beer halls and township bottle stores, and by the enforced closure of many shebeens by militant students and workers (Hirson 1981: 264-5; Scharf 1984). Students viewed shebeens as reactionary institutions 'serving to depress rather than heighten social consciousness' (La Hausse 1984: 314). Illustratively, in a statement issued in November 1976 by the Soweto Students' Representative Council urging that shebeens should cease operations, it was claimed that they constituted:

> a cause of unhappiness in the black man's life.... A number of lives have been lost because of the operation of these shebeens. Salaries have not reached homes because they were first opened in shebeens. Futures have been wrecked by the operation of these shebeens.... Nothing good has ever come out of them.

> Many of our fathers and brothers have been killed in or out of them. Hundreds of our colleagues have become delinquents, beggars, or orphans as shebeen kings and queens have become capitalists. We can no longer tolerate seeing our fathers' pay-packets emptied in shebeens.
>
> (cited in Kane-Berman 1978: 20)

With the diffusion of rioting to the townships of the Western Cape, once again shebeens were subject to attack (Scharf 1985). More than one hundred shebeens in the townships of the Cape Peninsula reportedly were destroyed in October 1976 by students, who additionally established pickets at bus stops to search adults bringing their liquor home (Kane-Berman 1978). Quite clearly, in the perceptions of the militants of 1976, the spread of consumerism and the existence of the shebeen as informal sector niche were a deflection of the Black struggle and divisive of Black solidarity (Scharf 1985: 57).

From shebeen to tavern

The immediate priority of the state after 1976 was to rebuild, as rapidly as possible, the hundreds of beer halls and township bottle stores which had been destroyed during the unrest. Design innovations were prompted for new beer halls, viz.,

> The walls and roofs will be made of fire-proof reinforced concrete; the buildings will have no windows, but will be fully air-conditioned for the comfort of the liquor dealers and their patrons; bullet-proof glass will separate cashiers from customers; steel security doors, and direct radio links to the cops will also be installed.
>
> (Financial Mail, 29 April 1977)

Although economic imperatives underpinned the programme of reconstruction, there was a growing realization that liquor profits could not be relied upon as a basis for future township revenues (Stein 1981). Moreover, the state could no longer afford to have its liquor outlets seen as part of the instruments of domination, nor did it wish to continue to endure financial losses which the township violence was exacting (Scharf 1984, 1985). Accordingly, from mid-1977 there began the initial faltering steps away from state monopolization of the township liquor trade and hints that suitably controlled shebeens might be legalized. The new strategy of legalizing shebeens is inextricably a part of the broader state drive to cultivate the support of the township Black middle class, a group committed to the principles of free enterprise and the ethos of consumerism (Hudson and Sarakinsky 1986).

The initial moves for transforming the illegal shebeen into a licensed tavern occurred with the enactment in 1977 of new liquor legislation, which facilitated the entry of Black entrepreneurs into the township liquor trade. Further, a major inquiry was mounted between 1977-80 on the whole question of liquor consumption and distribution in Black residential areas, with a strong focus on the question of the possible formalization of shebeens. It was during this period that the country's largest liquor concerns surfaced openly in their support for legalizing the informal shebeen trade. The stakes of the country's leading liquor companies in the shebeen trade were now enormous; for example, shebeens accounted in 1979 for R150 million, or one-third of South African Breweries Ltd national beer sales (Financial Mail, 2 November 1979). With the continuation of police-raiding on shebeens and the seizure of vehicles, goods, and refrigerators, it was manifestly in the interests of the liquor companies to pressurize the state in favour of formalization. In January 1980, South African Breweries made an urgent appeal to the state for the unconditional licensing of shebeens to neutralize police depredations (Ranaka 1980: 1). Nevertheless, the state relented only in May 1980, with the announcement that it would agree to a process of formalizing shebeens, thereby transforming them into legal township taverns (Rogerson and Hart 1986).

Struggles over the spoils

The road to 'formality' was not, however, an easy one for shebeeners to traverse. The issuance of licences to trade as legal taverns was couched in a complex mesh of regulations and controls. In order to comply with the basic requirements for licensing, township residential sites would have to be rezoned for business purposes and a dividing wall constructed to separate the tavern area from the rest of the dwelling (Simpson 1984). Further regulations specified, inter alia, a separate kitchen with lockable serving hatch, suitable storerooms for both full and empty liquor containers, separate toilets for males and females to those used by the licensee, a properly enclosed drinking area, and the erection of a 2-metre high wall around the perimeter of the whole premises. The cost of complying with what officials described as these 'practical and reasonable' (Simpson 1984: 28) regulations was conservatively estimated at R15,000 (US$7,200), a sum way beyond the reasonable means of the majority of shebeeners (Rogerson and Hart 1986). In addition to the above controls, the regulations further imposed strict trading hours of business upon the licensed shebeens. Once re-zoned, shebeeners were liable for additional rental payments as well as an annual licence fee, set initially at R600. Finally, it was made clear by officials that liquor licences would not be granted to persons with criminal convictions under existing liquor legislation.

Aspiring tavern licensees had to accept such controls and weave a pathway through a cumbersome licensing procedure involving at least three different government departments. Most critical in the approval of licences was the participation of the administration boards (later re-christened development boards) - institutions established in the early 1970s to run the affairs of urban Black townships (Bekker and Humphries 1985). These institutions, the progeny of apartheid planning, had a direct financial stake in retarding the granting of tavern licences because of their statutory monopoly on off-sale liquor retail outlets in the townships. Indeed, delays by officialdom in dealing with licence applications slowed the matter of shebeen licensing down to a glacial pace (Star, 4 September 1981, 2 March 1982). In response to what was seen as deliberate obduracy on the part of the West Rand Administration Board (WRAB), under whose aegis fell the issue of shebeen licensing in Soweto, an organization was formed in 1979 to represent shebeeners' interests and to hasten the progress of formalization (Star, 27 December 1979). The new organization, the Soweto Taverners' Association, initiated a mass boycott of all WRAB bottle stores in Soweto, committing itself to an open challenge of legislation which prohibited the conveyance of liquor from White areas into Black townships. Shebeen operators were to purchase their liquor requirements entirely from sources outside of Soweto, imposing massive financial losses upon WRAB liquor outlets (Star, 27 December 1979). The boycott called by the Soweto Taverners' Association ran for several weeks and was remarkably successful in its objective of effecting a streamlining of the licensing procedure and the concession of the first tavern licences. Indeed, again in 1981 the threat was made of a national boycott of administration board bottle stores if the authorities continued stalling as regards shebeen licensing (Sowetan, 2 September 1981).

Over the past five years, the number of shebeeners who have been formalized is relatively few as compared to the widespread nature of such activities in Soweto. Reliable statistics are sparse; however, it is estimated there are some 4,000 shebeens regularly functioning in Soweto, of which less than 40 had received the necessary re-zoning clearances by the beginning of 1985. Moreover, only three taverns had fulfilled the strict conditions with regards to the separation of the tavern from the licensee's living area (Sunday Times, 30 December 1984). Despite the energetic efforts mounted by the leading liquor companies to assist prospective taverners in submitting licence applications, there was little prospect that the mass of Soweto's small shebeens would ever match the stringent conditions and costs attached to a tavern licence. Indeed, the possibility emerged of the tavern trade being pre-empted by communities of traders and shopkeepers seeking to secure licences to convert their existing businesses into taverns (Star, 17 September 1981; Sowetan, 18 September 1981). The anticipated spoils from successfully obtaining a tavern licence in Soweto are captured by the poignant comment proffered by one legal taverner: 'A liquor licence is the most valuable licence in South Africa, better than a birth certificate and much harder to get than a firearm licence' (Sunday Times, 30 December 1984). For the apartheid state, the spoils of formalization are of an entirely different nature. By legalizing shebeens and extending some limited concessions to a Black petite bourgeoisie, the state strives to create the appearance that liquor distribution in South Africa's troubled townships has now become a depoliticized issue. Yet, ironically under South Africa's State of Emergency, shebeens are active participants in the resistance struggles launched by popular organizations, as illustrated by their commitment to early closure over the period of the 1986 'Christmas Against the Emergency' campaign called by the United Democratic Front (Weekly Mail, 12-18 December 1986).

CONCLUDING REMARKS

The saga of shebeen formalization in South Africa's Black townships represents a case study bridging the fledgling consumerism literature and the well-ploughed terrain of informal sector studies. Over the past three decades, the dramatic transformations wrought in the character of township shebeens mirror changes occurring in Black consumption patterns for alcohol. But, in many respects, it can be argued that the elite township shebeens led rather than merely reflected such changes, exhibiting consumption styles and drinking patterns to be emulated by the masses. At a broader level, the shift from traditional beers and illicit home concoctions to European beers and malt liquors can be viewed as a manifestation of wider social processes at work in the South African social formation. Most importantly, it evidences the historical forging of a mass market among Black consumers, the crystallization of a middle class and the penetration of consumerist ideologies.

The ramifications of these changed consumer preferences for alcohol have been wide-ranging. As has been shown elsewhere, the shift away from traditional sorghum beer to the consumption of 'White man's liquor' is a vital underpinning of the current fiscal crisis in the townships (Rogerson 1986d). Equally notable is the way in which the shebeen has been thrust into a key position in the South African liquor industry, providing fresh openings for capital to penetrate the mass market created in townships by the spread of the new drinking tastes. It was therefore as a consequence of the diffusion of consumerism and the acceptance of 'modern' over traditional patterns of alcohol consumption that pressures began to mount from liquor capital for the formalization of this informal income niche. After Soweto 1976, the apartheid state grudgingly acceded to the needs of liquor capital, sponsoring the legalization and transformation of shebeens to township taverns as part of its broader ploy to 'reform' apartheid and, in so doing, secure the support of an urban Black petit-bourgeois class.

Chapter Eleven

The last frontier: the emergence of the industrial palate in Hong Kong[1]

Scott MacLeod and Terry McGee

INTRODUCTION

On a global scale, increasing capital-intensive investment in food systems is rapidly turning food into an industrial commodity. Historically, food systems have been dominated by the tendency for production and consumption to be located within the same household, but with increasing urbanization and changes in the capital intensity and technology of production, processing, transportation, retailing, and preparation, a larger proportion of food is being consumed by non-producers. It is a combination of these developments that has led us to entitle the chapter 'The Last Frontier' because we view the food system - defined as production, processing, distribution, and consumption of food - as being one of the latest economic sectors to undergo increasing and thorough capital intensification (with its resultant industrialization).

However, it is not possible to predict precisely how the trend towards the industrialization of food systems is likely to develop in different milieux, or in various sub-markets or sectors. Indeed, it is the complexity and variance in the way the food industrialization process occurs what pose the most interesting research issues. The main purpose of this chapter therefore is to sketch out some of the many facets which conspire to produce the generalized trend of the industrialization of food systems on a world scale as they are played out within a local milieu. As such, this chapter is descriptive rather than theoretically inclined. We want to see if Hong Kong is experiencing an industrialization of its food system and if so, to identify the features of the process.

While this chapter is primarily concerned with the increasing industrialization of the Hong Kong food system, it is important to maintain a focus on the results of the process described - how they come to effect and reflect changes in the day-to-day lives of the people of Hong Kong.

Changes in consumption provide a touchstone for this investigation as they express the result of the interaction of various global generalizing forces and intricacies of local contingency in concrete space. One manifestation of this interaction seems to be an apparent convergence of consumption practices on a world scale (see Armstrong and McGee 1985). The concept of convergence is here used in a very broad way to describe the apparent gradual unification of global consumption norms towards an evolving global standard (i.e., a geographical homogenization of available products). The convergence of food consumption patterns is an indicator of the increasing pervasiveness of the world industrial food system on a local scale.

WHY FOOD?

> I think that it can be plausibly argued that changes in diet are more important than changes of dynasty or even religion.
>
> (George Orwell, cited in Barnett and Muller 1974)

The food sector seems an intriguing unit of analysis for a number of reasons. Perhaps the primary reason is that food consumption is so ubiquitous. It is very much a part of the 'everyday' in a truly Braudelian sense (Braudel 1973 or see Giddens 1984). Food consumption patterns are a fundamental part of the routines of life. Thus, changes in eating habits might be expected to illuminate other more sublimated social changes. Food consumption is at once a very social and a very individual activity. It is surrounded in a halo of social mores and driven by the basic necessity of hunger. It is a critical variable in social and individual reproduction (in both practical and sociological senses). Because dietary patterns are so common and so important, they can provide a good deal of information about a place and a people.

Local food consumption patterns were, until quite recently, circumscribed by the nature and dynamics of the local social formation and the local food system. These systems were in turn usually tightly and organically linked with one another and with local ecologies. Local isolated food systems were, and still are, often very idiosyncratic. Indeed, diet is one of the most distinguishing features of traditional societies and ethnic groups. It is an overt manifestation of the unity of a people, their links to one other and to a given place. Changes brought about by urbanization (particularly migration from rural areas), industrialization, and increased levels of global circulation have altered the nature of numerous social relations, among them the contexts of consumption - in this case local food systems. The shift away from locally generated parameters of consumption and production is perhaps the fundamental feature of development and increased global interdependence. Food systems, therefore, provide an important focus and a useful framework for examining the process of social change in situ.

WHY HONG KONG?

Hong Kong was selected as our case study for a number of reasons. First, Hong Kong is experiencing rapid social change, having evolved from a sleepy fishing village, through its role as colonial entrepot (see Faure et al. 1984), to that of a 'world city' (Hall 1984; Friedmann 1986). Hong Kong's growth rate puts it in the top ranks of the Asian NICs (Armstrong and McGee 1985: 91). Some reasons for this economic success are Hong Kong's aggressive internationalism, its laissez-faire ethos, and the adaptability of its small enterprises (Castells 1986: 113; Peattie 1983; K. Hopkins 1971: 160). These traits are usually discussed in terms of the production sphere, yet they also effect consumption patterns (see Filgueira 1981). As Hong Kong lacks a large politically integrated hinterland, and is 80 per cent dependent on imports for its food needs, the openness and suppleness of the Hong Kong food system should make it a useful bellwether of the pressures and constraints facing other developing nations.

Hong Kong also seems a particularly useful unit of analysis for examining changes in a food system and consumption practices because of the traditional strength of the Hong Kong dietary regime. Food is a central facet of Cantonese culture. The meal is the locus of much social activity; the food, the bearer of much social meaning. Cantonese cuisine is, no doubt at least partially, familiar to most readers.

Anderson (1977c) has strongly argued that Cantonese cuisine is closely integrated with local ecological parameters, that it presents an excellent example of the melding of social patterns and environmental requisites. It represents food consumption patterns closely linked to local place. Thus, shifts away from this dietary pattern may be indicative of the introduction of new, industrial or international patterns of regulation in the local milieu.

In this regard, Hong Kong is a fascinating amalgam of dietary consistency and change. Despite the apparent strength of the local diet, there are radical changes occurring in the daily dietary pattens of the people of Hong Kong. Hong Kong is an excellent example of a society (largely Cantonese) wedded to its cuisine to the point of chauvinism (Anderson 1977a: 13) yet increasingly flirting with the norms of the global industrial palate.

THE INDUSTRIAL PALATE

The phrase 'the industrial palate' refers to changes in both consumption and production which accompany and facilitate the increasingly manufactured nature of the foodstuffs consumed in much of the world (particularly in the cities). The processes of market expansion and capital intensification which are occurring, especially in the productive sphere (but downstream as well), have shifted the parameters of the food sector from local eco-cultural concerns (see Behar 1976; Croll 1983) to ones more closely linked to the requisites of a global industrialized marketplace.

The tendency towards the industrialization of the palate is the result of two sets of processes. These one may term the 'direct' and the 'indirect' routes. The direct route to changes in consumption and food systems is relatively confined to the dynamics of the particular market sector (i.e., food). It includes the structural dynamics of the local and world food systems and their interactions and articulations with each other and the consumer. This 'route' is the focus of most of this chapter. An example of this kind of change is found in frozen foods. Freezing is an important part of the industrialization of foodstuffs. It allows food to be transported great distances and gives it a longer shelf-life. It also requires changes in the local distribution system and (often) in household technologies. These changes reshape the very nature of the local food system and thus alter the context of consumption. Other changes via the 'direct' route are perhaps less tangible. They include the result of the competitive edge of industrial foods in industrial settings. Competition from mass-produced goods, goods produced by highly specialized producers, and 'scientific' distribution techniques and technologies (all discussed below) force local entrepreneurs to modify the way they address the market. In these ways the specific market sector, the direct outcome and context of consumption choices, is transformed.

The food system is also affected by forces largely outside the specific market sector. These forces form an 'indirect' route to changes in consumption patterns. In this grouping the generalized impacts of industrialization and urbanization (or, put more broadly, 'social change') are included.

Neither of these 'routes' stands alone as an explanation of observed changes in consumption patterns or market sectors. However, the focus of this chapter is on the direct route to changes in the food system and consumption because it reflects and shapes how the processes of industrialization and convergence are occurring within the broader parameters of the social formation as a whole.

The authors have examined the nature of the global food industry elsewhere[2] and as the subject of concern here is primarily with the Hong Kong case, those discussions need not be fully repeated here. However, it is necessary to identify some of the salient characteristics of food as a commodity, and the nature of large-scale food enterprises.

Food is not a 'typical' industrial good.[3] It can be distinguished from other commodities in a number of ways. As it is an essential good, it tends to be more recession-resistant than other commodities, due to its relative price and income inelasticity (this trait has made the food sector an attractive investment area). The inelasticity of foods also makes it difficult to sustain profits and market expansion. The share of staples in the total consumption profile is likely to decline as incomes rise (as outlined by Engel's Law) and as local markets expand (see the section on consumption). In order to maintain their return on investment, food companies must add value to the foods they produce and/or seek other markets. The result of these tendencies is that producers are increasingly forced to develop new products (or redefine and add value to old ones) and to create or find new markets. These two prerogatives underlie the capital intensification and internationalization trends. They escalate food production into the realm of the global capitalist market-place.

A recurring theme in this chapter is the manner in which the industrialization process is, to some extent, textured by the scale of operation. Through the economies of scale, the large producer needs a smaller 'margin' of profit per item sold; what is required is a mass market. The size of this market, and its continued growth, can lead to certain economies (e.g., the low margin required on goods which have moved through the 'product cycle') but it also limits the ability of the large-scale producer/distributor to service particular, geographically based market niches. For example, the transnational vendor is likely to be more interested in finding a niche that broadly serves all industrial societies than a specific national market. This kind of logic lies at the heart of the tendency towards standardization (see also Kaynak 1982, 1985; Buzell 1983; Levitt 1986) of product types and thus effects the convergence process.

This is the patterned canvas which lies beneath the unfolding scene. Capitalism's expansion in the contexts of industrialization and urbanization has resulted in the transformation of people's most basic requirement - food - from a part of their place to a placeless industrial commodity.

HONG KONG: THE INDUSTRIALIZATION OF A LOCAL FOOD PRODUCTION ECONOMY

It is useful to use the Hong Kong case to outline how complex and multi-layered is the process of industrialization of foodstuffs; for it is not simply a matter of external dictates, but also of a complex melange of local variables. In Hong Kong, the conjunction of international and local forces has led, in general, to an intensification of food production.[4] This intensification is characterized by shifting product types and increasing capital and technological inputs into existing foods. The intensification process permeates the food system as a whole - from the paddy bund to the food processing plant.

The industrial palate in Hong Kong

In the fields

Agriculture in Hong Kong has a tradition of being highly productive (Sit 1981; Wong 1983; Yeung 1985) and it is becoming more so. The process of agricultural intensification in Hong Kong is occurring in two ways. These are first, shifts in the types of products grown, and second, changes in production techniques (for example in the level of inputs).

The collapse of the rice-production sector very clearly illustrates the results of the impact of international competition and the move up-market in the Hong Kong food system as a whole. Hong Kong rice production was once famed for its technique and quality. However, as Schiffer's data have shown, by 1980 local production accounted for only 0.02 per cent of the level of imports (Schiffer 1984: 23). Clearly, one cannot overstate the shifts in the agricultural space economy, where agricultural land usage devoted to rice fell from 70.3 per cent to 0.4 per cent between 1954 and 1979 (Wong 1983). The corollary of the precipitous decline in rice production is the shift to more capital-intensive crops. At various periods, different crop regimes have become dominant as the agricultural sector has responded to a wide gamut of forces.

The first wave of change (c. 1945-75) was characterized by increasing land use for capital intensive crops like vegetables and fruits (Sit 1981; Wong 1983; Yeung 1985). The second wave of change (1975 to the present) is characterized by the production of high protein, higher intensity products such as chickens, pigs, and fish products (produced in fish ponds) (Hong Kong Government 1985a). These products exemplify the processes of intensification and industrialization as they require higher levels of capital inputs and are more closely linked to downstream industrial processing.

Finally, it is important to underline the fact that not only are the types of products grown changing, but so too are the techniques of growing. Hong Kong has experienced a shift of technique from hand labour, through primary mechanization (for example, power-tillers), to 'bio-technologies'. Changes in the technologies of food production have tended towards the capital intensive, and are increasingly bypassing the importance of local ecological factors through the introduction of advanced industrial food production systems. For example, hydroponics is becoming increasingly popular in Hong Kong (Yeung 1985: 33).

The reasons for the processes of specialization and intensification are various. They include: increasing local land values; monetization of the agrarian sector; increasing wage levels in Hong Kong; government policy; growing local demand for more expensive foods; and finally, the impact of new cultivators recently emigrated from China. None of these variables has pre-eminence, they act in toto, increasingly within the context of a broader world food system, for example through import competition or technology transfer.

The impact of Chinese immigrant agriculturists from Guandong illustrates the depth of interaction within this complex of forces, and gives some idea of how the intensification process occurs in situ through the example of one fragment of the larger dynamic. Most writers agree that the impact of an 'invasion' of vegetable and pond farmers from China is one factor in the changing structure of the Hong Kong food production sector (Sit 1981: 141). Yet these people were not independent agents freely altering the course of Hong Kong food production: they first had a need for capital, saw a potential market, and moved to fill it.

One reason for the need for capital was that the new immigrant tenants rented land on a monetary basis (a departure from the past). To the renters, their expertise and capital needs required a shift to more intensive vegetable and fish production. Vegetable farming increased cash flow nine times (Sit 1981: 127) while decreasing the amount of land needed for production. The drive behind this new production logic was the awareness of a decline in the value of rice due to competition from cheaper international sources and an increasing market for more expensive foodstuffs. Strauch, for example, has written that these immigrants were 'seeking fields suitable for markets' (Strauch 1984: 192). Competition from lower-priced rice imports, and the increased value of land, in conjunction with market shifts and local knowledge, thus led to a change in production priorities and a change in the local food supply systems.

In Hong Kong, therefore, one sees in microcosm the replaying of trends which continue around the world: the agricultural sector becomes increasingly capitalized. Faced with a global market of (often subsidized) competition, highly advanced agricultural techniques and transport modes, and high local land prices, the local rice producers lost much of their advantage of locale. As the relative price of rice declined, producers turned to more highly valued, capital-intensive products.

Although changes in the agriculture sector seem to be rather linear, such as intensification, the processes which combine to shape the resultant dynamic are complex and interactive. Yet, underlying this complexity, is the increasing monetization of social relations. As the level of capital input increases, so too does the importance of the (global) capitalist market-place. The result of this impingement is an increasing industrialization of the food system. In agriculture then, the combination of micro forces has led to an increasing intensification, ultimately to the industrialization of food production.[5]

THE HONG KONG FOOD INDUSTRY

The capital intensification process does not start or stop at the farmgate (or the paddy bund). Food processing is one of Hong Kong's fastest-growing industries. Although the number of establishments (and employees) dropped, gross output increased 380 per cent from 1976 to 1982. Further, gross fixed capital formation (or 'plant') increased 300 per cent and value addition was up 330 per cent (Data from United Nations various editions; and Hong Kong Government 1984a). These growth rates outstrip the rate of inflation by about 60 per cent over this period (World Bank 1983) and are indicative of the increasing scale of production in the food processing sector.

Table 11.1 illustrates the profile of the Hong Kong food processing industry, indicating that the larger establishments are the most productive. The biggest producers extract ten times more value addition per employee than the smallest grouping. However, the smaller establishments create more value addition per sales unit than the larger ones. It is in the nature of large production complexes to seek a small margin on each sale (this is their competitive edge). The larger establishments seek a mass market. They trade market 'fit' for volume of sales and standardization. They attempt to bring the market to themselves, through pricing, advertising, and other marketing techniques. Smaller establishments, on the other hand, need a smaller market. Thus they tend to 'go to' or to serve a particular market niche.

Table 11.1 Hong Kong: characteristics of the food manufacturing industry 1984

	Value of gross output ($,000)									
	<100	100-499	500-999	1,000-1,999	2,000-4,999	5,000 9,999	10,000 19,999	20,000 49,000	50,000 99,000	>100,000
No. of establishments	55	215	176	212	173	27	19	28	11	13
Percentage of total gross output	-	1.2	2.4	5.7	9.6	3.4	4.1	16.3	12.4	45.0
Percentage of total establishments	6.0	23.1	19.0	22.8	18.6	2.9	2.0	3.0	1.2	1.4
Percentage of total value added	0.1	0.1	2.9	6.5	10.1	3.7	5.7	16.9	16.4	36.9
Percentage of total employees	0.5	4.3	5.3	12.0	16.1	5.6	6.0	17.3	11.2	21.9
Value addition per sales unit	0.4	0.21	0.27	0.27	0.25	0.23	0.33	0.22	0.28	0.19
Value addition per employee	11.3	160	378	380	440	451	662	683	102.4	117.9

Source: Derived from Hong Kong Government(1984a)

The tendency in markets dominated by large establishments is towards standardization, and the inverse is likely to be true where smaller establishments predominate. The trend in the Hong Kong food processing sector seems to be to larger, more productive establishments. This implies a move towards mass-market production and away from the traditional Hong Kong norm of small establishments (Peattie 1983).

The increasing scale and intensity of production is likely to have an international bias because of technology transfer and foreign investment. The internationalizing nature of the food processing industry is most apparent in trends in foreign direct investment (FDI). FDI in the food and beverage 'manufacturing industry' increased from HK$5.4 million to HK$386.7 million[6] from 1971 to 1981. This increased the food and beverage sector's share of FDI from 0.79 per cent to 5.7 per cent (Hung 1984). In percentage terms, the increase in FDI in the Hong Kong food and beverage sector was 7,061. This rate of growth outstripped even the electronics sector.

This examination of the Hong Kong case has demonstrated how the process of capital intensification occurs *in situ*. It is the result of a wide range of factors, but is underlain by the permeation of the global capitalist marketplace into the local milieu. The importance of this relationship is not just confined to the production sector: as the following sections outline, it also greatly impacts on the realms of circulation and consumption.

CHANGES IN THE HONG KONG FOOD DISTRIBUTION SYSTEM

Changes in distribution networks are critical to the texture of the industrialization/convergence process because part of the nature of the expansion of the circulatory sphere is that 'a whole series of agents now intervened between the producer and consumer' (Goody 1982: 166). These interveners are important, for they act as the conduits through which foodstuffs find their way to the consumer. They are often very sensitive to the dynamics of the interplay of the requisites of the industrial palate and local communities. They are agents within the local milieu who by their actions define and redefine the nature of the discourse between the structure of the food system and the agency of consumption.

The Hong Kong food distribution system is, in fact, a very complex multitude of overlapping networks of distribution. However, one can identify two basic kinds of 'sectors' which are in some sense identifiable by the products they sell. The first is the more traditional Chinese produce and rice marketing system. This system tends to be government-regulated and deals mostly in foodstuffs from the New Territories or China.

A second 'sector' is that which handles mostly processed, preserved, and/or foreign foodstuffs. This grouping is more tightly articulated with the global food system. The degree of concentration of ownership is higher in this sector than in the former. The outlets of this sector tend to be grocers or supermarkets (and increasingly, of late, convenience stores such as 'Seven-Elevens').

Because shifts in the ecology and economy of food selling are most apparent at the retail level, we will focus our analysis on retail outlets. A number of surprising and at times apparently contradictory trends seem to be emerging in the Hong Kong food retailing sector (Fig. 11.1). Clearly, the Hong Kong food retailing sector is in a state of flux.

There are some rather interesting comparisons to be made with regard to the nature of different outlets.[7] Of particular note are the rather stark differences in the two growth areas of fresh fruit and vegetable (FFV) sellers and supermarkets. Few areas exemplify the divergence in the nature of these two kinds of establishments as much as the number of employees per establishment. Eighty-four per cent of supermarkets employ more than ten people, while only 4.1 per cent of FFV outlets do. Eighty-eight per cent of FFV outlets have fewer than five employees, while only 3.4 per cent of supermarkets do (Hong Kong Government1984b).

The difference between these two types of outlets go deeper than the number of employees (Table 11.2). The overall ratios of value addition and gross margin per establishment are far lower for FFV outlets than for supermarkets. This is part of their character: the FFV has a lower turnover and smaller market, but by directly serving its market, for example, by its location, it extracts the maximum value addition per unit of sales. The FFV goes to, and adapts to, the market. The supermarket, through its price and diversity, seeks to bring the consumer to it. This is a critical distinction.

Figure 11.1 Hong Kong: changing number of food retailing establishments 1974–85

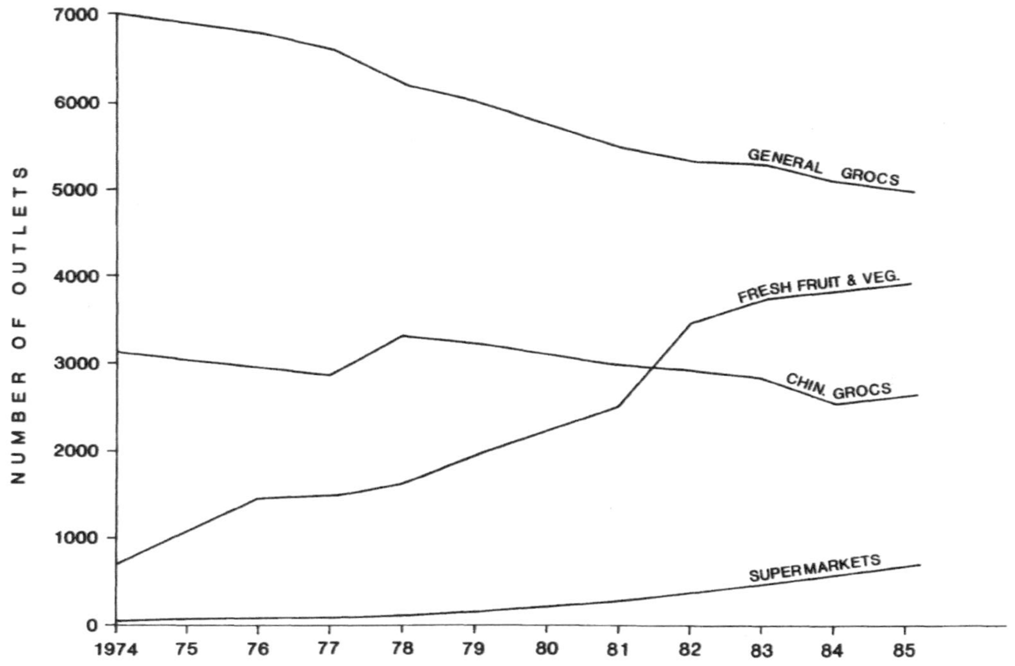

Table 11.2 Hong Kong sales, gross margin, and value addition for selected retail types (HK$s)

Retail types	VA/sales	VA/est.	GM/sales	GM/est.
Supermarkets	11.7	294.1	13.6	343.3
Fresh fruit and vegetable sellers	27.1	26.0	28.8	28.0

Key: VA - 'Census Value Added'
GM - Gross Margin
Est. - Establishment

Source: Derived from data in Hong Kong Government (1982)

THE SUPERMARKET IN HONG KONG

The most overt manifestation of the mass market in the food retailing sector is the supermarket. It represents what is to some the vanguard (Kaynak 1982: 246), and to others the dull homogenization, of the mass market. Newcombe, for example, writes that 'supermarkets are the same in Hong Kong as they are throughout the rest of the world' (Newcombe 1977: 336).

In Hong Kong, the number of supermarkets increased from sixty-two 1974-5 to 655 in 1984-5. Supermarkets have increased their share of the Hong Kong food retailing market from negligible levels in the late 1960s to approximately 55 per cent today (South China Morning Post (SCMP), 9 January 1986). The vibrancy of the supermarket sector is particularly evident if one focuses on a specific chain. The Wellcome chain, and its direct parent firm, Dairy Farm, seem the archetypal exemplars of the processes of growth, capitalization, and internationalization of this facet of the Hong Kong food system.

Dairy Farm had small beginnings. When it was founded in 1886 it was strictly a supplier of fresh milk. From that time it expanded into general dairy products (1896), into ice production (1918), then into a joint venture with Lane Crawford Ltd (UK) to open its first two supermarkets (1960). In 1964, Dairy Farm bought the Wellcome Company, then primarily a food wholesaler (with one supermarket). In 1972, Dairy Farm was acquired by Hong Kong Land. Since that time it has greatly expanded both within the Hong Kong market and internationally, acquiring (in 1979) the Franklins Food Chain in Australia. Dairy Farm has recently 'de-merged' from Hong Kong Land in order to give investors a chance to realize some of Dairy Farm's 'profitability and growth prospects which were submerged in the context of Hong Kong Land's large and assset-dominated balance sheet' (Hill and Knowlton Asia Inc. 1987). Dairy Farm had to subsidize Land to the tune of an estimated 33 per cent of Land's cash flow in the 1982-4 property squeeze (SCMP 9 January 1986).

Dairy Farm is now a very large and booming concern. Dairy Farm's direct parent firm is now a Bermuda holding company. Total international sales in 1986 were HK$10.2 billion: that was a 20 per cent increase over 1985. Its Hong Kong operations, while not at the scale of the Australian enterprise, exhibited markedly better growth, increasing from HK$1.1 billion in 1981 to HK$2.7 billion in 1985. The lion's share of this growth was a result of the success of the Wellcome chain. Table 11.3 illustrates the retailer's phenomenal growth. This growth shows few signs of abating. Wellcome plans to open twenty new outlets in Hong Kong during 1987 (Dairy Farm 1986).

Table 11.3 Hong Kong: growth of Wellcome supermarkets

Year	1981	1982	1983	1984	1985
Sales (HK$m)	872	1,083	1,302	1,785	2,201
No. of establishments	50	61	73	86	104
Percentage share of HK food sales	8	9.1	9.5	11.2	13.7

Source: Barclays de Zoete Wedd (1986)

The impact of supermarkets in Hong Kong

The rapidity of growth in the supermarket sector is indicative of the state of flux within the Hong Kong food system. It is also indicative of the market 'momentum' of supermarketing in Hong Kong. The two major chains in Hong Kong are 'Wellcome' and 'Park 'n' Shop' (the latter controlled by another 'Hong', i.e. Hutchison-Whampoa). Ian Wade, a managing director of Wellcome's parent company, claims that 'the two major chains are now so big that [they] virtually dictate the market' (The Bulletin, October 1986: 51). They have increased their market share through a number of aggressive marketing tactics. These include first predatory price competition: 'As in price competition anywhere else, small operations suffer. If you're bigger, you can afford to spread your problems,' Wade says (ibid). The second tactic is monopolization of selling rights; in Hong Kong the large retailers are establishing direct ties to producers in the international market-place and are even developing their own brand name. Third, lines of credit are also biased towards larger retailers, suppliers offer the big chains 120 days' credit, while the smaller operations only get a 90-day term. Large supermarkets can adopt these tactics because of their size; but it is also such tactics which increase their size.

Another source of market advantage for the chain supermarkets is through the technologies of retailing. The chains have introduced the latest in storage and inventory control technologies, for example, 'just-in-time' inventory control. The big chains are also importing new in-store production technologies, especially for their bakeries. For example, the Wellcome chain has just entered into a joint venture agreement with 'Mrs Field Inc.' to locally produce and distribute a variety of baked goods. 'Mrs Field' is a global corporation with 380 outlets worldwide (SCMP, 25 September 1986). Park 'n' Shop has signed a similar deal with Fuji Bakery, the largest in Japan with 2,000 retail outlets (The Standard, 13 April 1986). High levels of capital and technological inputs have made supermarkets more competitive and changed the nature of food retailing in Hong Kong.

A key feature of supermarkets is that they are parts of chains. Supermarkets adopt a marketing scheme (closely linked to the types of goods they sell) and use it in all outlets. In this way they facilitate the geographical dispersion of a given mode of marketing and enhance the process of standardization of product and the dissemination of the industrial palate. Kaynak feels that these outlets act in a manner similar to Schumpeter's 'innovating institutions'. He writes that: 'The competitive situation in the market is never the same after such institutions enter the marketplace; they are institutions that seem to differ not only in degree but in kind' (Kaynak 1982: 247).

This assertion seems particularly valid for supermarkets. Supermarkets require shifts in local consumption practices and they offer incentives to bring about these changes: by pricing of packaged goods, selection of produce, providing a wide range of goods and services, and by taking advantage of the marketing and advertising programmes of branded goods. Yet to succeed, the supermarket requires changes in consumer purchasing patterns. The extent to which the market is conducive to supermarketing is described by the level of 'outreach'.

'Outreach' describes a complex totality of social and infrastructural changes which are facilitated by, and facilitate, the growth of the supermarket sector. It covers such things as the decline in the role of local grocer, increased income levels, and the increasing pervasiveness of media exposure. Outreach is a reflection of the 'fit' between the nature of the supermarket as retailer and the consumer. Although both adjust, the dominant tendency is for the consumer to adapt to the nature of the supermarket, spurred on by convenience, cost, and marketing incentives. For example, as Goldman writes:

> The major factor limiting outreach is that in order for consumers to reap the benefits of low price and wide selection offered by the supermarket they have to change their basic shopping habits ... the supermarket shopper adjusts for this increase in travel distance and time by reducing shopping frequency.
>
> (Goldman 1974: 65)

Goldman's assertions inform this discussion in a number of ways. Centrally, they state the necessity of change in consumers' habits, household technologies, and spending patterns. In this way the supermarket effects changes via the direct route to consumption shifts, yet it requires changes via the indirect route to consumption shifts which are broadly analagous to supermarket 'outreach'. These include increased income levels, changes in household time economies, and spatial variables (such as transportation availability and the decline in the traditional grocer's role as community centre). All three areas are interrelated but income levels have primacy.

Outreach refers to the kinds of changes mentioned above, but it also refers to changes in the types of goods consumed. Consumers need to switch to packaged goods in order for the supermarket to fully exploit its market advantages. The linkage between the supermarket and the types of goods it sells is exemplified by the trait of 'selective adoption'. Research indicates that even when supermarket shopping becomes widely adopted, very few consumers in developing countries buy all their goods in them, and supermarkets tend to become sellers of groceries rather than produce (Goldman 1984: 25).

The FFV, selective adoption, and the competitive niches of retail types

'Selective adoption' also appears to be prevalent in Hong Kong. In Hong Kong, 'the wet markets' (as FFV outlets are called), have tenaciously held on to their share of the produce market and are growing in numbers. The viability of the traditional produce sector has a number of sources. These lie mainly in the areas of pricing, flexibility, quality control, location, and the Cantonese cultural predilection for fresh produce.

Smaller retailers purchase smaller lots and, therefore, wastage may be less of a factor in their price structure. As they have a smaller inventory, these traders can also go to the market. Small-produce vendors are much more responsive to shoppers' needs. They occupy a niche where it is critical to know precisely who needs what, and when. In Hong Kong the traditional cultural predilection for fresh produce has acted to enhance the viability of this sector. The booming growth of the FFV sector is indicative of the fact that change in the food system is not a simple unilinear process - local variables count.

The FFV, the convenience store, and the penetration of the world food system

The supermarket is not the only harbinger of the impingement of the global food system. Interestingly, it is possible to argue that the FFV sector is itself an excellent example of the depth of penetration of the world food system. In many ways, the FFV outlets illustrate the thesis of conservation - dissolution. Increasingly, these micro-scale outlets are themselves becoming enmeshed in the larger global food system. This is particularly true in the case of vegetables.

The share of imported versus local vegetables is increasing. Further, more and more of these vegetables are coming from 'core' sources, particularly the USA (Table 11.4). Fruits and vegetables from core countries tend to be highly capital-intensively produced and rely on advanced, equally capital-intense, transportation technologies. Yet one also sees a proliferation of small, fresh-fruit and vegetable sellers. In unison, these two trends exemplify the interaction of scales and the increasing depth of penetration of the world capitalist food system into the daily lives of the people of Hong Kong.

Table 11.4 Hong Kong: import levels of vegetables and fruit from China and USA

	(HK$m)			
	1983	1984	1985	% change
China	2,075	1,897	1,918	- 7.5%
USA	1,217	1,259	1,372	+12.7%

Source: Hong Kong Government (1986b)

Another example of the inaccuracy of viewing the supermarket as the only vanguard of the global industrial food system may be outlined in terms of the selling of processed and packaged foodstuffs. This area is no longer the bailiwick of supermarkets alone. They are no longer the sole vanguard of the mass market in processed foods.

In very recent times, Hong Kong has witnessed the growth in numbers of a third sort of retail outlet - the convenience store. One could, in fact, say that the growth of convenience stores has been explosive. In 1981 there were eight 'Seven-Elevens' in Hong Kong (the franchiser in Hong

Kong is, incidently, Jardine Marketing Services Ltd, a sister company of the Wellcome food chain); by 1987 there were 200 (Asian Business, August 1987: 54). This growth occurred alongside the boom in supermarkets and a general contraction in the number of food retail outlets from approximately 10,000 in 1983 to 8,000 in 1986.

The convenience store represents a hybrid of the large-scale, highly capitalized supermarket selling packaged branded goods, and the local small-scale enterprise (the Hong Kong norm). It differs from traditional grocers in a number of ways. The most central of these is its level of capital investment and use of retailing technologies, such as self-service drink dispensers and mini Hamburger machines. Linked to these characteristics is the fact that these outlets are franchises. Vendors buy an intact marketing and retailing programme which has been perfected abroad.

The advantage in franchises is in the science of selling. One franchisee sums it up thus: 'You do not really have to think about a lot of things. You just do what the system tells you to do' (Asian Business, August 1987: 54). Franchise outlets are prime sources of industrialization of diet and of convergence. Part of 'the package' is the standardization of produce. Such products are often those of international food corporations, such as Coke or Hostess. In this way, the global franchiser and local entrepreneur enhance the process of converging consumption patterns through the distribution of industrial food types.

The Hong Kong food distribution system is in a state of flux. Rapid and unprecedented growth rates are revolutionizing the way an increasing share of Hong Kong's food is being sold. The retail sector, one of the most direct contexts of consumption, is undergoing changes conducive to the emergence of the industrial palate. Changes in the retail sector effect and reflect changes in the most basic of daily routines - food consumption patterns.

CHANGING FOOD CONSUMPTION PATTERNS

In this section of the chapter the discussion centres on the focus of the various forces outlined above, i.e., the individual consumer. In fact, it is through consumption that the influence of the tangible and intangible forces outlined above become manifest. We turn first to the underlying social changes which permit consumption shifts (the indirect route) and then conclude with a description of changes in local food consumption patterns.

The parameters of the mass market in Hong Kong

A primary prerequisite for the 'industrialization' and convergence of food consumption patterns is the emergence of the mass market. This in turn requires a number of changes in the social formation. Drawing from the British experience, Frazer (1981) has outlined several features which he considers central to the emergence of the mass market. Table 11.5 lists these variables and presents relevant information for Hong Kong. Clearly, the prerequisites, as outlined by Frazer, are apparent in Hong Kong. However, present-day Hong Kong is not nineteenth-century Britain. It is experiencing the emergence of the mass market at a different rate and in a different geographical-historical conjuncture.

There are, therefore, some important variables which support the emergence of the industrial palate that Frazer neglects and that are important in Hong Kong. One such variable is the level of female labour force participation.[8] In Hong Kong, the percentage rate of female participation in the formal labour force increased from 43.6 in 1976 to 51.2 in 1985 (Hong Kong Government 1986b). This trend places greater stress on labour time in the home, making processed foods more attractive. It also increases the cash income of the family, facilitating the purchase of these foods.

Hong Kong has not really witnessed a transformation to the nuclear family accompanying industrialization as predicted by theorists such as Parsons. The tendency in Hong Kong may be even more conducive to increased levels of modernization of consumption, as the trend seems to be towards a family unit containing at least one other adult. Usually this extra member is an unmarried brother. This serves to increase the earning potential of the family unit, increasing available income for discretionary spending.

In terms of income, the distribution figures for Hong Kong are quite contentious, with Gini coefficients ranging from 0.41 to 0.74 (Chow and Papenek 1979). Despite the disparities of opinion, those data sets which have a temporal dimension tend to show similar trends. Notable gains have been made by the majority of the middle (especially upper-middle) income groupings at the expense of the very rich

Table 11.5 Hong Kong: indicators of a potential mass market

Variable	Hong Kong trends
A growing market	Population 1950 - 1.9 million 1985 - 5.43 million
Increasing disposable income	Using 1964 = 100 Index of nominal wages in manufacturing in 1980 = 525 Index of food prices in 1980 = 340
Growing literacy	Adult literacy 1960 - 70.4% 1979 - 90%
Organizational changes in retailing	Number of supermarkets 1974 - 62 1985 - 695
A stratified society	Share of ordinal group of income lowest 20% - 5.6% of total income highest 5% - 24% of total income
Increasing number of urban workers	Level of urbanization 90% Average density 1960 - 29,000/km 1980 - 48,000/km
Creation of a family type separate from the production process	Labour force in agriculture 1961 - 7.4% 1981 - 1.9%

Source: World Bank (1983); Chau (1983); Hong Kong Government (various years); Chow and Papenek (1979)

and the very poor (World Bank 1983; Chow and Papenek 1979).

There are, however, some variables unique to Hong Kong which adjust the income distribution picture somewhat and which have an impact on food consumption regimes. Although Hong Kong is often described as a laissez-faire free market economy (see Rabushka 1979; Chow and Papenek 1979), this is not quite true. The Hong Kong economy is not a truly free market - there are a number of factors which serve to subsidize the costs of the reproduction of labour power, to maintain consumption levels (see Schiffer 1984: 10-11), and to subsidize the price of staples in particular.

Basic food costs are subsidized in Hong Kong in two ways: first, through government intervention via marketing agencies; and second, through lower than world market prices on imports from China (40 per cent of total food consumption). Government intervention is most prevalent in the rice and vegetable sectors (the most traditional). Through the rice control scheme of 1955, for example, the government sets guidelines for the price of rice and strictly regulates and supervises thirty-eight registered rice stock holders.

Hong Kong is dependent on foreign sources for 80 per cent of its food supply, with half of this coming from China. However, China has clearly not abused its market dominance (Castells 1986). Chau has shown that 'prices of food imported from China would have been 29 per cent more expensive in 1979, had China adjusted its export price upward at the same rate as world export prices over the same period [1972-9]' (Chau 1983: 221). The impact of these lower prices is not felt equally across all income cohorts. Schiffer (1984: 4) has argued that there is an inverse relationship between income and dependency on foodstuffs from China. In this way the price of staples to the very poor is subsidized.

The picture which emerges is of a situation where the lowest income grouping has a declining share of income but where this shortfall is in part subsidized in other ways. The considerable growth of the middle class and the 'subsidization' of the incomes of the very poor support increased consumption levels of non-essentials. The stage is thus set for the emergence of the mass market and the convergence of food consumption patterns. As Goody has written, the industrialization of food 'led to a considerable degree of homogenization of food consumption and was dependent upon the effective increase in demand from the working class, which now desired access to foodstuffs' (Goody 1982: 170).

A second important feature of increasing real income levels is the way in which this affects what is purchased. As income levels increase, the percentage of expenditures devoted to food falls due to the basic inelasticity of demand for foodstuffs. This leaves more room for discretionary spending in general. More importantly (and correctly), the inelasticity of demand is particularly relevant in terms of the consumption of staple foods. Because staples are highly inelastic, increasing income allows for increasing discretionary spending within the food basket. It is largely (though not exclusively) in the realm of discretionary spending that new foods find an introduction into the local dietary regime. Due to various local factors (e.g., state subsidized housing) and international factors (e.g., cheap imports from China), and the general health of the economy, the people of Hong Kong have, in general experienced an increase in discretionary spending.

One could go on to portray other areas where it is evident that the Hong Kong social formation is undergoing a number of 'indirect' shifts conducive to the modernization of consumption, such as the levels of media density. Yet it seems clear from this brief consideration of some selected indicators of the potential mass market (or in supermarket terms, potential 'outreach') that Hong Kong is a society ripe for the embrace of the industrial palate. Hong Kong society is experiencing changes which both facilitate and require an increase in the purchase of value addition in foodstuffs.

There are two distinct areas of change in the consumption habits of the people of Hong Kong. The first is the changing locale of eating. The second is the change in the types of the foods found in the Home Shopping Basket (HSB).

Meals bought away from home

One of the major shifts in the food consumption patterns of the people of Hong Kong is in the locale of eating. Average expenditure on meals bought away from home (MBAH) increased from 7.5 per cent to 19 per cent of total expenditures between 1963-4 and 1984-5 (Hong Kong Government 1965; 1986a). The unadjusted growth rate for the restaurant sector as a whole from 1977-85 was 380 per cent (Hong Kong Government 1985b). A good deal of the buoyancy in this sector seems to be the result of growth in the Chinese Restaurant Sector (CRS). This trend is indicative of the fact that the process of modernizing consumption is not solely one of discarding traditional tastes.

The Cantonese of Hong Kong have traditionally placed

a great deal of emphasis on eating out (Anderson 1977a, b; Newcombe 1977; Chau 1980; Salaff 1981). This tradition is growing rapidly as the people of Hong Kong have, in general, experienced an increase in purchasing power. What lies at the base of the tendency to increased purchasing of meals away from home is the increased willingness of consumers to purchase more value addition in their foodstuffs, whether this is in the form of decor, the company of other patrons, the 'image', or the speed of service of the restaurant.

If the rates of growth and 'health' of different subsectors are examined, an intriguing picture emerges. The growth rate for the CRS (Chinese Restaurant Sector) from 1977-85 was 340 per cent, yet for the fast-food sector (FFS) it was 1,181 per cent. Moreover, sales per establishment are increasing faster in the FFS (30 per cent from 1980-3) than the CRS (10 per cent from 1980-3) (Hong Kong Government 1985b). Clearly, an analysis of the FFS reveals a great deal about trends in the Hong Kong restaurant sector: where and how people are increasingly eating out.

The fast-food sector

The three key areas for success in the fast-food sector in Hong Kong, and in general, are now examined: first, location. To Hong Kong 'the position is everything' (SCMP 7 April 1986). Prime positions are in the CBD, but as the market expands and competition deepens, franchises disperse geographically, bringing the industrial palate with them. Second, 'branding' is critical to the success of fast-food franchises, but linked to the international rather than the national image; to McDonalds rather than to the United States (Fujita 1986). The consumer must be able to identify the outlet and the products it purveys (they need to know what to expect). Third, and most important, using the 'production system' is the key to the expansion of the fast-food sector (Fujita 1986: 26; Hirst 1983). MacDonalds, for example, has 25,000 operating manuals (Fujita 1986). The production of the homogenized, guaranteed, standardized, and branded food item is the key to the fast-food outlet's origins and success (Hirst 1983).

In Hong Kong the leader in the fast-food industry is a local company called 'Cafe de Coral'. Though this company is ostensibly local, it fits quite smoothly into the mould of the globally franchised fast-food outlet (with whom it competes). The market is growing so fast that Cafe de Coral management feels it must expand at a rate of at least 20 per cent per annum to keep up (SCMP 18 May 1986). To do this, it will need to build seven new outlets, costing US$2-3 million each, every year. Clearly, maintaining market share is an expensive prosposition in a booming fast-food industry. This is one reason why franchising is so popular and why very large (often self-financing) transnational corporations are generally so successful. They provide the techniques and products which have already been developed and integrated elsewhere. Thus, the transnational corporation's cost of expansion may be very low and is often disproportionately devoted to marketing. The competitive edge, and production systems of the well-developed, fast-food franchises in some ways force the competitors to mimic certain facets of their marketing approach. The sheer capital intensity of the enterprise of entering the fast-food market requires a scientific approach to selling - comparable to the manufacture of food in a distinctly industrial manner. The fast-food outlet becomes a very small-scale food processor (as are all restaurants) and one which is very much integrated into the production and maintenance of the industrial palate. Its success marks the penetration of the process of the industrialization of food consumption patterns.

The Hong Kong fast-food sector exemplifies the interplay of local economies and ecologies with the imperatives of international capital. At one level, the nature of 'franchising' epitomizes the relationship of transnational capital and the local bourgeoisie. At another, the very introduction of highly capital-intensive and 'scientific' production and marketing techniques elicits a response from local competitors to do likewise.

Changing consumption patterns at home

The home shopping basket (HSB) is that section of the expenditure of foodstuffs which excludes expenditures on meals bought away from home (MBAH). The HSB is a useful category because it allows one to examine a differentiated totality within which shifting 'shares' may be discerned.

Table 11.6 lists the food groupings gaining an increasing share within the HSB between 1963-4 and 1979-80, ranked by nominal percentage change in share of the HSB. Few of the 'growth areas' in expenditure distribution could be termed traditional (for example, pork, rice products, and

Table 11.6 Hong Kong: shares of home shopping basket expenditure from 1963/64 to 1979/80

Rank	Foodstuff	Nominal change (%)	Rate of change (%)
1	Fruit, fresh	4.69	100.4
2	Meat/poultry, frozen	3.26	*
3	Fish, other	3.23	587.3
4	Confectionery	1.40	170.7
5	Meat/poultry, tinned	0.95	*
6	Meat, other	0.91	18.4
7	Foods, other	0.61	14.8
8	Fruit, other	0.57	203.6
9	Tea, coffee, soda, juices	0.48	25.0
10	Biscuits	0.27	33.0
11	Milk, fresh	0.20	36.4

* This grouping is not in the 1963/64 tables

Source: Hong Kong Government (1965; 1981)

fresh vegetables, though these are still important portions of the expenditure profile).

Indeed, the food types showing most gains have tended to be of the more processed, international and value-added types. The leading growth sector is fresh fruits: advances in super-cooling and other storage technologies have led to a trend for these fruits to increasingly come from further and further afield. Frozen meat and poultry are amongst the original industrial foods (see Horst 1974). 'Other' fish includes processed, canned, and frozen fish. 'Other' fruits are solely preserved and/or processed. Confectionery products are, of course, processed and exemplify a second underlying trait, that of branding. Most of the foodstuffs showing an increasing share of expenditures are subject to

Table 11.7 Hong Kong: decreasing shares of home shopping basket expenditures from 1963/64 to 1979/80

Rank	Foodstuff	Nominal change (%)	Rate of change (%)
1	Rice	8.4	56.0
2	Eggs	2.9	61.5
3	Beef, local	1.3	30.2
4	Dried fish produce	1.0	40.5
5	Pork, local	1.0	8.5
6	Milk, powder	0.8	32.8
7	Freshwater fish, fresh	0.6	9.9
8	Bread/cakes	0.5	10.0
9	Beans/peas	0.4	65.5
10	Sugar	0.3	36.6
11	Vegetables, other	0.2	19.5
12	Vegetables, fresh	0.1	1.0
13	Other cereals	0.1	4.9

Source: Hong Kong Government (1965; 1981)

branding: they are, in general, processed, 'international', and highly capitalized 'industrial-food' types.

The food types showing a declining share of expenditures in the population as a whole tend to be more traditional (Table 11.7). Clearly evident is the local, traditional low value-added and 'staple' nature of these foodstuffs; the listing seems almost an inventory of traditional food types. The consistency of the traditional nature of foods showing a decline is so marked that it raises some concerns about the trends that are revealed. Is the perceived drop in the share of expenditures due to relativities in the price of these goods or to an overall increase in real income levels? Such events might cause the proportion of expenditures given over to traditional staples to drop. This is because staples tend to be price and income inelastic. On the other hand, it has been shown that underpriced food from China has allowed Hong Kong a unique immunity to the food price inflation that so often accompanies rapid industrialization (Chau 1983: 197). Chinese policies have suppressed the price of staples in general, especially in comparison to other foodstuffs. For example, while rice prices rose about 2.5 times from 1963 to 1985, foodstuffs in general increased approximately 3.5 times. This could contribute to the relative decline in proportional expenditure on such staples.

This is not, however, to say that the data are of little use, quite the contrary is the case. The probability of uneven price changes fits within the parameters of the present argument for at least three reasons. First, the decline in the relative price of rice may be a reflection of a lack of demand (though it is more likely to be a result of a lack of demand and over-supply).

The second reason is perhaps more crucial to the present argument. The decline in the relative price of rice is only to be expected, as the rest of the food market contains more 'value-added'. Traditional staples like rice are not as subject to the pricing pressure which accompanies high levels of value addition in more processed foods.

Third, the relative decline in the price of staples is critical to the whole process of intensification/convergence. This is true in both the consumption and production spheres. An increase in income levels relative to the price of staples leaves more income available for discretionary spending. This in turn increases the potential for shifting consumption patterns. The increase in discretionary spending also creates an available market niche for the vendors of industrial foods. The generalized tendency for the relative value of staples to fall over the long term is a source of the continuing need of producers to add value, and of the ability of consumers to purchase that added value.

From the evidence on expenditure patterns, it appears that Hong Kong food consumption patterns are undergoing a radical transformation. The general nature of this change is away from traditional staples (exemplified by rice) and towards foods containing more value addition. These foods also tend to be rather more international in nature. It would seem that overall expenditure patterns do confirm the tendencies to increased levels of processing and internationalization characteristic of the industrial palate.

CONCLUSIONS

The aim of this chapter was to illuminate some of the intricacies of the process of the emergence of the industrial palate in a local setting. This discussion has suggested that this process is not as linear, nor as simple, as one might have expected. Hong Kong is experiencing a multi-faceted, multi-levelled industrialization of its food system. This industrialization affects and reflects changes in the realms of production, circulation, and consumption (as well as in the broader social formation as a whole). One can envisage the productive sphere (with its increasing capital intensity) as being the motor behind the emergence of industrial food; the circulatory sphere, including advertising, as being a context for, and a facilitator of, the emergence of a mass market for industrial foods; and the consumption sphere as shaping the specific foodstuffs required (within the constraints and possibilities of the overall market and local social formation). None of the variables acts alone. They exist and interact in an interdependent manner, shaping and reshaping themselves in interaction with the local social totality and world market forces.

In Hong Kong, these dynamics have worked in a synergistic manner to lead to an industrialization of the local food system. The increasing impact of international factors, such as competition, technology transfer, and investment, in combination with local potentialities, has led to an increasing level of capital input into the food sector. Such developments fit quite well into an expanding market. Hong Kong consumers, in general, appear ready and willing to purchase an increasing amount of processed foodstuffs, which their changing social niches seem to require.

However, through the dynamics of the 'direct route' to consumption changes, the industrialization of the food system has a momentum of its own. For example, the ongoing restructuring of the food system effects the consumption patterns of the very poor as well as the better off.[9] At a more fundamental level, the industrialization of the local food system changes its orientation. The increasing level of capital input and the increasing reliance on externally originating techniques and technologies shift the local food system away from parameters closely linked to the eco-cultural nature of 'place', and in turn ensconce it more deeply into the warp and weft of the global industrial food system. Thus one sees the emergence of the industrial palate, the crumbling of the last frontier.

NOTES

1. This research forms part of an SSHRC-funded grant to the Institute of Asian Research at UBC and the Centre for Development Studies in McGill University, Montreal.

2, A paper co-authored by the two present writers and Rex Casinader for the Annual Convention of the AAG at Portland in 1987 dealt more explicitly with the nature of the world food system, as does a forthcoming manuscript by S. MacLeod, published by the Institute of Asian Research, UBC.

3. It is a mistake to place too much emphasis on the intuitive division of agriculture and industry. Traditionally and presently the division is not that explicit. Also it seems important to clarify the distinction between food crops undergoing industrial processing and traditional industrial crops such as sisal or rubber. The focus of this discussion is on the industrialization of what are traditionally food crops.

4. A number of studies have pointed to the rapidly increasing level of capital intensity in the food sector (Burns et al. 1983; Conner et al. 1985; and the UN Center for Transnational Corporations 1981). The high level of capital intensity is at least in part due to the inelasticity of food as a staple. A great deal of the activity of transnational food companies is directed towards altering the inherent inelasticity of food products through increasing value addition. The producer/processor can add value by further processing a given foodstuff to meet a perceived potential market and/or value addition can be achieved through marketing techniques. The methods of achieving the goals of value addition and market expansion are usually capital intensive. Further, this value addition occurs not only on the farm, but also through industrial processing or marketing techniques.

5. There has also been a notable rise in the amount of abandoned land. Abandoned land was the second largest agricultural land use in 1979 (Wong 1983). The primary reason for this abandonment seem to lie largely in the pressures exerted on peri-urban land by rapidly expanding urban areas, exaggerated in Hong Kong by government regulations (Yeung 1985) and the development of the new

towns (Sit 1981). For the purposes of the present paper, this level of abandonment is particularly pertinent, in so far as it highlights the increasing level of intensification on other lands.

6. In 1988 the exchange rate was US$1.00 = HK$ 7.7.

7. Due to space constraints, the changes in the more traditional grocery store will be omitted. However, there has been a marked decline in the number of small grocers selling western goods, and a surprising stability in the number of grocers selling traditional Chinese products.

8. Other factors include the level of educational attainment, which is rapidly climbing in Hong Kong; the inceasing emphasis that Hong Kong's youth are placing on their peer groups rather than on their family; and the increasing tendency to put off the timing of the birth of the first child, thus increasing the period of greatest consumption.

9. One problem with the increasing industrialization and convergence of food consumption patterns is how they come to affect the very poor. The availability of food types becomes circumscribed as the depth of penetration of the world industrial food system grows. Clearly, not all income groups are experiencing the increases in income necessary to support the increased purchase of value addition. In some societies this group may in fact form the majority. In Hong Kong at present, the relatively widespread benefits of growth have enabled much of the populace to experience the fruits and the drawbacks of the industrial palate. However, the alteration of the local food system may come to affect the range of available food choices to the more marginal members of a given society. When combined with the power of 'status emulation' and the related impact of marketing and advertising schemes, these shifts in the local food system may have a negative effect on the food selection possibilities of the poor of a given society who cannot afford to purchase the high level of value addition found in industrial foodstuffs.

Bibliography

Abba, A. et al. (1986) 'Diagnostico de la situacion habitacional de los sectores mas carenciados de la Capital Federal' in N. Clichevsky, B. Cuenya, and S. Benalva (eds) Habitat Popular: Experiencias y Alternativas en Paises de America Latina, Cuadernos de CEUR 16, Buenos Aires.

Albers, G. (ed.), (1980) Protection and Cultural Animation of Monuments, Sites, and Historic Towns in Europe, German Commission for UNESCO, Bonn.

Alden, D. (1968) Royal Government in Colonial Brazil, Berkeley.

Almanach para a cidade da Bahia, Anno 1812 (1973), facsimile edition, Salvador.

Amin, S. (1974) Accumulation on a World Scale, Monthly Review Press, New York.

Amis, P. (1982) 'Squatters or tenants: the commercialization of unauthorized housing in Nairobi', World Development, 12(1): 87-96.

Anderson, E.N. Jr. (1977a) 'Eating at the Ngs', fugitive manuscript.

Anderson, E.N. Jr. (1977b) 'Changing food patterns in rural Hong Kong', unpublished paper, 4th Asian Conference of South California, Berkeley.

Anderson, E.N. Jr. (1977c) 'Rural to urban in Hong Kong foodways', fugitive manuscript.

Anon (1956) 'Growth of advertising industry in S. Africa reflects the healthy state of a country's economy', Industrial Review of Africa, 7 (12): 2-5.

Arias, P. and Roberts, B. (1985) 'The city in permanent transition: the consequences of a national system of industrial specialization', in J. Walton (ed.), op. cit.: 149-75.

Aries, P. (1962) Centuries of Childhood: A Social History of Family Life, Vintage Books, New York.

Armstrong, W. and McGee, T.G. (1985) Theatres of Accumulation: Studies in Asian and Latin American Urbanization, Methuen, London.

Arrighi, G. (1985) Semiperipheral Development, Sage, London.

Arrighi, G. and Drangel, J. (1986) 'The stratification of the world economy: an exploration of the semi-periphery zone', Review, 10(1): 9-74.

Augel, J. (1986) 'Loteamentos e especulacao imobiliaria como fatores determinantes da expansao urbana', in D.W. Benecke, K. Kohut, G. Mertins, S. Schneider, A. Schrader (eds), Desarrollo Demografico, Migraciones y Urbanizacion en America Latina, Friedrich Pustet, Regensburg: 369-81.

Augel, J. and Augel, M.P. (1984) 'Salvador: Historische Grosse-schmerzliche Erneuerung', in R.M. Ernst (ed.), Stadt in Afrika, Asien und Lateinamerika, Colloquium, Berlin: 100-13.

Augel, M.P. (1980) Visitantes Estrangeiros na Bahia Oitocentista, Cultrix, Sao Paulo.

Azevedo, Th. de (1969) Povoamento da Cidade do Salvador, Editora Itapua, Salvador.

Bahia, Governo (1978) A Grande Salvador: Posse e Uso da Terra, Salvador.

Bahia, SEPLANTEC - CPE (1978-80) A Insercao da Bahia na Evolucao Nacional: 1ª Etapa, 1850-89, Salvador.

Bailey, P. (1978) Leisure and Class in Victorian England, Routledge & Kegan Paul, London.

Baltzell, E.D. (1980) Puritan Boston and Quaker Philadelphia, Free Press, New York.

Barclays de Zoete Wedd Research (1986) Hong Kong Research: Dairy Farm International, unpublished consultant's report, Hong Kong.

Bariatua San Sebastian, J.M. (1977) Las Asociaciones de Vecinos, Instituto de Estudios de Administracion Local, Madrid.

Barnet, R.J. and Muller, R.E. (1974) Global Reach: the Power of Multinational Corporations, Simon & Schuster, New York.

Beavon, K.S.O. (1981) 'From hypermarkets to hawkers: changing foci of concern for human geographers', Environmental Studies, Occasional Paper No. 23, Department of Geography and Environmental Studies, University of the Witwatersrand, Johannesburg.

Beavon, K.S.O. (1982) 'Black townships in South Africa: terra incognita for urban geographers', South African Geographical Journal, 64: 3-20.

Beavon, K.S.O. and Rogerson, C.M. (1980) 'The persistence of the casual poor in Johannesburg', Contree: Journal of South African Urban and Regional History, 7: 15-21.

Beavon, K.S.O. and Rogerson, C.M. (1984) 'Aspects of hawking in the Johannesburg central business district',

Proceedings, Geographical Association of Zimbabwe, 15: 31-45.
Beavon, K.S.O. and Rogerson, C.M. (1986a) 'The council vs. the common people: the case of street trading in Johannesburg', Geoforum, 17: 201-216.
Beavon, K.S.O. and Rogerson, C.M. (1986b) 'The changing role of women in the urban informal sector of Johannesburg', in D.W. Drakakis-Smith (ed.), Urbanization in the Developing World, Croom Helm, London: 203-20.
Behar, M. (1976) 'European diets versus fast foods', Food Policy, November: 432-5.
Bekker, S. and Humphries, R. (1985) From Control to Confusion: The Changing Role of Administration Boards in South Africa, 1971-83, Shuter & Shooter, Pietermaritzburg.
Berkner, K. (1977) 'Peasant household organization and demographic change in Lower Saxony (1689-1766)', in Ronald Demos Lee (ed.), Population Patterns in the Past, Academic Press, New York: 19-51.
Blowers, A., Brook, C., Dunleavy, P., and McDowell, L. (1982) Urban Change and Conflict: An Interdisciplinary Reader, Harper & Row, London.
Blumberg, M. (1959) 'Durban explodes', Africa South, 4(1): 9-17.
Bohigas, O. (1963) Barcelona entre el Pla Cerda i el Barraquisme, Barcelona.
Booth, D. (1985) 'Marxism and development sociology: interpreting the impasse', World Development, 13: 761-88.
Borja, J. (1973) 'Planeamiento y crecimiento urbano de Barcelona 1939/58. El fet urba de Barcelona', Construccion, Arquitectura y Urbanismo, 46: 1-45.
Borja, J. (1977a) 'Popular movements and urban alternatives in post-Franco Spain', International Journal of Urban and Regional Research, 1: 151-60.
Borja, J. (1977b) 'Urban social movements in Spain', in M. Harloe (ed.), Captive Cities, Wiley, Chichester: 47-59.
Borja, J. and Serra, J. (1984) Notas sobre Urbanizacion y Sociedad en la Espana Actual, UIMP Pazo de Marinan, La Coruna.
Borja, J. et al. (1972) La Gran Barcelona, Los Libros de la Frontera, Barcelona.
Bornschier, V. and Chase-Dunn, C. (1985) Transnational Corporations in Underdevelopment, Praeger, New York.

Boserup, E. (1970) Women's Role in Economic Development, Allen & Unwin, London.

Boxer, C.R. (1965) Portuguese Society in the Tropics: the Municipal Councils in Goa, Macao, Bahia, and Luanda, 1510-1800, University of Wisconsin, Madison.

Bozzoli, B. (1979) 'Popular history and the Witwatersrand', in B. Bozzoli (ed.) Labour, Townships, and Protest, Ravan, Johannesburg: 1-15.

Bramwell, W. (1984) Pubs and Localised Communities in Mid-Victorian Birmingham, Occasional Paper No. 22, Department of Geography, Queen Mary College, University of London.

Brandao, M. de Azevedo R. (1978) 'Origens da Expansao Periferica de Salvador', in SEPLANTEC: Planejamento, 6(2): 155-72.

Braudel, F. (1973) Capitalism and Material Life 1400-1800, Harper & Row, New York.

Breiger, R. (1981) 'Structures of economic interdependence among nations', in P. Blan and R. Merton (eds.), Continuities in Structural Inquiry, Free Press, New York: 353-80.

Bridenbaugh, C. (1938) Cities in the Wilderness, Ronald Press, New York.

Bridenbaugh, C. (1955) Cities in Revolt, Alfred Knopf, New York.

Bromley, R. (ed.) (1979) The Urban Informal Sector, Pergamon, Oxford.

Bromley, F. and Gerry, C. (eds) (1979) Casual Work and Poverty in Third World Cities, Wiley, Chichester.

Browett, J. (1981) 'On the role of development in development geography', Tijdschrift voor Econ. en Soc. Geog., 72(3): 155-61.

Brundenius, J. (1981) 'Growth with equity: the Cuban experience', World Development, 9: 1083-96.

Bryant, J.J. and Khan, F. (forthcoming) 'Population and housing in Fiji', in R. Chandra and J.J. Bryant (eds), Population Monograph of Fiji, South Pacific Commission, Noumea.

Bunting, B. (1958) 'Liquor and the colour bar', Africa South, 2(4): 36-43.

Burgess, R. (1978) 'Petty commodity housing or dweller control? A critique of John Turner's views on housing policy', World Development, 6(9/10): 1105-33.

Burns, J., McInverney, J., and Swinbank, J. (eds) (1983) The Food Industry: Economics and Policies, Heinemann,

London.
Butterworth, D. and Chance, J.K. (1981) Latin American Urbanization, Cambridge University Press, Cambridge.
Buzell, R. (1983) 'Can you standardize multinational marketing?' in I. Dickson (ed.), Managing Effectively in the World Marketplace, John Wiley, Toronto: 273-93.
Calcagno, A.E. and Jakobowicz, J.M. (1981) 'Some aspects of the international distribution of industrial activity', CEPAL Review, 13: 7-34.
Calmon, F. Marques de Goes (1925) Vida Economico-Financeira da Bahia: Elementos para a Historia de 1808 a 1899, Bahia (reprint 1978).
Cameron, J. (1985) 'Destitute allowance vs. family assistance: conflict over welfare labels in Fiji', Development and Change, 16: 485-502.
Candel, F. (1964) Els Altres Catalans, Edicions 62, Barcelona.
Capel, H. (1974) 'Agentes y estrategias en la produccion del espacio urbano espanol', Revista de Geografia, 8(1-2): 19-56.
Capel, H. (1975) Capitalismo y Morfologia Urbana en Espana, Los Libros de la Frontera, Barcelona.
Cardoso, F. and Faletto, E. (1978) Dependency and Development in Latin America, University of California Press, Berkeley.
Carreras, C. (1980) Sants. Analisi del Proces de Produccio de l'Espai Urba de Barcelona, Serpa, Barcelona.
Carreras, J.M. and Margalef, J. (1977) 'La evolucion de las ciudades catalanas entre 1857-1975', Ciudad y Territorio, 2: 32-45.
Carvalho, E. (1985) 'O que se passa no "centro historico" de Salvador?', in Bahia, Secretaria do Planejamento, Ciencia e Tecnologia, SEPLANTEC/CENEPES (ed.), Debates, 2(6): 319-37.
Cassasas i Simo, Ll. (1977) Barcelona i l'Espai Catala. El Paper de Barcelona en la Formacio i en l'Ordenament del Territori de Catalunya, Curial, Barcelona.
Castells, M. (1977) The Urban Question, Edward Arnold, London.
Castells, M. (1978) 'Urban social movements and the struggle for democracy', Int. Jnl. of Urban and Regional Research, 1: 133-46.
Castells, M. (1983) The City and the Grassroots, Edward Arnold, London.
Castells, M. (1986) The Shek Kip Mei Syndrome: Public

Housing and Economic Development in Hong Kong, Working Paper No. 15, Centre for Urban Studies and Urban Planning, University of Hong Kong, Hong Kong.
Ceccarelli, P. (1982) 'Politics, parties, and urban movements: Western Europe', in N.I. and S.S. Fainstein (eds), Urban Policy under Capitalism, Sage, Beverly Hills.
Centre d'Estudis d'Urbanisme (1976), Movimientos Urbanos en Espana, EDICUSA, Madrid.
CEPAL (Comision Economica para America Latina y el Caribe) (1985) 'Preliminary overview of the Latin American economy 1985', Notas Sobre la Economia y el Desarrollo, 424-5.
Chan, L.H. (1980) 'The jetty dwellers of Penang: incorporation and marginalization of an urban clan community', unpublished M.Soc.Sc. dissertation, Universiti Sains Malaysia, Penang.
Chandra, R. (1981) 'Rural-urban population movement on Fiji, 1966-76: a macro-analysis' in G.W. Jones and H.V. Richter (eds), Population Mobility and Development: South East Asia and the Pacific, Monograph No. 27, Development Studies Centre, Australian National University, Canberra: 329-54.
Chase-Dunn, C. (1978) 'Core-periphery relations: the effects of core competition', in B. Kaplan (ed.), Social Change in Capitalist World-Economy, Sage, Beverly Hills: 159-76.
Chase-Dunn, C. (1982) 'Socialist states in the capitalist world economy', in C. Chase-Dunn (ed.) Socialist States in the World System, Sage, Beverly Hills.
Chase-Dunn, C. (1984) 'Urbanization in the world system: new directions for research', in M. Smith (ed.), Cities in Transformation: Class, Capital and State, Sage, Beverly Hills: 111-20.
Chase-Dunn, C. and Robinson, R. (1977) 'Toward a structural perspective on the world system', Politics and Society, 7(4): 453-76.
Chau, L.C. (1980) 'The size and profile of poverty in Hong Kong', in C.K. Leung, J.W. Cushman, and Wang Gungwu (eds), Hong Kong: Dilemmas of Growth, RSPACS, Australian National University, Canberra: 495-526.
Chau, L.C. (1983) 'Imports of consumer goods from China and the economic growth of Hong Kong', in A.J. Youngsen (ed.), China and Hong Kong: The Economic Nexus, Oxford University Press, Hong Kong Press: 184-

225.
Chirot, D. (1977) Social Change in the Twentieth Century, Harcourt Brace Jovanovitch, New York.
Chirot, D. (1986) Social Change in the Modern Era, Harcourt Brace Jovanovitch, New York.
Chow, S. and Papenek, G.F. (1979) 'Laissez-faire, growth, and equity: Hong Kong', unpublished paper, Ninth Annual Canadian Council of South East Asian Studies, Vancouver.
COAB (1971) 'Notas sobre la marginalidad urbanistica', Cuadernos de Arquitectura y Urbanismo, 86: 85-91.
Cockburn C. (1977) The Local State: Management of Cities and People, Pluto Press, London.
Cohen, M. (1974) Urban Policy and Political Conflict in Africa: A Study of the Ivory Coast, University of Chicago Press, Chicago.
Cohen, R.B. (1981) 'The new international division of labour: multinational corporations and urban hierarchy', in M. Dear and A.J. Scott (eds), Urbanization and Urban Planning in Capitalist Society, Methuen, London: 287-315.
Collier, D. (1976) Squatters and Oligarchs, Johns Hopkins Press, Baltimore.
CONDER (1977) Plano Metropolitano de Desenvolvimento: Cenario Atual, Salvador.
CONDER (1979) Estudo de Uso do Solo e Transportes para a Regiao Metropolitana de Salvador, vol. 1, Salvador.
Connell, J. (1985) Migration, Employment, and Development in the South Pacific: Fiji, South Pacific Commission, Noumea.
Connell, J. and Curtain, R. (1982) 'The political economy of urbanization in Melanesia', Singapore Journal of Tropical Geography, 3(2): 119-36.
Conner, J.M., Richard, T., Marion, B.W., and Mueller, W.F. (1985) The Food Manufacturing Industries: Structure, Strategies, Performance and Policies, Lexington Books, Toronto.
Connolly, P. (1985) 'Mexico: state investment in the built environment and the debt problem', Proceedings of the Sixth Bartlett International Summer School, Venice, 4.1-4.12.
Conway, D. (1982) 'Self-help housing, the commodity nature of housing and amelioration of the housing deficits: continuing the Turner-Burgess debate', Antipode, 14(2): 40-6.

Cooke, P. (1983) Theories of Planning and Spatial Development, Hutchinson, London.

Corbett, D.C. (1985) Housing in Darwin: Policies and Their Results, Northern Australia Research Unit, Australian National University, Darwin.

Corbridge, S. (1986) 'Capitalism, industrialization, and development', Progress in Human Geography, 10: 48-67.

Cornelius, W. (1985) 'The political economy of Mexico', Mexican Studies, 1: 83-123.

Courtenay, P.O. (1982) Northern Australia, Longman, Sydney.

Crankshaw, O. (1986) 'Theories of class and the African "middle class" in South Africa, 1969-83', Africa Perspective, 1: 3-33.

Croll, E. (1983) The Family Rice Bowl, Biblio Distribution (UNRISO), New Jersey.

Cumings, B. (1981) The Origins of the Korean War, Princeton University Press, Princeton, NJ.

Cumings, B. (1984) The origins and development of the Northeast Asian political economy: industrial sectors, product cycles, and political consequences', International Organization, 38: 1-40.

Da Camargo, C.P. et al. (1975) Sao Paulo: Cresciemento e Pobreza, Edicoes Loyola, Sao Paulo.

Dairy Farm International Holdings Ltd (1986) '1986 Annual Reports: Chairman's Statement', unpublished report, Hong Kong.

Daus, R. (1983) Die Erfindung des Kolonialismus, Peter Hammer, Wuppertal.

Davenport, T.R.H. (1960) 'African townsmen? South African Native (Urban Areas) legislation through the years', African Affairs, 68: 95-109.

Davies, R. (1979) 'Capital restructuring and the modification of the racial division of labour in South Africa', Journal of Southern African Studies, 5: 181-98.

Dear, M. and Clark, G. (1978) 'The state and geographical process: a critical review', Environment and Planning A, 10: 177-84.

Dear, M. and Scott, A.J. (eds.) (1981) Urbanization and Urban Planning in Capitalist Society, Methuen, London.

Dicken, P. (1985) Global Shift: Industrial Change in a Turbulent World, Harper & Row.

Dietz, H. (1987) 'Rent-seeking, rent-avoidance, and informality in analyses of Third World urban housing', paper delivered at Center of US-Mexican Studies, 7-10

April.
Doherty, J. (1983) 'Beyond dependency', Professional Geographer, 35 (1), 81-3.
Drakakis-Smith, D.W. (1980) 'Alice through the looking glass: marginalization in the town camps of Alice Springs', Environment and Planning A, 12(3): 427-48.
Drakakis-Smith, D.W. (1981a) Urbanization, Housing, and the Development Process, Croom Helm, London.
Drakakis-Smith, D.W. (1981b) 'Aboriginal access to housing in Alice Springs', Australian Geographer, 15(1): 39-51.
Drakakis-Smith, D.W. (1981c) 'Aboriginal underdevelopment in Australia', Antipode, 13(1): 35-44.
Drakakis-Smith, D.W. (1983) 'Advance Australia Fair: internal colonialism in the Antipodes', in D.W. Drakakis-Smith and S.W. Williams (eds), op. cit.: 79-102.
Drakakis-Smith, D.W. (ed.) (1984) Housing in the North: Policies and Markets, Northern Australia Research Unit, Australian National University, Darwin.
Drakakis-Smith, D.W. (1985) 'The changing role of women in urbanization: a preliminary analysis from Harare', International Migration Review, 18: 1278-92.
Drakakis-Smith, D.W. (1986) Urbanization in the Developing World, Croom Helm, London.
Drakakis-Smith, D.W. and Williams, S.W. (eds) (1983) Internal Colonialism: Essays Around a Theme, Monograph No. 3, Developing Areas Research Group, Institute of British Geographers, London.
Duncan, B. and Lieberson, S. (1970) Metropolis and Region in Transition, Sage, Beverly Hills.
Edwards, M. (1982) 'Cities of tenants: renting among the urban poor in Latin America', in A. Gilbert, et al. (eds) op. cit. 129-58.
Edwards, M. (1983) 'Residential mobility in a changing housing market: the case of Bucaramanga, Colombia', Urban Studies, 20: 131-45.
Elder, G.H. Jr. (1978) 'Family history and the life course', in T. Hareven (ed.), Transitions: The Family and the Life Course in Historical Perspective, Academic Press, New York: 17-64.
Espinheira, G. (1985a) 'Urbanizacao segregada - a expansao de Salvador', in SEPLANTEC, Debates, 2(6): 311-18.
Espinheira, G. (1985b) 'Um adeus aos bordeis', Panorama da Bahia, 42: 12-16.
Evans, P. (1979a) 'Beyond centre and periphery: a comment

on the world-system approach to the study of development', Sociological Inquiry, 49: 15-20.

Evans, P. (1979b) Dependent Development, Princeton University Press, Princeton.

Ewell, J. (1984) Venezuela: a Century of Change, Stanford University Press, Stanford.

Ewen, S. (1976) Captains of Consciousness: Advertising and the Social Roots of the Consumer Culture, McGraw-Hill, New York.

Faria, V.E. (1980) 'Divisao inter-regional do trabalho e pobreza urbana: o caso de Salvador', in G.A.A. de Souza and V.E. Faria (eds), Bahia de Todos os Pobres, Vozes/CEBRAP, Sao Paolo: 23-40.

Faure, D., Hayes, S., and Birch, A. (eds) (1984) From Village to City, University of Hong Kong Press, Hong Kong.

Felix, D. (1983) 'Income distribution and quality of life in Latin America: patterns, trends, and policy implications', Latin American Research Review, 18: 3-34.

Ferras, R. (1977) Barcelone, Croissance d'une Metropole, Anthropos, Paris.

Ferrer, R.M. and Precedo, L.A. (1981) El Sistema de Localizacion Urbano e Industrial, Banco de Bilbao, Bilbao.

Fiala, R. and Kamens, D. (1986) 'Urban growth and the world polity in the nineteenth and twentieth centuries', Studies in Comparative International Development, 21(1): 23-35.

Fielding, A.J. (1982) 'Counter-urbanization in Western Europe', Progress in Planning, 17: 1-52.

Fiji Central Planning Office (1980) Eighth Development Plan, 1981-5, Central Planning Office, Suva.

Fiji Employment and Development Mission (1984) Work and Income for the People of Fiji: A Strategy for More Than Just Survival, Final Report to the Government of Fiji, Parliamentary Paper No. 66, Suva.

Fiji Ministry for Lands, Local Government, and Housing (1982) 'Resettlement of squatters', Draft Cabinet Memorandum (P/82), Suva.

Filgueira, C. (1981) 'Consumption in the new Latin American models', CEPAL Review, 15: 71-110.

Flandrin, J.L. (1979) Families in Former Times: Kinship, Household, and Sexuality, Cambridge University Press, Cambridge.

Forbes, D. (1981) 'Petty commodity production and underdevelopment: the case of pedlars and trishaw

riders in Ujung Pandang, Indonesia', Progress in Planning, 16(2): 105-78.
Forbes, D. and Rimmer, P. (eds) (1984) Uneven Development and the Geographical Transfer of Value, Monograph HG16, RSPACS, Australian National University, Canberra.
Fox, R.W. (1975) Urban Population Growth Trends in Latin America, Inter-American Development Bank, Washington D.C.
Franco, A.M. de Almeida (1983) 'Habitacao popular e solo urbano em Salvador', unpublished M.A. dissertation, Salvador.
Frank, A.G. (1969) Latin America: Underdevelopment or Revolution?, Monthly Review Press, New York.
Frank, A.G. (1978) Dependent Accumulation and Under-development, Macmillan, London.
Frankel, P. (1979) 'Municipal transformation in Soweto: race, politics, and maladministration in Black Johannesburg', African Studies Review, 22: 49-63.
Frazer, H.W.C. (1981) The Coming of the Mass Market, 1850-1914, Archon Books, Hamdon.
Friedmann, J. (1972) 'The spatial organization of power in the development of urban systems', Development and Change, 4: 12-50.
Friedmann, J. (1978) 'The role of cities in national development', in L.S. Bourne and J.W. Simmon (eds), Systems of Cities, Oxford University Press, New York: 131-59.
Friedmann, J. (1986) 'The world city hypothesis', Development and Change, 17: 69-84.
Friedmann, J. and Wolff, G. (1982) 'World City', International Journal of Urban and Regional Research.
Frosch-Asshauer, A. (1986) Tradition im Umbruch: Zur Soziologie eines Brasilianischen Altstadtviertels, Munster.
Fujita, D. (1986) 'Golden arches on the Ginza: selling an American institution in Japan', Speaking of Japan, 7(66): 24-7.
Fundacao do Patrimonio Artistico e Cultural da Bahia (1979) 10 Anos de Fundacao, Salvador.
FUNDACOMUN (1985) 'Inventario de los barros en el Distrito Federal y Estado Mironda', Informe Preliminar, Caracas.
Furtado, C. (1970) The Economic Development of Latin America, Harvard University Press, Cambridge, Mass.

Furtado, C. (1972/3) 'The post-1964 Brazilian "model" of development', Studies in Comparative International Development, 8: 115-27.

Galera, M., Roca, F., and Tarrago, M. (1973) Atlas de Barcelona (Siglos XVI al XX), COAB, Barcelona.

Galtung, J. (1971) 'A structural theory of imperialism', Journal of Peace Research, 8: 81-117.

Gaspar, J. (1984) 'Urbanization' in A.M. Williams (ed.), The Transformation of Southern Europe, Harper & Row, London: 208-35.

Gereffi, G. and Evans, P. (1981) 'Transnational corporations, dependent developments, and state policy in the semi-periphery: a comparison of Brazil and Mexico', Latin American Research Review, 16: 31-64.

Gerritsen, R. (1985) 'Left theorizing on the Northern Territory political economy: an introductory note', in P. Loveday and D. Wade-Marshall (eds), Northern Australia: Process and Prospects, Northern Australia Research Unit, Australian National University, Darwin: 258-84.

Giddens, A. (1984) The Constitution of Society, Polity Press, Cambridge.

Gidengil, E.L. (1978) 'Centres and peripheries: an empirical testing of Galtung's theory of imperialism', Journal of Peace Research, 15(1): 51-66.

Gilbert, A.G. (1974) Latin American Development, Penguin, Harmondsworth.

Gilbert, A.G. (1981) 'Pirates and invaders: land acquisition in urban Colombia and Venezuela', World Development, 9: 657-78.

Gilbert, A.G. (1983) 'The tenants of self-help housing: choice and constraint in the housing markets of less developed countries', Development and Change, 14: 449-77.

Gilbert, A.G. and Gugler, J. (1982) Cities, Poverty, and Development, Oxford University Press, Oxford.

Gilbert, A.G. and Healey, P. (1985) The Political Economy of Land: Urban Development in an Oil Economy, Gower, Aldershot.

Gilbert, A.G. and Ward, P.M. (1985) Housing, the State, and the Poor: Policy and Practice in Latin American Cities, Cambridge University Press, Cambridge.

Gilbert, A.G., Hardoy, J.E., and Ramirez, R. (eds) (1982) Urbanization in Contemporary Latin America, John Wiley, Chichester.

Ginatempo, N. (1985) 'Social reproduction and the structure of marginal areas in southern Italy: some remarks on the role of the family in the present crisis', International Journal of Urban and Regional Research, 9: 99-111.

Giner, S. (1985) 'The social structure of Catalonia', Iberian Studies, 14(1-2): 43-68.

Godfrey, M. and Langdon, S. (1976) 'Partners in underdevelopment?: the transnationalization thesis in a Kenyan context', Journal of Commonwealth and Comparative Politics, 14: 42-63.

Goldman, A. (1974) 'Outreach of consumers and the modernization of urban food retailing in developing countries', Journal of Marketing, 38: 8-16.

Goldman, A. (1984) 'Adoption of supermarket shopping in a developing country: the selective adoption phenomenon', European Journal of Marketing, 16: 17-26.

Gonzalez-Berenquer y Urrutia, J.L. (1977) 'Diez encrucijadas para el urbanismo espanol', Ciudad y Territorio, 1: 94-101.

Goody, J. (1982) Cooking, Cuisine, and Class: A Study in Comparative Sociology, Cambridge University Press, Cambridge.

Gormsen, E. (1984) 'Repercusiones del "boom" de los anos sesenta en el urbanismo espanol', Estudios Geograficos, 45(176): 303-27.

Grupo 2c (1972) 'Los planes de Barcelona', Construccion de la Ciudad, 0: 3-55.

Gugler, J. (1982) 'Over-urbanization reconsidered', Third World Planning Review: 173-89.

Gugler, J. and Flanagan, W. (1977) 'On the political economy of urbanization in the Third World: the case of West Africa', International Journal of Urban and Regional Research, 1: 272-92.

Gugler, J. and Flanagan, W. (1978) Urbanization and Social Change in West Africa, Cambridge University Press, Cambridge.

Gwynne, R.N. (1985) Industrialization and Urbanization in Latin America, Croom Helm, London.

Habitat (1987) Global Report on Human Settlements 1986, Oxford University Press, Oxford.

Hadjimichaelis, C. (1980) 'The geographical transfer of value: a comparative analysis of regional development in Southern Europe', doctoral dissertation, School of Urban Planning, UCLA, Los Angeles.

Hall, P. (1984) The World Cities, Weidenfeld & Nicolson, London.

Hance, W. (1970) Population, Migration, and Urbanization in Africa, Columbia University Press, New York.

Hanley, S.B. (1977) 'The influence of economic and social variables on marriage and fertility in eighteenth- and nineteenth-century Japanese villages', in R.D. Lee (ed.), Population Patterns in the Past, Academic Press, New York: 165-200.

Hareven, T. (1977) 'Family time and historical time', Daedalus, 106(2): 59-70.

Hareven, T. (ed.) (1978a) Transitions: The Family and the Life Course in Historical Perspective, Academic Press, New York.

Hareven, T. (1978b) 'The historical study of the life course' in T. Hareven (ed.), Transitions: The Family and the Life Course in Historical Perspective, Academic Press, New York: 1-16.

Hareven, T. (1982) Family Time and Industrial Time, Cambridge University Press, Cambridge.

Hareven, T. and Langenbach, R. (1978) Amoskeag: Life and Work in an American Factory - City, Pantheon Books, New York.

Harris, C.C. (1983) The Family and Industrial Society, George Allen & Unwin, London.

Harrison, B. (1973) 'Pubs' in H.J. Dyos and M. Wolff (eds), The Victorian City, Vol. 1, Routledge & Kegan Paul, London: 161-90.

Hart, D. (1984) 'South African literature and Johannesburg's black townships', unpublished M.A. dissertation, University of the Witwatersrand, Johannesburg.

Hart, D.M. and Rogerson, C.M. (1985) 'Literary geography and the informal sector', Geography Research Forum, 8: in press.

Hart, G.P. (1972) 'Some socio-economic aspects of African entrepreneurship', Occasional Paper No. 16, Institute of Social and Economic Research, Rhodes Univesity, Grahamstown.

Harvey, D. (1973) Social Justice and the City, Edward Arnold, London.

Harvey, D. (1981) 'The urban process under capitalism: a framework for analysis' in M. Dear and A.J. Scott (eds), Urbanization and Urban Planning in Capitalist Society, Methuen, London.

Harvey, D. (1982) The Limits to Capital, Blackwell, Oxford.

Harvey, D. (1985) The Urbanization of Capital, Blackwell, Oxford.
Hawley, A. (1971) Urban Society, John Wiley, New York.
Henderson, J. (1986) 'The new international division of labour and urban development in the world system', in D. Drakakis-Smith (ed.), Urbanization in the Developing World, Croom Helm, London : 63-82.
Henderson, J. and Castells, M. (1987) Global Restructuring and Territorial Development, Sage, London.
Hendler, P. (1985) 'Urban Foundation strategies for change: a black future for free enterprise', Work in Progress, 39: 35-7.
Hendler, P. (1986) 'Capital accumulation, the state, and the the housing question: the private allocation of residences in African townships on the Witwatersrand, 1980-5', unpublished M.A. dissertation, University of the Witwatersrand, Johannesburg.
Heppel, M. (ed.) (1979) A Black Reality: Aboriginal Camps and Housing in Remote Australia, Australian Institute for Aboriginal Studies, Canberra.
Hermalin, A.I. and van de Walle, E. (1977) 'The civil code and nuptiality: empirical investigations of a hypothesis' in D. Lee (ed.), Population Patterns in the Past, Academic Press, New York: 71-111.
Higueras Arnal, A. (1980) 'Los desequilibrios regionales en Espana' in Asociacion de Geografos Espanoles, La Region y la Geografia Espanola, AGE, Vallodolid.
Hill and Knowlton Asia Inc. (1987) 'Dairy Farm announces profits of HK$288 million', unpublished report, Hong Kong.
Hinderink, J. and Sterkenburg, J. (1978) 'Spatial inequality in underdeveloped countries and the role of government policy', Tijdschrift voor Economisch en Sociale Geografie, 69(1-2): 5-16.
Hirson, B. (1981) Year of Fire, Year of Ash: The Soweto Revolt; Roots of a Revolution?, Zed, London.
Hirst, M.E. (1983) 'Fast-food merchandising - its origins and significance', in Food in Motion: The Migration of Foodstuffs and Cookery Techniques, Proceedings of the Oxford Symposium, Vol. 11, Oxford: 70-86.
Hollnsteiner, M. and Lopez, M.E. (1976) 'Manila: the face of poverty', in Social Science Research Institution, International Christian University (ed.), Asia Urbanizing, Simul Press, Tokyo: 69-86.
Hong Kong Government (1965) The Household Expenditure

Survey, 1963/64, and Consumer Price Index, Hong Kong: Statistics Branch, Commerce and Industry Department.

Hong Kong Government (1981) Report of the Household Expenditure Survey, 1979-80, and the New Consumer Price Index System, Census and Statistics Department, Hong Kong.

Hong Kong Government (1982) Survey of Wholesale, Retail and Import/Export Traders, Restaurants and Hotels, Census and Statistics Department, Hong Kong.

Hong Kong Government (1984a) Survey of Industrial Production, Census and Statistics Department, Hong Kong.

Hong Kong Government (1984b) Survey of Wholesale, Retail and Import/Export Traders, Restaurants and Hotels, Census and Statistics Department, Hong Kong.

Hong Kong Government (1985a) Hong Kong Annual Digest of Statistics, Census and Statistics Department, Hong Kong.

Hong Kong Government (1985b) Survey of Wholesale, Retail and Import/Export Traders, Restaurants and Hotels, Census and Statistics Department, Hong Kong.

Hong Kong Government (1986a) Report of the Household Expenditure Survey, 1984-5, and Consumer Price Index, Census and Statistics Department, Hong Kong.

Hong Kong Government (1986b) Hong Kong Annual Digest of Statistics, Census and Statistics Department, Hong Kong.

Hong Kong Government (various years) Survey of Wholesale, Retail and Import/Export Traders, Restaurants and Hotels, Census and Statistics Department, Hong Kong.

Hopkins, A. (1973) An Economic History of West Africa, Columbia University Press, New York.

Hopkins, K. (ed.) (1971) Hong Kong: The Industrial Colony. A Political, Social, and Economic Survey, Oxford University Press, Hong Kong.

Horst, T. (1974) At Home Abroad: A Study of the Domestic and Foreign Operation of the American Food-Processing Industry, Ballinger, Cambridge, Mass.

Hoselitz, B. (1954) 'Generative and parasitic cities', Economic Development and Cultural Change, 3: 278-94.

Hudson, P. and Sarakinsky, M. (1986) 'Class interests and politics: the case of the urban African bourgeoisie', in SARS (South African Research Service) (eds), South African Review, III, Ravan, Johannesburg: 169-85.

Hung, C.L. (1984) 'Foreign investments', in D. Lethbridge

(ed.), The Business Environment in Hong Kong, Oxford University Press, Hong Kong.
Inikori, A. (1982) Forced Migration, Academic Press, New York.
IPAC (1983) 'Microcensus of 1983', unpublished report, Salvador.
Jacobowicz, A. (1984) 'The green ban movement: urban struggle and class politics', in J. Holligan and C. Paris (eds), Australian Urban Politics, Longman, Cheshire, Sydney: 149-66.
Jenkins, R. (1984) 'Divisions over the international division of labour', Capital and Class, 22: 28-57.
Jimenez, E. (1982) 'The value of squatter dwellings in developing countries', Economic Development and Cultural Change, 30(4): 739-52.
Johnston, R.J. and Taylor, P.J. (1986) A World in Crisis: Geographical Perspectives, Basil Blackwell, Oxford.
Jouet, J. (1984) 'Advertising and transnational corporations in Kenya', Development and Change, 15: 435-56.
Kane-Berman, J. (1978) Soweto: Black Revolt, White Reaction, Ravan, Johannesburg.
Kaplinsky, R. (1979) 'Inappropriate products and techniques: breakfast food in Kenya', Review of African Political Economy, 14: 90-6.
Kaplinsky, R. (1984) 'The international context for industrialization in the coming decade', Journal of Development Studies, 21: 75-95.
Kaynak, E. (1982) Marketing in the Third World, Praeger, New York.
Kaynak, E. (1985) Global Perspectives in Marketing, Praeger, Toronto.
Keare, D.H. and Parris, S. (1982) Evaluation of Shelter Programs for the Urban Poor: Principal Findings, World Bank Staff Working Paper No. 547, World Bank, Washington.
Kegan, N. (1978) 'African settlements in the Johannesburg area 1903-23', unpublished M.A. dissertation, University of the Witwatersrand, Johannesburg.
Kick, E. (1986) 'The form and operation of the world system: a multiple network analysis', unpublished paper, Sun Belt Network Conference, Santa Barbara.
Kies, C.W. (1982) Problems Relating to the Use of Leisure in Soweto - Preliminary Survey, Pretoria, Human Sciences Research Council, Institute for Sociological and Demographic Research, Report S-78.

Kim, D.S. and Roemer, M. (1970) Growth and Structural Transformation, Harvard University Press, Cambridge, Mass.

King, J.C. (1971) 'Housing in Spain', Town Planning Review, 4(4): 381-403.

Kirk-Greene, A. and Rimmer, D. (1981) Nigeria Since 1970: A Political and Economic Outline, Africans, New York.

Kitchin, A.K (1978) 'A strategy for retail activity in a Black urban area', unpublished B.Sc. (TRP) dissertation, University of the Witwatersrand, Johannesburg.

Knox, P.L. (1988) 'The social production of the built environment', Progress in Human Geography, 13(2): 354-77.

Koch, E. (1983a) 'Without visible means of subsistence: slumyard culture in Johannesburg 1918-40', in B. Bozzoli (ed.), Town and Countryside in the Transvaal, Ravan, Johannesburg: 151-75.

Koch, E. (1983b) 'Doornfontein and its African World Class, 1914 to 1935: a Study of Popular Culture in Johannesburg', unpublished M.A. dissertation, University of the Witwatersrand, Johannesburg.

Koo, H. (1982) 'A conceptual framework for a political economic analysis of dependent development: a preliminary analysis of South Korea and Taiwanese cases', unpublished paper, Annual Meeting of the American Scoiological Association, San Francisco.

Kowarick, L. and Campanario, M. (1986) 'Sao Paulo: the price of world city status', Development and Change, 17: 159-74.

Kraychete, G. (1986) 'Regiao Metropolitana de Salvador: os deserdados do crescimento', in Cadernos do CEAS, 106: 22-35.

Kruger, H.-J. (1978) 'Industrialisierung und Stadtentwicklung in Salvador (Nordostbrasilien)', in Ibero-Amerikanisches Archiv, Neue Folge, 4(3): 185-216.

Kuper, L. (1965) An African Bourgeoisie: Race, Class, and Politics in South Africa, Yale University Press, New Haven.

Lagboa, F.Y. and Pe, R.E. (1981) 'Transnationals and consumerism: impact on Filipino tastes and values', State and Society, 2: 91-9.

La Hausse, P. (1984) 'The struggle of the city: alcohol, the Ematcheni, and popular culture in Durban, 1902-36', unpublished M.A. dissertation, University of Cape Town.

Langdon, S. (1974) 'The political economy of dependence: note toward analysis of multinational corporations in Kenya', Journal of East African Research and Development, 4: 123-59.
Langdon, S. (1975) 'Multinational corporations, taste transfer, and underdevelopment: a case study from Kenya', Review of African Political Economy, 2: 12-35.
Laquian, A.A. (1983) Basic Housing: Policies for Urban Sites, Services and Shelter in Developing Countries, International Development Research Center, Ottawa.
Laslett, P. (1969) 'Size and structure of the household in England over three centuries', Population Studies, 23: 199-223.
Laslett, P. and Wall, R. (eds) (1972) Household and Family in Past Time, Cambridge University Press, Cambridge.
Lasuen, J.R. (1972) 'La politica del suelo urbano', Arquitectura, 162, COAM, Madrid: 16-28.
Lasuen, J.R. (1974) Ensayos sobre Economia Regional y Urbana, Arial, Barcelona.
Lasuen, J.R. and Racionero, L. (1971) Las Areas Metropolitanas Espanolas, Madrid.
Laubscher, P. (1977) 'The magnificent obsession - the literature of the urban Black and the liquor question', unpublished B.A. dissertation, University of the Witwatersrand, Johannesburg.
Lea, J.P. and Zehner, R. (1986) Yellowcake and Crocodiles: Town Planning, Government, and Society in Northern Australia, Allen & Unwin, Sydney.
Lee, R.D. (ed.) (1977) Population Patterns in the Past, Academic Press, New York.
Leeming, R. (1977) Street Studies in Hong Kong: Localities in a Chinese City, Oxford University Press, New York.
Lefebvre (1976) The Survival of Capitalism, Allison and Busby, London.
Leira, E., Gago, J., and Solana, I. (1976) 'Cuarenta anos de crecimiento urbano', Ciudad y Territorio, 2-3, 43-66.
Lenski, G. (1976) 'History and social change', American Journal of Sociology, 82(1): 549-63.
Leopold, M. (1985) 'TNC food companies and their global strategies', International Social Science Journal, 37(3): 315-30.
Levine, D. (1977) Family Formation in an Age of Nascent Capitalism, Academic Press, New York.
Levitt, T. (1986) The Marketing Imagination, Free Press, New York.

Lewis, R. (1966) 'A 'city' within a city - the creation of Soweto', South African Geographical Journal, 48: 45-85.
Lewis, W.A. (1954) 'Economic development with unlimited supplies of labour', Manchester School of Economics and Social Studies, 20: 139-92.
Linn, J. (1983) Policies for Efficient and Equitable Growth of Cities in Developing Countries, Oxford University Press, Oxford.
Lipton, M. (1977) Why Poor People Stay Poor: Urban Bias in World Development, Australian National University Press, Canberra.
Llarch, E. et al. (1987) Dimensio Economica i Territorial del Barcelones, Caixa d'Estalvis de Catalunya, Barcelona.
Lloyd, P. (1979) Slums of Hope, Penguin, Harmondsworth.
London, B. (1980) Metropolis and Nation in Thailand: Political Economy of Uneven Developments, Westview Press, Boulder, Colo.
London, B. (1985) 'City-hinterland relationships in an international context: development as social control in Northern Thailand', in M. Timberlake (ed.), Urbanization in the World Economy, Academic Press, New York: 207-30.
Lotter, J.M. (1981) 'The structure and function of shebeens in Black and Coloured communities' in South Africa, Republic of, Report of the Conference: Alcohol in Perspective, Director-General of Health and Welfare, Pretoria: 223-36.
Lotter, J.M. and Schmidt, J.J. (1974) 'The nature and function of the shebeen in an urban Bantu community' in Proceedings on Alcoholism and Drug Dependence, Department of Social Welfare and Pensions, Pretoria: 41-8.
Lotter, J.M. and Schmidt, J.J. (1975) 'The shebeen in an urban Bantu community', Humanities, 3: 59-65.
Lottman, H.R. (1976) How Cities are Saved, Universe Books, New York.
Loveday, P. and Wade-Marshall, D. (1985a) Economy and People in the North, Northern Australia Research Unit, Australian National University, Darwin.
Loveday, P. and Wade-Marshall, D. (eds) (1985b) Northern Australia: Process and Prospects, Northern Australia Research Unit, Australian National University, Darwin, 258-84.
Lowder, S. (1980) The Evolution and Identity of Urban Social Areas: The Case of Barcelona, Occasional Papers Series

No. 4, Geography Department, University of Glasgow, Glasgow.
Lowe, S. (1986) Urban Social Movements: The City After Castells, Macmillan, London.
Lubeck, P. (1977) 'Contrasts and continuity in a dependent city: Kano, Nigeria', in J. Abu-Lughod and R. Hay (eds), Third World Urbanization, Maaroufa Press, Chicago: 281-90.
Lugar, C. (1980) 'The merchant community of Salvador, Bahia, 1780-1830', unpublished doctoral dissertation, State University of New York, New York.
Lunday, J. (1984) The Movement Toward North America: States and Capitalist Competition in the World Economy, unpublished doctoral dissertation, Johns Hopkins University, Baltimore.
Mabin, A. (1986) 'Labour, capital, class struggle and the origins of residential segregation in Kimberley, 1800-1920', Journal of Historical Geography, 12: 4-26.
Mabin, A. and Parnell, S. (1983) 'Recommodification and working-class home ownership: new directions for South African cities?' South African Geographical Journal, 65: 148-66.
Mabogunje, A. (1968) Urbanization in Nigeria, University of London, London.
McCallum, D. and Benjamin, S. (1985) 'Low-income urban housing in the Third World: broadening the economic perspective', Urban Studies, 22: 277-87.
McGee, T.G. (1967) The Southeast Asian City, Praeger, New York.
McGee, T.G. (1973) Hawkers in Hong Kong: A Study of Policy and Planning in a Third World City, University of Hong Kong Press, Hong Kong.
McGee, T.G. (1976) 'The persistence of the proto-proletariat: occupational structures and planning of the future of Third World cities', Progress in Geography, 9: 3-38.
McGee, T.G. (1978) 'Rural - urban mobility of South and Southeast Asia: different formulations ... different answers', in J. Abu-Lughod and R. Hay (eds), Third World Urbanization, Maaroufa Press, Chicago.
McGee, T.G. (1984) 'Middle-class households and the creation of a mass market in Malaysia: a proposal', unpublished paper, Research Seminar on Third World Urbanization and the Household Economy, Universiti Sains Malaysia, Penang.

McGee, T.G. (1985) 'Mass markets, little markets: some preliminary thoughts on the growth of consumption and its relation to urbanization: a case study of Malaysia', in S. Platner (ed.), Monographs in Economic Anthropology, No.4, University Press of America, Lanham.

McInnes, R.M. (1977) 'Childbearing and land availability: some evidence from individual household data' in R. Lee (ed.), Population Patterns in the Past, Academic Press, New York: 201-27.

McKelvey, E. (1969) The City in American History, George Allen & Unwin, London.

Malaysia, Government of (1970) Second Malaysia Plan 1971-5, Government Press, Kuala Lumpur.

Mandy, N. (1984) A City Divided, Macmillan, Johannesburg.

Mars, G. (1982) Cheats at Work, George Allen & Unwin, London.

Mashile, G.G. and Pirie, G.H. (1977) 'Aspects of housing allocation in Soweto', South African Geographical Journal, 59: 139-49.

Massey, D. (1979) 'In what sense a regional problem?', Regional Studies, 13: 233-43.

Mather, C.T. (1985) 'Racial zoning in Kimberley: 1950-59', unpublished report, University of the Witwatersrand, Johannesburg.

Matos Mar, J. (1985) Desborde Popular y Crisis del Estado: el Nuevo Rostro de Peru en la Decada de 1980, Instituto de Estudios Peruanos, Lima.

Mattedi, M.R.M. (1979) 'As invasoes em Salvador: uma alternativa habitacional', unpublished M.A. dissertation, Federal University, Salvador.

Mattedi, M.R.M., de Brito, M.R., and Barreto, S.S. (1979) 'O processo de urbanizacao', in Bahia SEPLANTEC - CPE, Habitacao e Urbanismo em Salvador, Salvador: 337-64.

Mattos, W. (1978) Evolucao historica e cultural do Pelourinho, Barbero, Rio de Janeiro.

Medhurst, K.N. (1973a) 'The central-local axis in Spain', Iberian Studies, 2(2): 81-7.

Medhurst, K.N. (1973b) Government in Spain, Pergamon, Oxford.

Meyer, D. (1986) 'The world system of cities: relations between international financial metropolises and South American cities', Social Forces, 64(3): 553-81.

Mills, E. and Song, B.N. (1979) Studies in the Modernization of the Republic of Korea: 1945-75, Harvard University Press, Cambridge, Mass.

Mingione, E. (1981) Social Conflict and the City, Basil Blackwell, Oxford.

Mingione, E. (1985) 'Social reproduction of the surplus labour force: the case of Southern Italy' in N. Redclift and E. Mingione (eds), Beyond Employment: Household, Gender, and Subsistence, Basil Blackwell, Oxford: 14-54.

Miro, J. et al. (1974) La Catalunya Pobra, Nova Terra, Barcelona.

Mohamad, M. (1984) 'Gender, class, and the sexual division of labour in a rural community in Kedah', Kajian Malaysia, II(2): 101-22.

Montero, J. (1972) 'La planificacion parcial en la comarca de Barcelona', Cuadernos de Arquitectura y Urbanismo, 87: 23-47.

Morales, R. (1984) 'Transitional labor: undocumented workers in the Los Angeles automobile industry', International Migration Review, 17: 570-96.

Morawetz, D. (1981) Why the Emperor's New Clothes are Not Made in Colombia, Oxford University Press, Oxford.

Morris, P. (1980) Soweto: A Review of Existing Conditions and Some Guidelines for Change, The Urban Foundation, Johannesburg.

Morris, P. (1981) A History of Black Housing in South Africa, South Africa Foundation, Johannesburg.

Morse, R.M. (1971) 'Trends and issues in Latin American urban research, 1965-70', Latin American Research Review, 6: 3-52, 9-75.

Moser, C. (1978) 'Informal sector or petty-commodity production: dualism or dependence in urban development?', World Development, 6: 1041-64.

Moser, C. (1984) 'The informal sector reworked: viability and vulnerability in urban development', Regional Development Dialogue, 5: 135-78.

Motumi, M. (1979) 'Draft report on survey on shebeens in Moroka', unpublished report for the Urban Foundation, Johannesburg.

Nakasa, N. (1975) The World of Nat Nakasa, Ravan, Johannesburg.

Narli, A.N. (1984) 'Development, Malay women, and Islam in Malaysia', Kajian Malaysia, II(2): 123-35.

Naylon, J. (1987) 'Iberia' in H.D. Clout (ed.), Regional Development in Western Europe, David Fulton, London: 383-418.

Neary, J.P. and van Wijnbergen, S. (eds) (1986) Natural Resources and the Macroeconomy, Basil Blackwell, Oxford.

Nemeth, R. and Smith, D. (1983) 'Divergent patterns of urbanization in the Philippines and South Korea: an historical structural approach', Comparative Urban Research, 10: 21-45.

Nemeth, R. and Smith, D. (1985) 'The political economy of contrasting urban hierarchies in South Korea and the Philippines', in M. Timberlake (ed.), Urbanization in the World Economy, Academic Press, New York: 183-206.

Nettels, C. (1952) 'British mercantilism and the development of the thirteen colonies', Journal of Economic History, 12: 105-14.

Netting, R.M., Wilk, R.R., and Arnould, E.J. (eds) (1984) Households: Comparative and Historical Studies of the Domestic Group, University of California Press, Berkeley.

Newcombe, K. (1977) 'From hawkers to supermarkets: patterns of food distribution in Hong Kong', Ekistics, 259: 336-41.

Ngubane, J.K. (1946) 'Market for 8,000,000 natives', South African Industry and Trade, 42(17): 133-9.

Ntshangase, D. (1961) 'A study of a sample of Johannesburg shebeens', File 11/1, Public Health Department Archives, Johannesburg.

Nun, J. (1969) 'Sobrepoblacion relativa, ejercito industrial de reserva y masa marginal', Revista Latinoamericana de Sociologia, 4: 178-237.

O'Donnell, G. (1978) 'Reflections on the bureaucratic-authoritarian state', Latin American Research Review, 13: 3-38.

Oliveira, F. (1972) 'A economia Brasileira: critica a razao dualista', Estudos CEBRAP, 2: 5-82.

OMPU (Oficina de Planeamiento Urbano) (1972) 'Diagnostico parcial y politicas generales en relacion a las areas de ranchos de Caracas', Informe Preliminar, Caracas.

Onselen, C. van (1976) Chibaro: African Mine Labour in Southern Rhodesia, 1900-33, Pluto, London.

Onselen, C. van (1982) Studies in the Social and Economic History of the Witwatersrand, 1886-1914, Ravan, Johannesburg.

Pahl, R.E. (1984) Divisions of Labour, Basil Blackwell, Oxford.

Pahl, R.E. and Wallace, C. (1985) 'Household work strategies in economic recession', in N. Redclift and E. Mingione (eds), Beyond Employment: Household, Gender, and Subsistence, Basil Blackwell, Oxford: 189-227.

Palloix, C. (1975) 'The internationalization of capital and the circuit of social capital', in H. Radice (ed.), International Firms and Modern Imperialism, Penguin, Harmondsworth: 47-61.

Palloix, C. (1977) 'The self-expansion of capital on a world scale', Review of Radical Political Economics, 9(2): 3-27.

Paris, C. et al. (1982) 'From public to welfare housing', unpublished paper, Australian Housing Conference, Sydney.

Parkes, D. (1984) Northern Australia, Academic Press, Sydney.

Parliament of Fiji (1977) Report on the Census of the Population 1976: Volume 1, Parliamentary Paper No. 13, Fiji Government Printing Service, Suva.

Parliament of Fiji (1982) The Housing Authority: Report for the Year 1980, Parliamentary Paper No. 26, Suva.

Parliament of Fiji (1988) Report on Fiji Population Census: Volume 1 - General Tables, Parliamentary Paper No. 4, Fiji Government Printing Service, Suva.

Pease, W. and Pease, J. (1985) The Web of Progress: Private Values and Public Styles in Boston and Charleston: 1828-43, Oxford University Press, New York.

Peattie, L. (1983) Small Enterprises in the Development Process, Working Paper No. 2, Centre for Urban Studies and Urban Planning, University of Hong Kong, Hong Kong.

Peres, F. da Rocha (1974) Memoria da Se, Salvador.

Perry, R., Dean, R., and Brown, B. (eds) (1986) Counter-urbanization: Case Studies of Urban to Rural Movement, Geo Books, Norwich.

Petshek, K.R. (1973) The Challenge of Urban Reform, Temple University Press, Philadelphia.

Pickvance, C. (1976) 'On the study of urban social movements', in C. Pickvance (ed.), Urban Sociology, Tavistock, London: 128-218.

Pickvance, C. (1983) 'What has become of urban social movements?', conference paper, University of Paris X, Nanterre.

Pirie, G.H. (1977) 'Commuter mode choice amongst Sowetans', unpublished M.A. dissertation, University of the Witwatersrand, Johannesburg.

Pirie, G.H. (1982) 'Mostly "Jubek": urbanism in some South African English literature', South African Geographical Journal, 64: 63-71.
Pirie, G.H. (1984a) 'Ethno-linguistic zoning in South African Black townships', Area, 16: 291-8.
Pirie, G.H. (1984b) 'Letters, words, worlds: the naming of Soweto', African Studies, 43, 43-51.
Pirie, G.H. and da Silva, M. (1986) 'Hostels for African migrants in Greater Johannesburg', Geo Journal, 12: 173-82.
Population Reference Bureau (1985) World Population Data Sheet, Population Reference Bureau, Washington D.C.
Portes, A. (1983) 'The informal sector: definition, controversy, and relation to national development', Review, 7: 151-74.
Portes, A. (1985a) 'The informal sector and the world economy: notes on the structure of subsidized labour', in M. Timberlake (ed.), Urbanization in the World Economy, Academic Press, New York: 53-62.
Portes, A. (1985b) 'Latin American class structures: their composition and change during the last decades', Latin American Research Review, 20: 7-39.
Portes, A. and Walton, J. (1976) Urban Latin America: The Political Conditions from Above and Below, University of Texas Press, Austin.
Prebisch, R. (1976) 'A critique of peripheral capitalism', CEPAL Review, 1: 9-76.
Prebisch, R. (1978) 'Socio-economic structure and crisis of peripheral capitalism', CEPAL Review, 5: 159-252.
Prebisch, R. (1981) 'The Latin American periphery in the global system of capitalism', CEPAL Review, 13: 143-50.
Precedo Ledo, A. (1976) 'El proceso de urbanizacion en Espana y sus relaciones con la industrializacion y la terciarizacion', Boletin de la Real Sociedad Geografica, 112(2): 457-75.
Puig, J.O. (1974) 'La conflictualidad urbana: algunas reflexiones sobre el reciente movimiento de barrios en Barcelona', Papers, 3: 275-323.
Quijano, A. (1974) 'The marginal pole of the economy and the marginalized labour force', Economy and Society, 3: 393-428.
Rabushka, A. (1979) Hong Kong: a Study in Economic Freedom, University of Chicago, Chicago.
Ranaka, C. (1980) 'SAB recommends legalizing of shebeens',

eSpotini, 1(4): 1-2.
Redclift, N. (1985) 'The contested domain: gender, accumulation, and the labour process', in N. Redclift and E. Mingione (eds), Beyond Employment: Household, Gender and Subsistence, Basil Blackwell, Oxford: 92-125.
Redclift, N. and Mingione, E. (eds) (1985) Beyond Employment: Household, Gender and Subsistence, Basil Blackwell, Oxford.
Reitsma, H.A. (1982a) 'Development geography, dependency relations, and the capitalist scapegoat', Professional Geographer, 34(2): 125-30.
Reitsma, H.A. (1982b) 'Geography and dependency: a rejoinder', Professional Geographer, 34(3): 337-42.
Reitsma, H.A. (1983) 'A conceptual model of dependency', Professional Geographer, 35(3): 330-1.
Richardson, A. (1977) 'City-slum in national spatial strategies in developing countries', World Bank Working Paper 252, Washington D.C.
Richardson, A. (1980) 'Polarization reversal in developing countries', Paper of the Regional Science Association, 45: 67-85.
Riddell, J.B. (1978) 'The migration to the cities of West Africa: some policy considerations', The Journal of Modern African Studies, 16(2): 241-60.
Rimmer, P.J. and Forbes, D. (1982) 'Underdevelopment theory: a geographical review', Australian Geographer, 15(4): 197-211.
Rimmer, P.J. and Forbes, D. (eds) (1983) Uneven Development and the Geographical Transfer of Value, Monograph HG16, Research School of Pacific Studies, Australian National University, Canberra.
Roberts, B. (1978) Cities of Peasants: The Political Economy of Urbanization in the Third World, Sage, Beverly Hills.
Roch, F. and Guerra, F. (1981) Especulacion del Suelo? Notas para una Politica del Suelo Urbano, Nuestra Cultura, Madrid.
Rodriguez, A. and D'Alos Moner, R. (1978) Economia y Territorio en Catalunya: Los Centros de Gravedad de Pobalcion, Industria y Renta, Banca Mas Sarda, Barcelona.
Rodriguez, J. and Sala, M. (1976) 'La construccion como mecanismo de acumulacion de capital', Construccion, Arquitectura y Urbanismo, 49: 43-63.
Rodriguez, O. (1981) 'On peripheral capitalism and its

transformation', CEPAL Review, 13: 151-9.

Rofman, A. (1974) Dependencia, Estructura de Poder y Formacion Regional en America Latina, Editorial Suramericana, Buenos Aires.

Rogerson, C.M. (1983) 'The casual poor of Johannesburg, South Africa: the rise and fall of coffee-cart trading', unpublished doctoral dissertation, Queen's University, Kingston, Ontario.

Rogerson, C.M. (1984) 'The spatial concentration of corporate control in South Africa, 1965-80', South African Geographical Journal, 66: 97-100.

Rogerson, C.M. (1985) 'The first decade of informal sector studies: review and synthesis', Environmental Studies, Occasional Paper No. 25, Department of Geography and Environmental Studies, University of the Witwatersrand, Johannesburg.

Rogerson, C.M. (1986a) 'Feeding the common people of Johannesburg, 1930-62', Journal of Historical Geography, 12: 56-73.

Rogerson, C.M. (1986b) 'The apartheid informal sector: from problem to solution?', unpublished paper, Vereinigeng von Afrikanisten in Deutschland, West Berlin.

Rogerson, C.M. (1986c) 'Johannesburg's informal sector: historical continuity and change', African Urban Quarterly, 1: 139-51.

Rogerson, C.M. (1986d) 'A strange case of beer: the state and sorghum beer manufacture in South Africa', Area, 18: 15-24.

Rogerson, C.M. (1987) 'Late apartheid and the urban informal sector', in J. Suckling and L. White (eds), After Apartheid: Renewing the South African Economy, James Currey, London.

Rogerson, C.M. and Beavon, K.S.O. (1980) 'The awakening of "informal sector" studies in Southern Africa', South African Geographical Journal, 62: 175-90.

Rogerson, C.M. and Beavon, K.S.O. (1982) 'Getting by in the "informal sector" of Soweto', Tijdschrift voor Economische en Sociale Geografie, 73: 250-65.

Rogerson, C.M. and Beavon, K.S.O. (1985) 'A tradition of repression: the street traders of Johannesburg' in R. Bromley (ed.), Planning for Small Enterprises in Third World Cities, Pergamon, Oxford: 223-45.

Rogerson, C.M. and da Silva, M. (1986) 'From backyard manufacture to factory flat: the industrialization of

South Africa's Black townships', unpublished paper, International Geographical Union, Commission on Industrial Change, Madrid.

Rogerson, C.M. and Hart, D.M. (1986) 'The survival of the "informal sector": the shebeens of Black Johannesburg', Geo Journal, 12: 153-66.

Romaguera Amat, R. and Dot Palleres, J.M. (1972) Barcelona: Genesis y Problematica del Area Metropolitana, Banco Urquijo/Editorial Moneda y Credito, Madrid.

Rosenthal, St. T. (1980) The Politics of Dependency: Urban Reform in Instanbul, Greenwood Press, London.

Rosser, C. (1973) Urbanization in Tropical Africa, International Urbanization Survey, Ford Foundation, New York.

Rowley, C.D. (1973) Outcasts in White Australia, Penguin, Ringwood.

Rowley, C.D. (1976) The Remote Aborigines, Penguin, Ringwood.

Rowley, C.D. (1978) The Destruction of Aboriginal Society, Penguin, Ringwood.

Russell-Wood, A.J.R. (1968) Fidalgos and Philanthropists: The Santa Casa da Misericordia of Bahia, 1550-1755, Berkeley.

Russell-Wood, A.J.R. (1982) The Black Man in Slavery and Freedom in Colonial Brazil, St Martin's Press, New York.

Sabater Cheliz, S. (1977) 'El proceso de urbanizacion en Barcelona y su traspais', Ciudad y Territorio, 3: 3-19.

Sachs, C. (1983) 'The growth of squatter settlements in Sao Paulo: a study of the perverse effects of the state housing policies', Social Science Information, 4/5: 749-75.

Sala Schnorkowski, M. (1977) 'El sector inmobiliario en Cataluna', Ciudad y Territorio, 3: 59-68.

Salaff, J.W. (1981) Working Daughters of Hong Kong: Filial Piety or Power in the Family?, Cambridge University Press, Cambridge.

Salih, K. (1981) 'Malaysia and the world system: a perspective essay on incorporation, social groups, and the state', unpublished paper, Universiti Sains Malaysia, Penang.

Salih, K. and Lo, F-C. (1975) Industrialization Strategy and the Growth Pole Approach: Case Study of Peninsular Malaysia, UNCRD, Nagoya.

Salih, K. and Tan, J. (1985) 'Impact of industrial development corporations and industrial estates on regional development: case of Malaysia', unpublished paper, Seminar on the Role of Public Enterprises in Developing Countries Workshop, New Delhi.

Salih, K. and Young, M.L. (1985) 'Employment, unemployment, and retrenchment in Malaysia: the outlook and what is to be done about it?', unpublished paper, MTUC Conference on 'Unemployment and Retrenchment in Malaysia', Penang.

Salih, K. and Young, M.L. (1987) 'Malaysia in the new international division of labour', in J. Henderson and M. Castells (eds), Global Restructuring and Territorial Development, Sage Publications, London: 168-202.

Salih, K., Young, M.L., Chan, L.H., Chan, C.K., Loh, K.W., and Tan, L.C. (1985) Young Workers and Urban Services: A Case Study of Penang, Malaysia, Final Report, Participatory Urban Services Project, Universiti Sains Malaysia, Penang.

Salih, K., Young, M.L., and Rasiah, R. (forthcoming) 'Transnational capital and local conjuncture: the semiconductor industry in Penang, Malaysia', Conference on Transnational Capital and Urbanization on the Pacific Rim, UCLA, Los Angeles.

Salvador, Prefeitura, OCEPLAN (1976) Evolucao Demografica de Salvador (1940-2000), Salvador.

Salvador, Prefeitura, SEPLAM, OCEPLAN (1985) Plano Diretor de Desenvolvimento Urbano, Salvador.

Sampson, A. (1957) 'Life in a South African slum', The Listener, 57: 703-4.

Sanchez-Leon, A. et al. (1979) Tugurizacion en Lima Metropolitana, DESCO, Lima.

Sandbrook, R. (1982) The Politics of Basic Needs, Heinemann, London.

Santos, M. (1959) O Centro da Cidade do Salvador: Estudo de Geografia Urbana, Salvador.

Sarda Dexeus, J. et al. (1984) La Economia de Cataluna Hoy y Manana, Banco de Bilbao, Barcelona.

Sarmento, W. Morais (1984) Nordeste: A Urbanizacao do Subdesenvolvimento, Mercado Aberto, Porto Alegre.

Sassen-Koob, S. (1986) 'New York City: economic restructuring and immigration', Development and Change, 17: 85-120.

Saunders, P.C. (1981) Social Theory and the Urban Question, Hutchinson, London.

Scharf, W. (1984) 'The impact of liquor on the working class (with particular focus on the Western Cape): the implications of the structure of the liquor industry and the role of the State in this regard', unpublished doctoral dissertation, University of Cape Town.

Scharf, W. (1985) 'Liquor, the State, and urban blacks' in D. Davis and M. Slabbert (eds), Crime and Power in South Africa: Critical Studies in Criminology, David Philip, Cape Town: 48-59.

Scheiner, S. (1983) 'Occupational mobility among blacks in South Africa', Working Paper No. 58, South African Labour and Development Research Unit, University of Cape Town.

Schiffer, J.R. (1984) 'Anatomy of a laissez-faire government: the Hong Kong growth model reconsidered', in P. Hills (ed.) State Policy, Urbanization, and the Development Process, Centre for Urban Studies and Urban Planning, University of Hong Kong, Hong Kong: 1-29.

Schmidt, E. (1980) 'The Sullivan principles: decoding corporate camouflage', United Nations Centre Against Apartheid, Notes and Documents, 4/80.

Schmitz, H. (1984) 'Industrialization strategies in less developed countries: some lessons of historical experience', The Journal of Development Studies, 21: 1-21.

Schwartz, St. B. (1973) Sovereignty and Society in Colonial Brazil: The High Court of Bahia and its Judges, 1609-1751, University of California Press, Berkeley.

Serrahima, F. and Marcos, J.A. (1970) 'Lo que cuesta una vivienda en terminos de arte especulativo', Construccion, Arquitectura y Urbanismo, 43: 36-41.

Sethuraman, S. (1977) 'The informal sector in Africa', International Labour Review, 126(3): 343-52.

Shepherd, J. and Walton, G. (1972) Shipping, Maritime Trade, and Economic Development of Colonial North America, Cambridge University Press, Cambridge.

Sibley, D. (1983) 'Government policies for peripheral regions: a case of categorical conflict', in D.W. Drakakis-Smith and S.W. Williams (eds), op. cit.: 64-75.

Simas, A. (1978) 'Evolucao fisica-urbana de Salvador', unpublished report, PLANDURB, Salvador.

Simkins, C. and Hindson, D. (1979) 'The division of labour in South Africa', Social Dynamics, 5: 1-12.

Simpson, R.W.J. (1984) 'The origin and future of shebeens', Fidelitas, (June-August): 28-29.

Sindicato da Industria da Construcao Civil da Cidade de Salvador (1980) Industria da Construcao Civil da Bahia. Atualizacao da Analise Comparativa do Diagnostico 77/78, Salvador.
Singer, P.I. (1980) 'A economia urbana de um ponto de vista estrutural: O caso de Salvador', in G.A.A. de Souza and V.E. Faria (eds), Bahia de Todos: os Pobres, Vozes/ CEBRAP, Sao Paulo: 41-69.
Singer, P. (1985) 'Capital and the National States: a historical interpretation', in J. Walton, (ed.), Capital and Labour in the Urbanized World, Sage, London: 17-42.
Sit, V.F.S. (1981) 'Agriculture under the urban shadow', in F.S. Sit (ed.), Urban Hong Kong, Summerson, Hong Kong: 128-40.
Slater, D. (1978) 'Towards a political economy of urbanization in peripheral capitalist societies', International Journal of Urban and Regional Research, 2: 26-52.
Slater, D. (1985a) 'The state and issues of regional analysis in Latin America', in J. Walton (ed.), Capital and Labour in the Urbanized World, Sage, London: 76-108.
Slater, D. (1985b) 'Social movements and a recasting of the political' in D. Slater (ed.), New Social Movements and the State in Latin America, CEDLA, Amsterdam: 1-21.
Slater, D. (1986) 'Capitalism and urbanization at the periphery', in D.W. Drakakis-Smith (ed.), Urbanization in the Developing World, Croom Helm, London: 7-22.
Smit, P. (1985) 'The process of Black urbanization' in H. Gilionee and L. Schlemmer (eds), Up Against the Fences: Poverty, Passes, and Privilege in South Africa, David Philip, Cape Town: 114-25.
Smith, C. (1985) 'Class relations and urbanization in Guatemala: toward an alternative theory of urban primacy', in M. Timberlake (ed.), Urbanization in the World Economy, Academic Press, New York: 121-67.
Smith, D. (1984) 'Urbanization in world economy: a cross-national and historical-structural analysis', unpublished doctoral dissertation, University of North Carolina, Chapel Hill.
Smith, D. (1985) 'International dependence and urbanization in East Asia: implications for planning', Population Research and Policy Review, 4: 203-33.
Smith, D. (1987) 'Dependent urbanization in colonial America: the case of Charleston, South Carolina', Social Forces, 66(1): 1-28.

Smith, D. and Nemeth, R. (1986) 'Urban development in Southeast Asia', in D.W. Drakakis-Smith (ed.), Urbanization in the Developing World, Croom Helm, London: 121-40.
Smith, D. and White, D. (1986) 'A dynamic analysis of international trade and world system structures: 1965-80', unpublished paper, International Studies Association, Anaheim.
Smith, M.A. (1983) 'Social usages of the public drinking house: changing aspects of class and leisure', British Journal of Sociology, 34: 367-85.
Smith, N. (1982) 'Theories of underdevelopment: a response to Reitsma', Professional Geographer, 34(3): 332-7.
Smith, T.C. (1977) Nakahara: Family Farming and Population in a Japanese Village, 1717-1830, Stanford University Press, Stanford.
Snyder, D. and Kick, E. (1979) 'Structural position in the world system and economic growth', American Journal of Sociology, 84: 1096-1126.
Soja, E. (1984) 'A materialist interpretation of spatiality', in D. Forbes and P. Rimmer (eds), op. cit.: 43-78.
Soussan, J. (1984) 'Recent trends in South African housing policy', Area, 16: 201-7.
South Africa (1913) Report of the Commission Appointed to Enquire into Assaults on Women 13/1913, Government Printer, Cape Town.
South Africa (1960) Report of the Commission of Inquiry into the General Distribution and Selling Prices of Intoxicating Liquor, UG 55/1960, Government Printer, Pretoria.
South Africa (1980) Report of the Commission of Inquiry into the Riots at Soweto and Elsewhere, June 1976 to February 1977, RP 55/1980, Government Printer, Pretoria.
South Africa (1982) Report of the Commission of Inquiry into Private Sector Involvement in Solving the Housing Backlog in Soweto, RP 14/1982, Government Printer, Pretoria.
Southall, R. (1980) 'African capitalism in contemporary South Africa', Journal of Southern African Studies, 7: 38-70.
Stadler, A. (1979) 'Birds in cornfields: squatter movements in Johannesburg 1944-7', in B. Bozzoli (ed.), Labour, Townships and Protest, Ravan, Johannesburg: 19-48.
Steiber, S. 'The world system and world trade', Sociological

Quarterly, 20: 23-6.
Stein, M. (1981) 'State liquor policy since 1880', unpublished paper, History Workshop, University of the Witwatersrand, Johannesburg.
Stinchcombe, A. (1979) Theoretical Methods in Social History, The Free Press, New York.
Stone, L. (1979) The Family, Sex, and Marriage in England 1500-1800, Harper Colophon, New York.
Storper, M. (1984) 'Who benefits from industrial decentralization? Social power in the labour market, income distribution and spatial policy in Brazil', Regional Studies, 18: 143-64.
Strauch, J. (1984) 'Middle peasant and market gardeners, the social context of the vegetable revolution in a small agricultural community in the New Territories, Hong Kong', in F. David, J. Hayes, and A. Birch (eds), From Village to City, University of Hong Kong Press, Hong Kong: 122-39.
Struyk, R.J. and Lynn, R. (1983) 'Determinants of housing investments in slum areas: Tondo and other locations in metro Manila', Land Economics, 59(4): 444-54.
Sunkel, O. (1973) 'Transnational capitalism and national disintegration in Latin America', Social and Economic Studies, 22: 132-76.
Sunkel, O. and Fuenzalida, E. (1977) The Transnationalization of Capitalism and National Development, Institute of Development Studies, University of Sussex, Brighton.
Suva City Council (1983) 'Report of a survey of unauthorized structures (squatter settlements) within the city of Suva', unpublished report, Suva City Council, Suva.
Tarrago, M. (1971) 'Els Tres Turons', Cuadernos de Arquitectura y Urbanismo, 86: 32-6.
Tarrago, M. (1976) Politica Urbana y Luchas Sociales, Avance, Barcelona.
Taylor, J. and Lea J.R. (1988) 'Darwin: city profile', Cities, August: 226-34.
Teedon, P. and Drakakis-Smith, D.W. (1986) 'Urbanization and socialism in Zimbabwe: the case of low-cost urban housing', Geoforum, 17(2): 309-24.
Teran, F. de (1978) El Planeamiento Urbano en la Espana Contemporanea. Historia de un Proceso Imposible, Gustavo Gili, Barcelona.
Teran, F. de (1982a) El Problema Urbano, Salvat, Barcelona.
Teran, F. de (1982b) Planeamiento Urbano en la Espana

Contemporanea (1900-80), Alianza, Madrid.

Themba, C. (1961) 'Special report on shebeens', F11/1, vol. 4, Public Health Department Archives, Johannesburg.

Thrift, N. (1986) 'The geography of international economic disorder', in R.J. Johnston and P.J. Taylor (eds), op. cit.: 12-67.

Tilly, A. and Scott, J.W. (1978) Women, Work and Family, Holt, Rinehart & Winston, New York.

Timberlake, M. (ed.) (1985) Urbanization in the World Economy, Academic Press, New York.

Todaro, M. and Stilkind, J. (1981) City Bias and Rural Neglect, Population Council, New York.

Tokman, V. (1982) 'Unequal development and the absorption of labour: Latin America 1950-80', CEPAL Review, 17: 121-33.

Tomaselli, R.E. (1983) 'Indian flower sellers of Johannesburg: a history of people on the street', in B. Bozzoli (ed.) Town and Countryside in the Transvaal, Ravan, Johannesburg: 215-39.

Townroe, P.M. and Hamer, A.M. (1984) 'Who benefits from industrial decentralization? Response to Storper', Regional Studies, 18: 339-44.

Townroe, P.M. and Keen, D. (1984) 'Polarization reversal in the state of Sao Paulo, Brazil', Regional Studies, 10: 45-54.

Turner, J.F.C. (1967) 'Barriers and channels for housing development in modernizing countries', Journal of the American Institute of Planners, 33: 167-81.

Turner, J.F.C. (1978) 'Housing in three dimensions: terms of reference for the housing question redefined', World Development, 6(9/10): 1135-45.

Tuswa, P. (1983) 'Changes in the residential landscape of Soweto', unpublished paper, University of the Witwatersrand, Johannesburg.

UFBa, Centro de Estudos da Arquitetura na Bahia (1980) Evolucao Fisica de Salvador, Salvador.

United Nations (1979) Commission on Human Settlements, Documents HS/C/2/3, United Nations, New York.

United Nations (1980) Patterns of Urban and Rural Development Growth, United Nations, New York.

United Nations Centre on Transnational Corporations (1981) Transnational Corporations in Food and Beverage Processing, United Nations, New York.

United Nations (1986) Demographic Yearbook 1985, United Nations, New York.

United Nations (various editions) Yearbook of Industrial Statistics, Vol. I., United Nations, New York.
Vayrynen, R. (1983) 'Semi-peripheral countries in the global economic and military order', in H. Tuomi and R. Vayrynen (eds), Militarization and Arms Production, Croom Helm, London: 163-92.
VerSteeg, C. (1975) Origins of the Southern Mosaic, University of Georgia Press, Athens.
Vidal Bendito, T. et al. (1980) Atlas Socio-Economico de Catalunya, Caixa d'Estalvis de Catalunya/Banco Occidental/Banco Urquijo/Department de Geografia de la Universidad de Barcelona, Barcelona.
Walker, R. and Storper, M. (1981) 'Capital and industrial location', Progress in Human Geography, 5: 473-509.
Wallerstein, I. (1974a) 'The rise and future demise of the world capitalist system: concepts for comparative analysis', Comparative Studies in Society and History, 16: 387-415.
Wallerstein, I. (1974b) The Modern World-System I, Academic Press, New York.
Wallerstein, I. (1976) 'Semi-peripheral countries and the contemporary world crisis', Theory and Society, 3: 461-84.
Wallerstein, I. (1980) The Modern World-System II, Academic Press, New York.
Wallerstein, I. (1982) 'Household structures and labour force formation', unpublished paper, Seminar on Households and the World Economy, University of Bielefeld.
Wallerstein, I. (1985) 'The relevance of the concepts of semi-periphery to Southern Europe' in G. Arrighi (ed.), Semi-Peripheral Development, Sage, London.
Wallerstein, I. and Hopkins, T. (1977) 'Patterns of development of the modern world-system', Review, 1(2): 11-45.
Walsh, A.C. (1979) 'Have we been squatting too long? Some lessons from Fiji', unpublished paper, Ninth Waigani Seminar, University of Papua New Guinea, Port Moresby.
Walsh, A.C. (1984) 'The search for an appropriate housing policy in Fiji', Third World Planning Review, 6(2): 185-200.
Walsh, R. (1970) 'The mechanics of revolutionary Charleston: 1760-76', in R. Mohl and N. Betten (eds), Urban America in Historical Perspective, Weywright and Talley, New York.

Walton, J. (1977) 'Accumulation and comparative urban systems: theory and some tentative contrasts of Latin America and Africa', Comparative Urban Research, 5: 5-18.
Walton, J. (1981) 'The new urban sociology', International Social Science Journal, 33(2): 374-90.
Walton, J. (1982) 'The international economy and peripheral urbanization', in N. and S. Faintstein (eds), Urban Policy Under Capitalism, Sage, Beverly Hills: 119-36.
Walton, J. (ed.) (1985a) Capital and Labour in an Industrializing World, Sage, Beverly Hills.
Walton, J. (1985b) 'The third "new" international division of labour' in J. Walton (ed.), op. cit.: 3-14.
Ward, P.M. (ed.) (1982) Self-help Housing: a Critique, Mansell, London.
Warren, W. (1973) 'Imperialism and capitalist industrialization', New Left Review, 81: 3-44.
Warren, W. (1980) Imperialism: Pioneer of Capitalism, New Left Books, London.
Wellings, P. and Sutcliffe, M. (1984) 'Developing the urban informal sector in South Africa: the reformist paradigm and its fallacies', Development and Change, 15: 517-50.
Wells, J. (1983) 'Industrial accumulation and living standards in the long-run: the Sao Paulo industrial working class, 1930-75, Part 1', Journal of Development Studies, 19: 145-69.
Western, J. (1981) Outcast Cape Town, University of Minnesota Press, Minneapolis.
Wilkie, J.W. and Perkal, A. (eds) (1983) Statistical Abstract of Latin America 23, Latin American Center, University of California, Los Angeles.
Wilkie, J.W. and Perkal, A. (eds) (1985) Statistical Abstract of Latin America 24, Latin American Center, University of California, Los Angeles.
Wilkinson, P. (1983) 'Housing' in SARS (South African Research Services), South African Review I, Ravan, Johannesburg: 270-8.
Wilkinson, P. (1984) 'The sale of the century? A critical review of recent developments in African housing policy', Carnegie Conference Paper No. 160, South African Labour and Development and Research Unit, University of Cape Town.
Williams, L.S. and Griffin, E.C. (1978) 'Rural and small town depopulation in Colombia', Geographical Review, 68: 13-30.

Williams, P. (1984) 'Economic processes and urban change: an analysis of contemporary patterns of residential structuring', Australian Geographical Studies, 22: 39-57.
Wolpe, H. (1977) 'The changing class structure of South Africa', in P. Zarembka (ed.), Research in Political Economy, Jai Press, Greenwich, Conn.: 143-74.
Wong, C.T. (1983) 'Land use in agriculture', in T.N. Chiu and C.L. So (eds), A Geography of Hong Kong, Oxford University Press, Hong Kong.
World Bank (1982) Atlas: Gross National Product. Population and Growth Rates, World Bank, Washington D.C.
World Bank (1983) World Tables, World Bank, Washington D.C.
World Bank (1985) World Development Report 1985, World Bank, Washington D.C.
Wynn, M. (1979a) 'Peripheral urban growth of Barcelona in the Franco era', Iberian Studies, 8(1): 13-28.
Wynn, M. (1979b) 'Barcelona: planning and change 1854-1977', Town Planning Review, 50(2): 185-203.
Wynn, M. (1980) 'San Cosme, Spain: planning and renewal of a state housing area', Journal of the American Planning Association, 16: 3-23.
Wynn, M. (1981) 'The residential development process in Spain - a case study', Planning Outlook, 24(1).
Wynn, M. (1984a) 'Spain' in M. Wynn (ed.) Housing in Europe, Croom Helm, London: 173-201.
Wynn, M. (1984b) 'Spain' in M. Wynn (ed.) Planning and Urban Growth in Southern Europe, Mansell, London: 111-63.
Yang, J. and Stone, R. (1985) 'Investment dependence, economic growth, and status in the world system', Studies in Comparative International Development, 2: 98-120.
Yeung, Y.-M.(1985) Urban Agriculture in Asia: A Substantive and Policy Review, Occasional Paper No. 81, The Chinese University of Hong Kong, Hong Kong.
Young, M.L. (1983) 'A survey of the bibliographic material on the Malay family in Malaysia', unpublished paper, Workshop on Family Research in Asia, East-West Center Population Institute, Honolulu, Hawaii.
Young, M.L. and Salih, K. (1986) 'International labour migration in the world system' in D.W. Drakakis-Smith (ed.), Urbanization and the Developing World, Croom Helm, London: 83-97.
Young, M.L. and Salih, K. (forthcoming) 'Industrialization,

retrenchment, and household processes: implications of the recession', Himpunan Sains Sosial IV, Persatuan Sains Sosial Malaysia. Universiti Malaya, Kuala Lumpur.

Zweig, S. (1984) Brasilien: Ein Land der Zukunft, Suhrkamp, Frankfurt.

Index

Index

Index